Proceedings in Technology Transfer

The book series "Proceedings in Technology Transfer" provides a valuable platform for researchers, practitioners, and policymakers to share their experiences and insights related to technology transfer. Our aim is to facilitate knowledge exchange between academic researchers and industry professionals, accelerating the introduction of new technologies and driving economic growth.

Our series publishes books and conference proceedings that focus on original research findings, best practice recommendations, emerging trends, and technology evaluations. The series covers all topics which explore the transfer of technology in various industry sectors that require innovations and economic growth.

Join us in advancing the technology transfer and its impact on innovation. Publish your proceedings with us and offer insights and expertise to professionals in your field.

Mohamed Hamdaoui · Saber Ben Abdessalam ·
Adel Ghith · Ayda Baffoun · Imed Feki
Editors

Proceedings of the Third International Conference of Innovative Textiles and Developed Materials

ITDM'3; 09–10 May 2025, Monastir, Tunisia

INNOVATIVE TEXTILES
AND
DEVELOPED MATERIALS

Springer

Editors
Mohamed Hamdaoui
Textile Materials and Processes Research Laboratory
Monastir National School of Engineers, ENIM
Monastir, Tunisia

Saber Ben Abdessalam
Textile Materials and Processes Research Laboratory
Monastir National School of Engineers, ENIM
Monastir, Tunisia

Adel Ghith
Textile Materials and Processes Research Laboratory
Monastir National School of Engineers, ENIM
Monastir, Tunisia

Ayda Baffoun
Textile Materials and Processes Research Laboratory
Monastir National School of Engineers, ENIM
Monastir, Tunisia

Imed Feki
Textile Materials and Processes Research Laboratory
Monastir National School of Engineers, ENIM
Monastir, Tunisia

ISSN 2948-2321 ISSN 2948-233X (electronic)
Proceedings in Technology Transfer
ISBN 978-981-95-7703-3 ISBN 978-981-95-7704-0 (eBook)
https://doi.org/10.1007/978-981-95-7704-0

This Springer imprint is published by the registered company Springer Nature Singapore Pte Ltd.
The registered company address is: 152 Beach Road, #21-01/04 Gateway East, Singapore 189721, Singapore

Preface

We are very pleased to present the Conference Proceedings Book of the Third International Conference on Innovative Textiles and Developed Materials (ITDM'3), held at the Serail Skanes Hotel in Monastir, Tunisia, from 09 to 10 May 2025.

We extend our sincere gratitude to all the distinguished participants for their valuable contributions to the success of the conference. The event brought together researchers and experts from both Europe and Africa to present and discuss their work within the framework of four main thematic areas:

- Innovative materials and polymers
- Technical and Smart textiles
- Circularity and Sustainability in Textiles
- Dyeing and Finishing
- Textile Industry 4.0
- Management, Quality, and Textile Processes

This conference served as a platform for academics, researchers, institutions, and national and international socio-economic stakeholders to exchange insights. It provided a valuable opportunity to share research findings and practical experiences, identify current challenges, and propose forward-looking perspectives.

We are deeply grateful to all speakers and participants for their insightful contributions. We also extend our warmest thanks to the keynote speakers and guest presenters for their thought-provoking plenary lectures and presentations.

It was an honor to welcome our invited keynote speakers: Prof. Xavier Legrand (ENSAIT, University of Lille, France), Prof. Mohamed Hassen V. Baouab (IPEIM, University of Monastir, Tunisia), Prof. Saber Ben Abdessalem (ENIM, University of Monastir, Tunisia), Prof. Marcjanna Augustyn and Associate Prof. Kaouther Kooli (Bournemouth University, England), Alexander Ross (CEO, Green Threads DPP Ltd, England), Shah Siddiqui (CEO, Time Research & Innovation (TRI) – COO, Green Threads DPP Ltd, England), and Tim Almond (Heartland Industries, England). A total of 41 accepted communications were presented and discussed during the event. Following a rigorous peer-review process, 24 full papers were selected and included in this volume.

The editors hope that this book will provide readers with a broad overview of the latest research and developments in innovative textiles and advanced materials, and that it will serve as a valuable reference for further studies.

We would also like to express our heartfelt appreciation to the members of the Scientific Committee for their thorough reviews, constructive feedback, and invaluable guidance to the authors.

Special thanks are due to Dr. Imed Feki and his organizing team for their outstanding efforts in preparing the conference venue, managing databases, addressing delegate inquiries, securing sponsorships, arranging the fashion show, and coordinating the many other essential tasks that contributed to the success of this event.

Saber Ben Abdessalam
Conference General Chairman

Mohamed Hamdaoui
Conference General Co-chairman

Adel Ghith
Scientific Chairman

Ayda Baffoun
Scientific Co-chairman

Imed Feki
Steering Committee Chairman

Organization

Conference General Chairman

Saber Ben Abdessalem	UR-MPTex, ENIM, University of Monastir, Tunisia

Conference General Co-chairman

Mohamed Hamdaoui	UR-MPTex, ENIM, University of Monastir, Tunisia

Steering Committee

Imed Feki	MPTex, ENIM, University of Monastir, Tunisia
Naoufel Bhouri	MPTex, ENIM, University of Monastir, Tunisia
Sofien Mokhtar	MPTex, ENIM, University of Monastir, Tunisia
Adel Elamri	MPTex, ENIM, University of Monastir, Tunisia
Khmaies Zdiri	MPTex, University of Monastir, Tunisia
Bouthaina Missaoui	MPTex, ENIM, University of Monastir, Tunisia
Nada Hajri	MPTex, ENIM, University of Monastir, Tunisia
Sirine Guedouar	MPTex, ENIM, University of Monastir, Tunisia
Amel Ben Abdallah	MPTex, ENIM, University of Monastir, Tunisia
Nouha Sebia	MPTex, ENIM, University of Monastir, Tunisia
Sirine Ltaief	MPTex, ENIM, University of Monastir, Tunisia

Scientific Commitee

Adel Ghith (Scientific Chairman)	UR-MPTex, ENIM, University of Monastir, Tunisia
Saber ben abdessalem	MPTex, ENIM, University of Monastir, Tunisia
Mohamed Hamdaoui	UR MPTex, ENIM, University of Monastir, Tunisia
Fatma Göktepe	Çorlu Engineering Faculty, Tekirdağ Namık Kemal University, Turkey
Paul Kickens	Textile Department, Ghent University, Belgium

Editors

Mohamed Hamdaoui	MPTex, ENIM, University of Monastir, Tunisia
Saber Ben Abdessalem	MPTex, ENIM, University of Monastir, Tunisia
Adel Ghith	MPTex, ENIM, University of Monastir, Tunisia
Ayda Baffoun	MPTex, ENIM, University of Monastir, Tunisia
Imed Feki	MPTex, ENIM, University of Monastir, Tunisia

Organizers

MPTex

UNIVERSITE DE MONASTIR

ENIM
ECOLE NATIONALE D'INGÉNIEURS DE MONASTIR

Partners

Association Tunisienne de Mécanique
الجمعية التونسية للميكانيك

AS ENIM

Sponsors

Contents

Textile Industry 4.0

Management, Quality, and Textile Processes

Innovative Materials and Polymers

Novel Braiding Structure and its Mechanical Properties

Georges Cahuzac[1], Peng Wang[2], and Xavier Legrand[1(✉)]

[1] ENSAIT, ULR 2461 - GEMTEX - Génie et Matériaux Textiles, Univ. Lille, F-59000 Lille, France
xavier.legrand@ensait.fr
[2] ENSISA, LPMT, University of Upper Alsace, F-68093 Mulhouse, France

Abstract. The textile structure known as braid, obtained using Maypole technology, has one major drawback: its lack of homogeneity. We are therefore proposing a new structure. Our initial estimates of the mechanical properties of this new and innovative weave are detailed. For this weave, it is necessary to develop a demonstrator machine, so a few dimensioning criteria are proposed.

Keywords: new braid pattern · estimated mechanical properties · machine design

1 Introduction

Mr. Georges Cahuzac, in his patent [1], defines an elementary mesh showing 6 wires (or strands) in the thickness at any given point. In comparison, and as a reminder, a conventional braid shows 0, 1, 2 or 3 wires, depending on the location. This feature (6 wires at any point in the braid plane) is a major innovation, as it leads to much greater homogeneity of mechanical properties in the braid plane.

However, it requires a special machine and, above all, involves a much more complex pattern. Consequently, a new pattern will be presented in this article, along with an estimation of its mechanical properties and the design of the first device.

2 A New Braid Pattern

2.1 Why a New Braid Pattern

A conventional braid shows 0, 1, 2 or 3 wires, depending on the location [2, 3]. Indeed, in the Fig. 1, at point A, this is 0 wire, at B, there is 1 wire, only 2 wires at point C and 3 wires at point D. This observation at meso-scale shows a great heterogeneity of a such material. Such heterogeneity can lead to significant local fragility, a major limitation for braided structures.

A new braided structure is proposed to avoid this defect. It has 6 wires in its thickness, at every point of the surface.

M. Hamdaoui et al. (Eds.): ITDM'3 2025, PTT, pp. 3–13, 2026.
https://doi.org/10.1007/978-981-95-7704-0_1

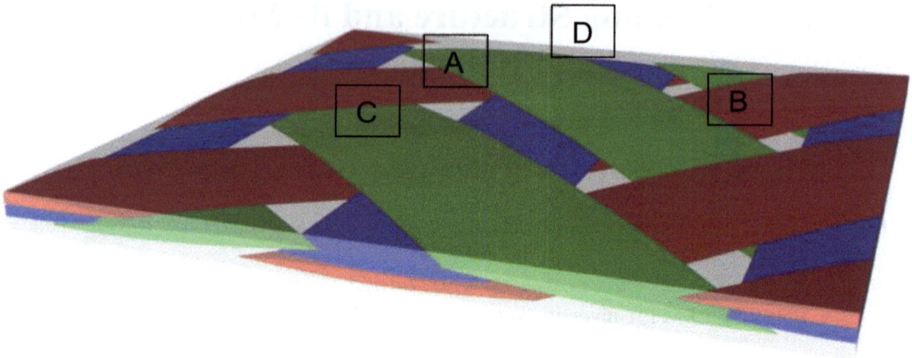

Fig. 1. A 3D representation of a tri-axial maypole classic braid.

2.2 The New Braid Pattern

Once again based on Maypole technology, the new pattern is shown in Fig. 2. Such figure is made using Texgen [4] and a Python script [5].

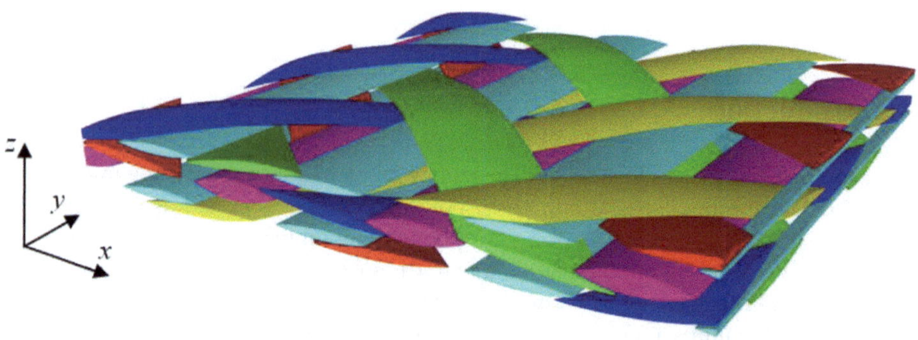

Fig. 2. A 3D representation of the new tri-axial maypole braid.

In Fig. 2, the straight pink wires are double: there are two wires represented in one large wire. all other wires are single. There are therefore 6 wires in the thickness, everywhere.

This representation is based on a sample cut, made by hand on a frame and then resin injected. The sections observed under the microscope confirmed the structure imagined in the initial patent, as illustrated in Fig. 3.

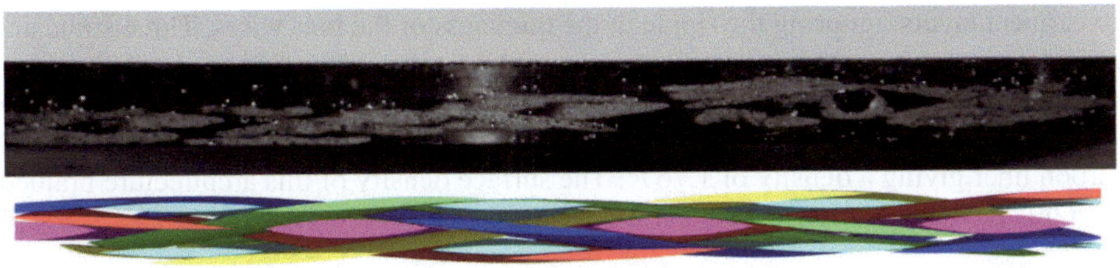

Fig. 3. *C*omparison between our model and a sample of hand-made new braid.

3 An Estimation of Mechanical Properties

3.1 Method

In order to estimate the in-plane stiffness of the new braided reinforcement composite, the elastic moduli of the equivalent laminate are calculated based on classical laminate theory (CLT). In a laminated composite, the reinforcing fibers are held in a laminar plane without undulation through the thickness. New braid architecture consists of two superimposed layers of interwoven axial wires with two sets of bias wires ($-\theta°$ and $+\theta°$). Thus, a low level of waviness can be considered for the wires making up the braid. This enables laminate theory to be applied to obtain an estimate of the stiffness modulus of this composite reinforced with an innovative braid, as determined by Cox [6] for a similar structure.

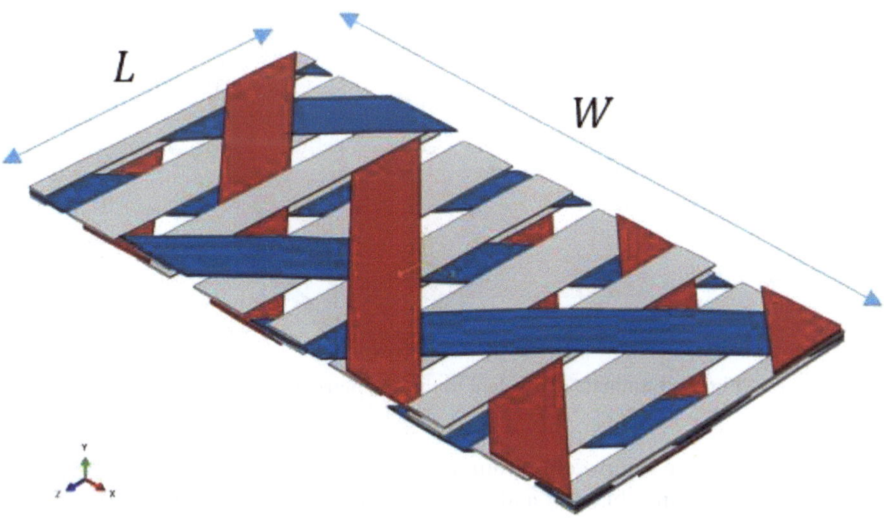

Fig. 4. Geometric model of the new braid

The approach followed to obtain the equivalent laminate is based on considering the same VER dimensions and the same number, orientation and stacking sequence of the constituent layers, ignoring the ripple in the thickness of the bias wires. Three strata are considered to correspond to 0° axial yarns, −° bias yarns and +° bias yarns. The fiber volume for each wire group is obtained from the geometric model established for the given length (= 20) and width (= 40) of the VER. Fiber mass is obtained for T300 carbon fiber giving a density of 1.76 /^{3}. The surface density of this architecture braided with this type of fiber is estimated at 367 /^{2} (Table 1).

Table 1. Fiber volume and mass in the VER of new braid based on the geometric model

Fibre set	Axial	Bias $-\theta^0$	Bias $+\theta^0$
$V_k^{f/C}$ in RVE [mm^3]	98.595	34.867	34.867
$m_k^{f/C}$ in RVE [g]	0.173	0.061	0.061
$V_k^{f/C}/V^{f/C}$	0.585	0.207	0.207

The thickness of each stratum (h^L), defined as (*zk* - *zk*-1) in Eq. (1) is obtained by (Eq. 1) from the composite thickness (H) considering a partition factor related to the ratio of each fiber group ($V^{f/C}$) to the total fiber content of the composite ($V^{f/C}$) (Fig. 5).

$$h_k^L = \frac{V_k^{f/C}}{V^{f/C}} H; k = 0, -\theta^0, +\theta^0 \quad (1)$$

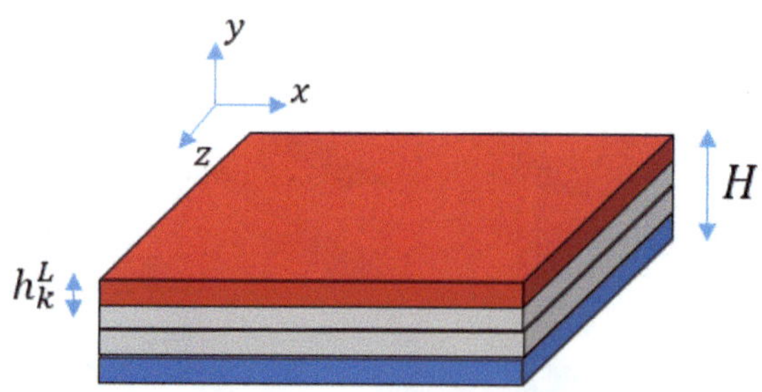

Fig. 5. Laminate equivalent to the geometric model in Fig. 4.

The volume of the equivalent stratum for each fiber set (V^L) is obtained by the thickness of the stratum obtained and the same dimensions in the VER plane (L and w). Thus, the volume of each equivalent stratum can be expressed by the volume of the composite (V^C) as a function of the fraction of a fiber set to the total volume of fibers in the composite ($V^{f/C}$) as expressed in (Eq. 2).

$$V_k^L = \frac{V_k^{f/C}}{V^{f/C}} V^C = \frac{V_k^{f/C}}{V^{f/C}} L\, w\, H; k = 0, -\theta^0, +\theta^0 \quad (2)$$

Considering that the volume of a set of fibers in the equivalent lamina ($V^{f/L}$) is equal to the volume of this set of fibers in the composite ($V^{f/L} = V^{f/C}$), the volume fraction of fibers in each lamella ($v^{f/L}$) can be obtained from Eq. (3) and is equal to the total fraction of fibers in the composite ($v^{f/C}$) (Fig. 6).

$$V_k^{\frac{f}{L}} = \frac{V_k^{\frac{f}{C}}}{V_k^L} = \frac{V_k^{\frac{f}{C}}}{\frac{V_k^{\frac{f}{C}}}{v^{\frac{f}{C}}} V^C} = \frac{V^{\frac{f}{C}}}{V^C} = v^{\frac{f}{C}}; k = 0, -\theta^0, +\theta^0$$

$$v_k^{\frac{f}{C}=} = v^{f/C} = \frac{V_k^{f/C}}{L\,w\,H} \tag{3}$$

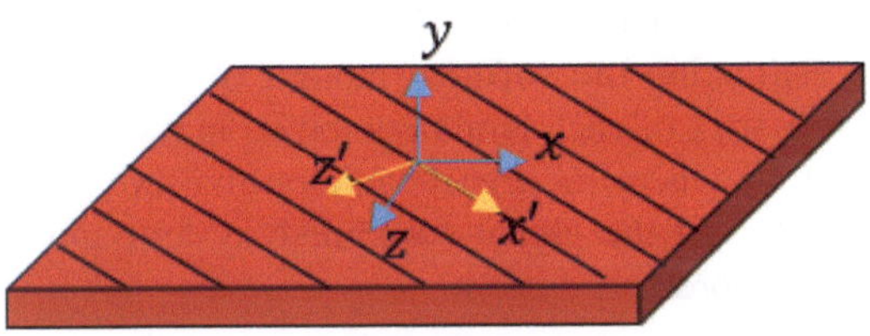

Fig. 6. Modules are calculated for each state

The stiffness matrix of each strate [C′]k with respect to the strate's local coordinate system associated with that of the corresponding wire is deduced based on the analytical micromechanical model consisting of the Reuss model and the Chamis model. The Reuss model (also called the first mixing rule) is often used to predict longitudinal modulus (1, 12 = 13) as it provides a good estimate for design [7]. This model is based on the iso-constraint assumption, where the strain in the fiber and in the matrix is assumed to be the same = = . For a representative volume V of fiber and matrix (with fiber volume fraction and matrix volume fraction ; + = 1, no void and matrix and fiber are perfectly bonded) subjected to an external load on its outer surface with consideration of volume-average stress-.

$$E1 = vf.Ef1 + vm.Em \tag{4}$$

$$v12 = vf.vf12 + vm.vm12 \tag{5}$$

Chamis [8, 9] proposed simplified SME micromechanical equations for transverse modulus and shear modulus (2, 12). In this model 2, 12 are calculated by:

$$E_2 = \frac{E_m}{1 - \sqrt{v_f\left(1 - \frac{E_m}{E_{f2}}\right)}} \qquad G_{12} = \frac{G_m}{1 - \sqrt{v_f\left(1 - \frac{G_m}{G_{f2}}\right)}} \tag{6}$$

Chamis and Caruso [10] have shown that good agreement is obtained for 2 and 12 calculated by EMS compared with 3D finite element models. The modulus of elasticity

Table 2. Modulus of elasticity of materials used.

Carbon T300 (transversally isotropic)	E_l[GPa]	250
	E_t[GPa]	8
	G_{lt}[GPa]	273
	V_{lt}	0.256
	ν_{tt}	0.3
Resin Epoxy (isotropic)	E[GPa]	3.5
	ν	0.35

of T300 carbon fiber is considered as reported by Miyagawa et al. [11] for transverse modulus, with epoxy resin as matrix, Table 2.

The next step in estimating the stiffness moduli of the equivalent laminate is to transform the stiffness matrix of each lamina obtained in the global coordinate system of the laminate by (Eq. 7) where the transformation matrix [T] is obtained based on the angle between axis "$_1$" of the global coordinate system and axis "' "of the local coordinate system. local coordinate system.

$$[C] = [\bar{T}][C'][\bar{T}]^T \tag{3.14}$$

$$[T] = \begin{bmatrix} a_{11}^2 & a_{12}^2 & a_{13}^2 & 2a_{12}a_{13} & 2a_{11}a_{13} & 2a_{11}a_{12} \\ a_{21}^2 & a_{22}^2 & a_{23}^2 & 2a_{22}a_{23} & 2a_{21}a_{23} & 2a_{21}a_{22} \\ a_{31}^2 & a_{32}^2 & a_{33}^2 & 2a_{32}a_{33} & 2a_{31}a_{33} & 2a_{31}a_{32} \\ a_{21}a_{31} & a_{22}a_{32} & a_{23}a_{33} & a_{22}a_{33}+a_{23}a_{32} & a_{21}a_{33}+a_{23}a_{31} & a_{21}a_{32}+a_{22}a_{31} \\ a_{11}a_{31} & a_{21}a_{32} & a_{13}a_{33} & a_{12}a_{33}+a_{13}a_{32} & a_{11}a_{33}+a_{13}a_{31} & a_{11}a_{32}+a_{12}a_{31} \\ a_{11}a_{21} & a_{12}a_{22} & a_{13}a_{23} & a_{12}a_{23}+a_{13}a_{22} & a_{11}a_{23}+a_{13}a_{21} & a_{11}a_{22}+a_{12}a_{21} \end{bmatrix}$$

$$a_{ij} = \cos(e_i', e_j) \qquad i = 1, 2, 3 \ : \tag{7}$$

The classical general theory of laminates is implemented to deduce the matrix [ ] (defining the elastic membrane and bending properties of the laminate) by Eq. (7). Next, the in-plane stiffness matrix (considering the in-plane stress assumption) and the elastic modulus of the laminate are obtained from Eq. (8).

$$E_2 = \frac{E_m}{1 - \sqrt{v_f\left(1 - \frac{E_m}{E_{f2}}\right)}} \qquad G_{12} = \frac{G_m}{1 - \sqrt{v_f\left(1 - \frac{G_m}{G_{f2}}\right)}} \tag{8}$$

This approach is encoded in MATLAB

3.2 Results

The geometric model cannot accurately predict composite thickness. Consequently, fiber volume fraction and elastic modulus are estimated by the stratified theory approach as a function of composite thickness (H) and presented in Fig. 7.

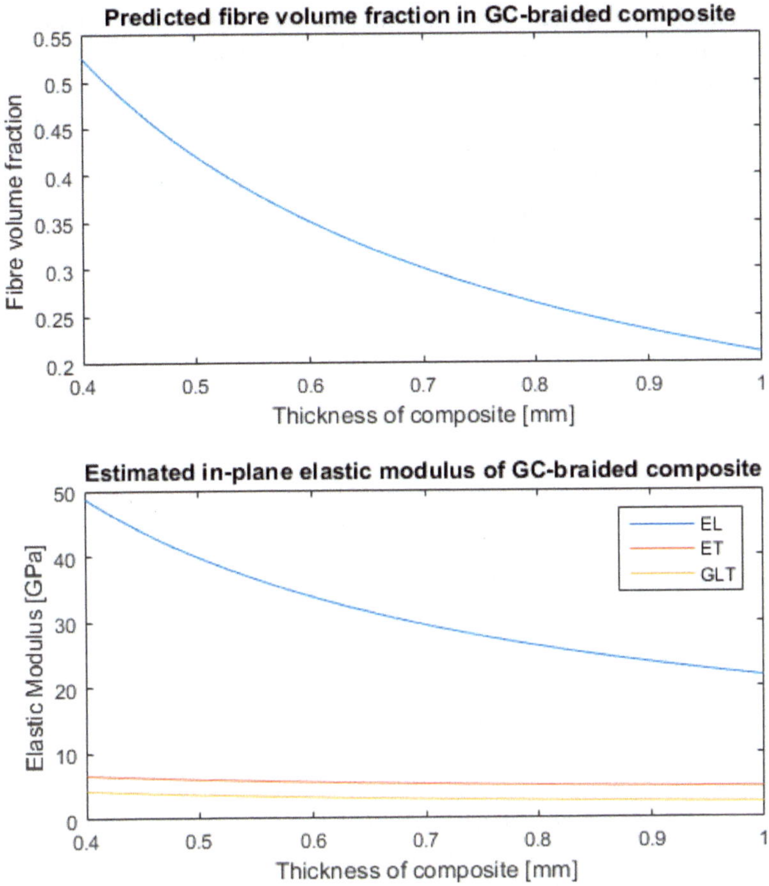

Fig. 7. Estimation of fiber volume fraction (top) and modulus of elasticity (bottom) in new braided composite as a function of composite thickness

The first curve in Fig. 7 shows a maximum fiber volume fraction of between 50 and 55% when the composite has a thickness of 0.4 mm. As the thickness of the reinforcement is determined, as the thickness of the composite (matrix and reinforcement) increases, the fiber volume fraction falls. The curves in the graph at the bottom of Figs. 3 and 4 show almost constant shear and transverse moduli as a function of composite thickness. The longitudinal modulus of elasticity, on the other hand, tends to decrease with increasing composite thickness.

4 A Design of the First Device

4.1 Reminder of the Principle

A possible machine arrangement to realize this braiding architecture corresponds to an axial braider with two circular rows of notched wheels. As a result, a geometrical problem has to be faced, implying that the notched and toothed wheels in the inner circular row are inevitably smaller than those in the outer circular row. This solution, although favored by most manufacturers, has both advantages and disadvantages. First, the factor of scale: the diameters of industrial axial braiding machines are generally several meters high, for products a few centimeters in diameter. This dimensional concern is compensated for by the provision of one or more braiding rings, forcing the wires to be removed. Second, more annoying, however, is the difference in diameter between the gears. For two gears to mesh perfectly, their module must be identical.

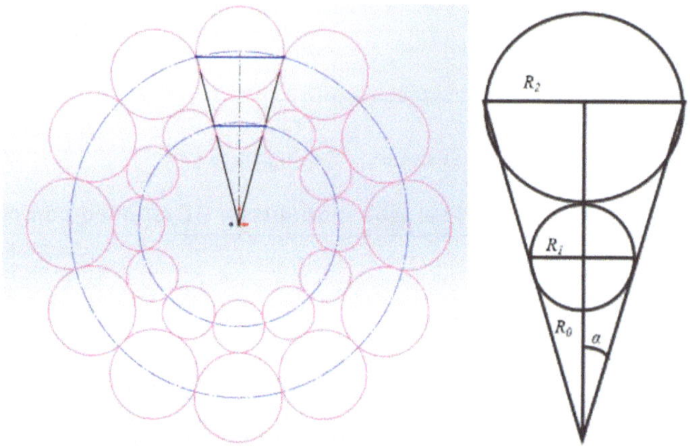

Fig. 8. Estimation of fiber volume fraction (top) and modulus of elasticity (bottom) in new braided composite as a function of composite thickness

Given the need for gears with the same module, the module is assumed to be fixed. The relationship between the diameter (d in mm), the module (m in mm) and the number of teeth (Z) of a gearwheel is:

$$d = mZ \tag{9}$$

Gear dimensions are standardized. A parametric study was carried out to determine the ideal combination of these parameters (see Fig. 8). As the number of gear teeth must be an integer, these limits or even blocks viable solutions when the R0 parameter is fixed. In addition to d, m and Z, the influence of angle α is also studied according to the following equations (Table 3):

$$\tan\alpha = \frac{R_2 - R_1}{R_1 + R_2} \tag{10}$$

$$\alpha = \tan^{-1}\left(\frac{R_1}{R_0}\right) \tag{11}$$

$$\mathrm{R_1} = \mathrm{R_0} \cdot \tan\alpha \tag{12}$$

It is then necessary to study configurations where the module is fixed, as well as the number of gearwheels in the assembly (parameter 360/2α). More cogwheels mean higher cost when purchasing or possibly custom-building, if no standardized dimensions are applicable.

Table 3. Parametric study table

360/2.α	2.α	α	tan(α)	N1	N2	R0	R1	R2	Z1	Z2	m(module)
6	60,00	30,00	0,58	3	11,20	100	57,74	215,47	86,60	323,21	0,33
12	30,00	15,00	0,27	3	5,20	100	26,79	46,41	4,47	7,74	3,00
18	20,00	10,00	0,18	3	4,28	100	17,63	25,18	2,20	3,15	4,00
24	15,00	7,50	0,13	3	3,91	100	13,17	17,16	1,32	1,72	5,00
6	60,00	30,00	0,58	4	14,93	100	57,74	215,47	4,81	17,96	6,00
12	30,00	15,00	0,27	4	6,93	100	26,79	46,41	1,91	3,32	7,00
18	20,00	10,00	0,18	4	5,71	100	17,63	25,18	1,10	1,57	8,00
24	15,00	7,50	0,13	4	5,21	100	13,17	17,16	0,73	0,95	9,00

4.1.1 Axial Braiding Machine and Associated 3D Prototype

To confirm the design, a prototype was designed and produced in 3D printing. Innovative 3D-printed braider prototypes are based on models designed using SolidWorks geometric modeling software (Dassault Systèmes). Initially printed on a reduced scale, they are printed on a full scale once several design stages have been validated. The first prototype of an innovative braider to be printed and presented is the axial braider (see Fig. 9).

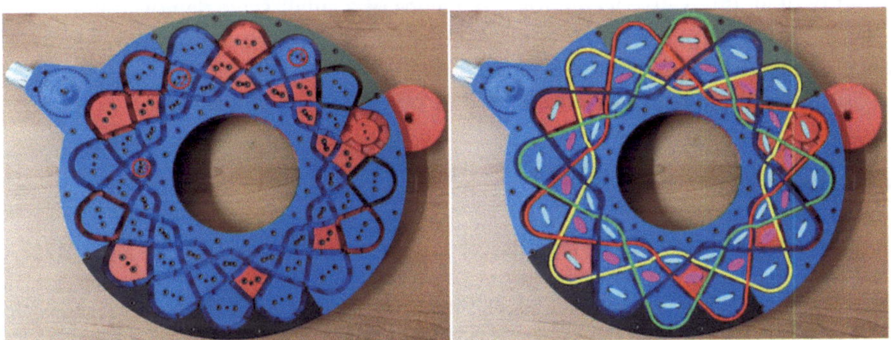

(a) (b)

Fig. 9. 3D-printed prototype of the new axial braiding machine

This prototype is designed to be driven by an electric motor, visible in the top left of Fig. 9a, to show the kinematics of the notched wheels. It is possible to insert a spindle element and observe its travel through the cam track. The travels of the spindle groups are highlighted in Fig. 9b below, corresponding to the oriented wires. Axial threads, oriented at 0° in the production direction, are shown (exaggeratedly) in cyan and magenta as illustrated in Fig. 10.

Fig. 10. 3D new tri-axial braid machine vs. Fig. 9b

5 Conclusion

In this article, we present a new braiding weave that solves a major problem of conventional braiding: its lack of homogeneity. Our initial estimates of the mechanical properties of this new structure show just how interesting it is. So, we set out to design the ad hoc machine: one capable of structuring the material according to the weaves we had imagined. Although more complex than a conventional Maypole braider, with its 4 cam paths, 2 braiding rings and 4×16 axial wires, we were able to demonstrate the feasibility of an innovative technology by means of a small-scale demonstrator.

Future work will focus on designing and building a full-scale prototype to produce braided structures using this new weave.

This production will enable us to validate our estimates of mechanical properties.

References

1. G. Cahuzac, "Triaxial multilayer armature and its manufacturing process," EP3771757A1, 2021
2. B. Duchamp, "Thèse Contribution à l' élaboration de préformes textiles pour le renforcement de réservoirs souples. PhD thesis. Université Lille. 2016
3. Kyosev, Y.: Introduction: The main types of braided structure using maypole braiding technology. In: Kyosev, Y. (ed.) Braiding Technology for Textiles, pp. 1–25. Woodhead Publishing, Cambridge (2015). https://doi.org/10.1533/9780857099211.1
4. TexGen (Version 3.5.2). University of Nottingham. 2012. http://texgen.sourceforge.net Accessed 5 Nov 2020
5. J.-V. Risicato, "Optimisation de l'architecture des fils dans une structure textile 3D pour le renforcement de pièces composites," PhD thesis. Université de Lille. p. 193, 2012
6. Cox, B.N., Dadkhah, M.S., Morris, W.L.: On the tensile composites failure of 3D woven. Composite: Part A. **27**, 447–458 (1995)
7. Gibson, R.F.: Principles Of Composite Material Mechanics. CRC Press, Boca Raton (2007). https://doi.org/10.1201/9781420014242
8. Chamis, C.C.: Simplified composite micromechanics equations for hygral, thermal and mechanical properties. SAMPE Q. **15**(3), 14–23 (1984)
9. Chamis, C.C.: Simplified composite micromechanics equations for mechanical, thermal and moisture-related properties. In: Weeton, J.W., Peters, D.M., Thomas, K.L. (eds.) Engineer's guide to composite materials. ASM International Materials Park, Materials Park, OH (1987)

10. Caruso, J.J., Chamis, C.C.: Assessment of simplified composite micromechanics using three-dimensional finite element analysis. J. Compos. Technol. Res. **8**(3), 77–83 (1986)
11. Miyagawa, H., Sato, C., Mase, T., Drown, E., Drzal, L.T., Ikegami, K.: Transverse elastic modulus of carbon fibers measured by Raman spectroscopy. Mater. Sci. Eng. A. **412**, 88–92 (2005). https://doi.org/10.1016/j.msea.2005.08.037

Rethinking Hospital Design: A Prevention Strategy Through Anti-Microbial Materials

Donia Maalej(✉)

Higher institute of Arts and Crafts of Sfax, University of Sfax, Sfax, Tunisia
dony.maalej@hotmail.com

Abstract. There is an increased risk of microbial transmission in hospital settings. Indeed, the French public health has conducted a national survey on the evolution of risk of contamination within sanitary areas. And this showed that 5.71% of hospitalized patients have a nosocomial infection in 2022. Comparing this value to 2017, which had 4.98%, it indicates a 0.73% increase in patients who have been infected with nosocomial infections. And this is due to the transmission through the surfaces of materials. However, the materials used are often chosen without a systematic assessment of their impact on infection prevention, patient safety, or sustainability. This study aims to identify and classify the most commonly used finishing materials in healthcare facilities, with the aim of supporting safer and more hygienic design. To achieve this objective, a mixed methodology was adopted: it combines a review of specialized literature, the analysis of data from hospital design guides and specialized books, as well as site visits and exchanges with field professionals. This approach has resulted in an inventory of materials by location. The materials were then grouped by family, taking into account specific selection criteria such as ease of cleaning, resistance to chemical agents, durability, cost and environmental impact. The identification of these materials is an essential first step towards creating a dedicated database. This database will then be enriched with quantitative and qualitative data from comparative evaluations according to the defined criteria. In the long term, this database will be integrated into the database of a digital design aid tool to support designers and architects in their choice of materials in hospitals.

Keywords: Antimicrobial Materials · User Safety · Contamination · nosocomial infections

1 Introduction

Until the end of the 20th century, health facilities were considered as non-complex care institutions. However, they now face major challenges in preventing nosocomial infections, specifically during the post-pandemic period. Users; patients, visitors and hospital staff are largely favored by the transmission of these infections due to direct interactions with soiled objects such as door handles, reception counter, medical equipment [1].

Several recent studies confirm this. Among them, the study by King et al. [2] states that the duration of surface contamination depends essentially on the type of material

M. Hamdaoui et al. (Eds.): ITDM'3 2025, PTT, pp. 14–25, 2026.
https://doi.org/10.1007/978-981-95-7704-0_2

and its maintenance. Indeed, some surfaces can house pathogens for longer than others, which poses an indirect risk of contamination. This reflection highlights the importance of choice of materials within care facilities.

In this context, the integration of materials with antimicrobial properties in sanitary spaces appears to be the most suitable solution. It not only reduces the contamination of pathogens, but also prevents the safety of users [3]. The objective of this research is to examine the different characteristics of materials in a strategy for prevention of nosocomial infections. Based on scientific, technical and epidemiological data, this study proposes to compare materials according to their effectiveness against the risks of contamination, in order to guide designers towards healthier choices. Thus, the selection of materials is no longer limited to a question of functionality or aesthetics, but becomes an essential lever for health safety.

2 Materials and Methods

This study adopted a mixed methodological approach, integrating a systematic literature review with the analysis of existing databases and technical documentation. The objective was to identify, characterize, and classify finishing materials commonly used in healthcare facilities, emphasizing their role in infection prevention and control. Our study is based on a literature review. These data are collected from scientific, technical and epidemiological sources.

The first step was to identify the importance of antimicrobial materials in preventing user safety through systematic theoretical research. The second step is to identify the evaluation criteria. This information was cross-checked with data on the materials present in each subspace: rooms, corridors, operating rooms, waiting rooms, etc.

As a complement, our methodology is based on a preliminary field study. During this in situ investigation, we identified around thirty materials used in various healthcare facilities. Among them, we selected the five materials most frequently in contact with users, and therefore the most exposed to daily contamination risks.

We use a qualitative multi-criteria approach. The latter will allow us to analyze and classify materials according to several elements, namely:

2.1 Aesthetics and Ergonomics

Visual and functional comfort are essential in the design of care spaces. In this sense, the use of aesthetic and suitable materials contributes to the psychological comfort of users. Also, taking into consideration the ergonomics of the space promotes safety, accessibility and better adaptation of users within these establishments [16].

2.2 Antimicrobial Properties

Taking antimicrobial properties into account is a central element in health environments. Their applications reduce the transmission of pathogens to contact surfaces. It is therefore an anticipation of integration of these materials presents itself as a strategy to combat the spread of infections.

2.3 Hygiene

The objective of our research is to preserve the health safety of users. Therefore, exposed surfaces and susceptible to transmit nosocomial infections. Following this logic these surfaces must pass through a hygienic protocol. Taking this protocol into consideration directly influences the choice of textures of different materials in a hospital project. The smooth, non-porous and easy-to-clean surfaces will further limit the accumulation of germs.

2.4 Sustainability

In this hygiene logic, exposed surfaces are subjected to frequent cleaning. Indeed, ensuring a fairly high sustainability for the various materials without rapid deterioration, further reduces the cost of maintenance or replacement of objects. This criterion ensures an economic and ecological solution for sanitary spaces.

2.5 Cost and Availability

For a successful architectural project, the cost of the material is a significant factor in government-funded projects such as hospitals. The materials must be available on the market and adapted to the budgets of public or private institutions. This allows the architects to ensure a successful completion of the architectural project.

2.6 Decontamination Time

In sanitary environments with a high risk of contamination and frequentation, the cleaning and decontamination time of surfaces are two essential criteria in the choice of materials, without compromising the effectiveness of the disinfection protocol.

2.7 Reduction of Infections

The criterion of reduction of infections allows contact surfaces to reduce the transmission of nosocomial infections. In an overall design approach, this criterion allows the designer to adapt their use in space with respect to their location, their exposure to contact.

Using these criteria, the materials were classified according to a criteria grid, their level of risk of contamination. This classification is based on their use in each space and the frequency of contact with users. Data collection was carried out from scientific articles, statistics and technical sites.

This work lays the foundations for a future research project which aims to create a design aid tool. This tool is intended for architects and designers. It aims to facilitate the choice of materials in health care facilities. It integrates infectious risks and helps design safer spaces for all.

3 Literature Review

The transmission of nosocomial infections is a major problem in health facilities, requiring a study on materials and their characteristics. And this is done in order to facilitate the choice of these materials when designing the space concerned. The use of these materials aims to limit the spread of pathogens, ensuring selection with antimicrobial properties and facilitating hospital hygiene, and subsequently ensures the safety of users of nosocomial infections.

3.1 The Importance of Materials in the Prevention of Nosocomial Infections

In order to base our research on a sufficiently solid study, we begin by researching the different contact surfaces most exposed to microbial transmissions. Indeed, hospital surfaces in frequent contact with patients and medical staff represent a major reservoir of contamination. According to the study by Baranovsky, door handles, ramps, elevator knobs and examination tables have high levels of microbial contamination, out of 5,329 samples collected the study shows 567 contaminated samples which represents 10,63% excluding pandemic [4]. This percentage is significant.

In addition, the manual disinfection of various surfaces in sanitary spaces is insufficient. It does not limit the persistence and transmission of pathogens. The lack of strict surface disinfection protocols and regular maintenance of equipment creates an environment conducive to cross-contamination between patients and staff, contributing to the spread of infections. This observation reinforces the importance of hygiene procedures and raising awareness among all staff about good cleaning practices to reduce the risk of transmission of nosocomial infections [5]. This leads us to identify the different surfaces that can transmit nosocomial infections.

The identification of these surfaces makes it possible to classify materials according to the rate of transmission of nosocomial infections. And it is with the aim of ensuring a better selection of materials to offer a more safe and secure hospital environment. The use of this classification method makes it possible to facilitate the choice of materials during design and subsequently reduce the spread of contamination, and effectively protect all users (hospital staff, visitors and patients). It is in this logic that the designer ensures a sustainable and safe prevention strategy.

3.2 Theoretical Approaches to Material Classification

The classification of materials is based on different theoretical approaches to organize and structure their selection according to various criteria. These classification models are fundamental to optimize the use of materials according to their intrinsic properties and their suitability for specific application requirements. In our case we chose the qualitative multi-criteria decision method. Our choice is justified by the requirement to take into account several criteria simultaneously allowing both an assessment of different materials and a comparison between them. Indeed, this method is put to ensure the solution of complex decision problems, thanks to the different criteria that takes into consideration. Among the approaches to this method, we find the Weighted Sum Model (WSM), the Analytic Hierarchy Process (AHP), TOPSIS, PROMETHEE, and VIKOR. Using these approaches offers the possibility of classifying materials according to different properties chosen according to the preferences of designers. In this context where the safety of users is at stake, the choice of classification method is essential. By adapting hybrid approaches, these methods improve the accuracy and reliability of decisions when using the design aid tool [6]. They present the best solution to classify materials according to criteria of sustainability, safety, reduction of nosocomial infections... However, there are some challenges such as the selection of appropriate design criteria. To avoid any kind of confusion and complexity in the choice of material criteria we have chosen the Extended PROMETHEE II method. The latter presents a qualitative multi-criteria decision approach. Its application allows designers to enrich the different evaluation and classification criteria. It is particularly suitable for situations where several criteria must be taken into account to evaluate materials, such as those used in the design of different elements within hospital spaces. This approach is based on the logic that each material can be classified following a comparison of different criteria according to the specification of choice of designers. And thanks to this directed classification, the tool can propose a final classification of different design alternatives. This approach is based on the following process (Fig. 1):

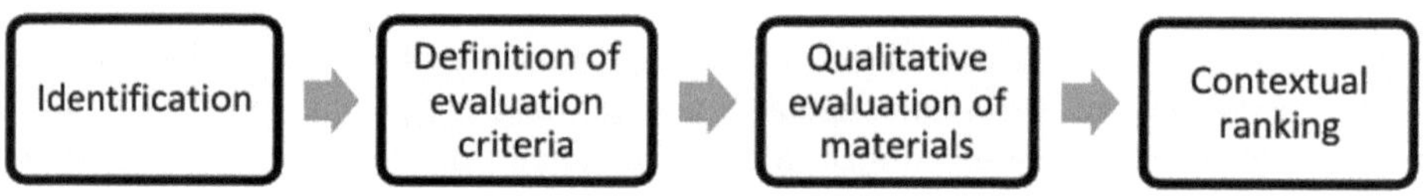

Fig. 1. Process of qualitative multi-criteria decision method

- Identification of alternatives

The materials used in the various hospital spaces (rooms, corridors, operating rooms, etc.) were first identified according to their actual presence on the ground and their current use in the sanitary design.

- Definition of evaluation criteria

Criteria were selected from scientific research and a field study. It aims to define materials according to their effectiveness against the risks of contamination taking into account the comfort of the user, the aesthetics of the space, its sustainability, its cost and its functionality

- Qualitative evaluation of materials

Each material was evaluated qualitatively, based on a cross-literature analysis, based on the seven criteria defined. The evaluation is based on comparative descriptions between them, in order to reflect the specificities of use in each subspace.

- Contextual ranking of alternatives

The classification of materials is based on their suitability for different hospital environments and their level of risk of contamination. Materials are grouped according to their overall performance, frequency of contact with users and ability to limit the spread of infections. We will focus on the first two steps of the process to create a classification table that will compare different material criteria.

3.3 Key Criteria for the Classification of Materials

To further guide this method, we have chosen to propose the various key criteria for classification of materials such as antimicrobial properties, sustainability and resistance to disinfectants, cost and market availability, recycling, aesthetics, ergonomics, decontamination time, reductions in nosocomial infections.

3.3.1 Antimicrobial Properties

Antimicrobial properties are the ability of a material to inhibit growth or eliminate microorganisms such as bacteria and viruses. Several research studies have investigated antimicrobial properties and their effectiveness of materials. Taking these properties into account in the classification is essential to help architects and designers when designing sanitary spaces but also to ensure improved safety for patients, visitors and hospital staff, and to help control the spread of infections. Taking the example of copper, its surface area can limit the spread of nosocomial infections from 70% to 99.9% in a few hours. While bacteriostatic polymers reduce contamination rates by 30% to 60% [7]. In addition, antimicrobial coatings on frequently touched surfaces have shown a reduction of nosocomial infections by up to 50% [8].

Antimicrobial surfaces can reduce decontamination time. Following this idea, the integration of these surfaces within hospitals ensures a better disinfection protocol [9]. It is in this reflection; the study of decontamination time is essential when choosing different materials within hospital spaces. Thanks to this decontamination character, these materials can thus guarantee the elimination of pathogens and ensures, subsequently, the safety of users.

In addition to decontamination time, there is the reduction of nosocomial infections. This criterion contributes to the reduction of pathogen transmission. It allows the materials to act as an active barrier against these agents. Therefore, several clinical trials have been conducted to identify the efficacy of this antimicrobial criterion among them the study conducted by Sharpe and Schmidt discusses the integration of these materials and the evaluation of their effectiveness in real-life situations [10].

3.3.2 Sustainability and Resistance to Disinfectant Protocols

In order to ensure the safety of users within hospital spaces, maintenance and disinfection are an important act. Their role is to reduce the risk of nosocomial infections. On the one hand, several studies show that there are materials such as stainless steel, which requires frequent disinfection protocols. While some materials such as treated polymers, whose surfaces are smooth and non-porous, are easier to maintain [11]. Another study highlights the effectiveness of ceramic materials when applied in surgical environments [12]. On the other hand, some materials are not resistant to frequent disinfection protocols. Following this reflection, the sustainability of materials and resistance to frequent disinfection protocols are presented as an essential criterion that ensures the longevity of the material, maintain disinfection effectiveness and ensure safety against infection transmission.

3.3.3 Cost and Availability

The choice of materials is also related to different economic constraints and market availability. In this sense, a study estimated that by integrating antimicrobial surfaces, nosocomial infections could be reduced by 23% of the maintenance rate [13]. However, some materials that validate all the criteria chosen by the designers are sometimes unavailable on the market. It is therefore important to take into account the cost and availability of the chosen material. Indeed, these factors directly influence the implementation of the large-scale project, thus ensuring an optimal design for the fight against nosocomial infections.

3.3.4 Recyclability

In an ecological approach, being aware of the issues that non-recyclable material can cause after a period allows us to move towards a more ecological choice strategy in hospital environments. Indeed, the study by Freitas et al. shows that the use of recycled materials can promote a sustainable approach without compromising space hygiene [14]. In addition, durable medical textiles can be compatible with disinfection requirements while improving the overall recyclability of hospital equipment [15]. It is in this thinking that taking recyclability into account contributes to a more sustainable management of resources by adopting a more ecological approach without sacrificing the quality of care or hygiene (Table 1).

4 Application of Criteria on a Few Materials in Relation to a Corpus Previously Studied

Table 1. Criteria for material identification

Materials	Hygiene	Sustainability	Recyclability	Antimicrobial properties	% Reduction infections	Virus Survival time (approximate.)	Surfaces characters
Stainless steel	High	Excellent	Average	Low (but easy to clean)	50–60%	Up to 48 h	Non-porous surface
PVC	Good	High	Weak	Moderate according to treatment	Up to 40%	Up to 72 h	Includes rigid plastics
Glass	High	Average	Excellent	Weak (but not porous)	Up to 30%	Up to 96 h	Smooth, non-porous
Copper surfaces	Very high	Good	Average	Naturally antimicrobial	58%	Approximately 4 h	Naturally antimicrobial properties
Linoleum	Average	Good	High	Low to moderate	Unknown	Less than 4 h	Some antiviral properties according to study

(continued)

Table 1. *(continued)*

Materials	Hygiene	Sustainability	Recyclability	Antimicrobial properties	% Reduction infections	Virus Survival time (approximate.)	Surfaces characters
Vinyl	Good	Good	Weak	According to treatment	Up to 30%	Up to 72 h (estimate)	Similar to PVC
Antimicrobial textiles†	Average to good	Variable	Low.to medium	Fibres treated with zinc,	silver and copper 50–70%	Variable according to treatment	Depends on the active agents used

5 Results and Discussion

Our research makes evident the identification of materials with regard to their hygiene, their costs and their availability on the market, their sustainability and recycling, their aesthetic and ergonomic characteristics and their antimicrobial properties. These materials studied have specific characteristics that influence their integration in sanitary spaces.

- Stainless steel has a non-porous surface, which limits the growth of microbes and facilitates maintenance and disinfection. This material has better resistance to cleaning agents than polymers and aluminum [17]. Thanks to its strength, stainless steel is generally used for surgical equipment and medical furniture [18, 19]. However, its cold aesthetic appearance may be perceived as unsuitable for patient care spaces [20].
- The polyvinyl chloride known by the abbreviation PVC. This material is applied mainly as a wall and floor covering. And it is thanks to its easy maintenance and resistance to disinfection protocols. Its smooth surface limits the accumulation of dust and pathogens. However, some concerns remain due to recycling difficulties [21]. The results presented by Advanced Plastiform show an average reduction of 35–40% for PVC depending on the type of treatment applied.
- Glass is frequently used in hospitals because of its non-porous surface, which facilitates cleaning and disinfection. It is mainly used for partitions and windows. Its recycling makes it a sustainable material [22].
- Stainless steel is characterized by a non-porous surface, limiting the antimicrobial properties.
- Antimicrobial surfaces reduce contamination within hospital spaces. This idea was confirmed in the research of Salgado et al. Which explains that the use of these materials has decreased nosocomial infections [23]. However, the high cost of these materials remains a crucial factor in material selection [24].
- The textile materials, although less present in the areas of the hospital space (furniture, curtains, cloths, separation...), their antimicrobial treatments limit microbial transmission [25]. However, regular maintenance is essential to ensure their effectiveness. Some synthetic fiber textiles are designed to be resistant to repeated washing and disinfectants [26].
- Copper surfaces: naturally antimicrobial, observed reduction of 58% in 4 h [23], confirming expected efficacy.

- Linoleum is a material with natural aesthetics and high recyclability. It has average hygiene and durability. Its antimicrobial properties range from low to moderate, with a viral survival of less than 4 h. Some studies report antiviral properties, but data is limited. Percentage of infection reduction is not specified.
- Vinyl, similar to PVC, has variable hygiene, ergonomics and aesthetics. It offers a reduction in infections of up to 30% depending on the treatments applied. Its recyclability is low and viral survival can reach 72 h (estimate).

The application of technical criteria to our corpus of six antimicrobial materials resulted in a detailed table for each of them detailing their basic attributes (hygiene, cost, antimicrobial properties, durability, etc.). This first step confirms the relevance of a database organized around these common characteristics, facilitating comparison, classification and targeted query of materials according to their technical performance.

However, this table is still insufficient to guide the interior designer in the spatial integration of these solutions: it does not provide information on the modes of use (installation, formats, supports) or their contribution in terms of design and quality of life. Take the example of a modulable antimicrobial textile, enriched with silver and copper nanoparticles, available in rolls of 1.4 x 25 m or ready-to-install panels of 60 x 120 cm. Its matt finish, water-repellent and stain-resistant, can be dyed according to the Pantone colour chart and retains a slight curly texture. This composition reduces the microbial load by more than 50%. Designed for intensive use, the fabric is washable at 60 °C and compatible with disinfection. It contributes to the improvement of visual comfort and indoor acoustics, while being hypoallergenic and easy to maintain. This material contributes to a better quality of life for users by combining hygiene, well-being and serenity. It is particularly suitable for box curtains; mattress covers or wall coverings outside medical imaging areas. However, its use is not recommended in the direct radiology beam so as not to alter the quality of the images.

Nevertheless, the criteria identified provide a solid basis for structuring knowledge of antimicrobial materials and guiding the first phases of selection in health settings. They allow the architect to understand the issues – hygiene, cost, surface treatment – without providing him with the final decision keys, which require a conceptual dimension.

This is why the rest of our work will aim at enriching this reference by a second categorisation, centred on:

1. Its application (wall, floor, partition, etc.)
2. The aesthetic and tactile surface treatment (colors and texture),
3. The format and mode of marketing (large or small outlets),
4. Impact on improving quality of life,
5. Safety (does the material pose any risks, like being slippery).

This approach, based on a multi-criteria method, will pave the way for a more contextual selection of materials based on categorization based on conceptual characteristics and categorization based on material characteristics, and will encourage creativity and innovation in their architectural application.

6 Conclusion

Faced with the spread of nosocomial infections, the design of health facilities can no longer be limited to a functional and aesthetic approach only. Therefore, it must adapt a more comprehensive strategy to prevent the transmission of these infections. It is in this sense that the choice of materials during a design becomes an essential step. It will ensure the safety of users against contaminations related to pathogens.

Following this reflection, our research focused on the issues that users face in health facilities. From the analysis we have all design within these spaces must be focused on preserving the safety of users, ensure the ergonomics of space, take into account the impact of space on the comfort of users. In addition, adapted design not adequate in a sanitary space generates major risk. In a post-pandemic period, this research makes sense. It aims to anticipate health safety failures from the earliest stages of design. And it is through the identification of inappropriate materials that can be applied within these sanitary spaces. It is in this perspective that the present research lays the foundations for identifying the characteristics of materials from a technical point of view, with the aim of creating a database intended to create a design aid tool. However, this identification remains partial in terms of architecture. Indeed, the categories defined according to the applications of the materials should integrate both technical and conceptual criteria, in order to better support architects and designers in the design of adapted sanitary spaces. This dimension will be further developed in our future research.

By placing scientific research at the heart of the architectural decision, this study paves the way for a safer and more sustainable hospital design. The future of health design lies in its ability to combine scientific rigor, material innovation and human commitment. The future tool will not be a simple technical support: it aims to be an actor of change, serving public health and collective well-being.

References

1. McFee, R.B.: Nosocomial or hospital-acquired infections: An overview. Dis. Mon. **55**(7), 422–438 (2009). https://doi.org/10.1016/j.disamonth.2009.03.014
2. Kong, C., Wang, X., Li, H., Zhou, H., Yang, L.: Advanced antibacterial materials for the prevention of nosocomial infections. Nanomed. Res. J. **3**(2), 97–108 (2024). https://doi.org/10.1097/NR9.0000000000000056
3. Dey, N. et al.: Graphene materials: Armor against nosocomial infections and biofilm formation – A review. Environ. Res. **214**(Part 2), 113867 (2022). https://doi.org/10.1016/j.envres.2022.113867
4. Baranovsky, S.: Circulation et persistance de pathogènes nosocomiaux multirésistants et hautement résistants émergents dans l'environnement hospitalier: Complexité des unités de transmission. Thesis, Université Montpellier (2020). https://theses.hal.science/tel 03118483/
5. Cissé, M.A.O. Évaluation des mesures de prévention des infections nosocomiales au CHU de Kati. Thèse de doctorat, Université des Sciences, des Techniques et des Technologies de Bamako, (2014) https://www.bibliosante.ml/bitstream/handle/123456789/539/14M121.pdf?sequence=1&is Allowed=y
6. Zavadskas, E.K., Turskis, Z., Kildienė, S.: State of art surveys of overviews on MCDM/MADM methods. Technol. Econ. Dev. Econ. **20**(1), 165–179 (2014). https://doi.org/10.3846/20294913.2014.892037

7. Monneret, C.: Le cuivre contre les infections nosocomiales. Société Chimique de France. https://new.societechimiquedefrance.fr/wp-content/uploads/2019/12/2015-392 janvier-p7-monneret-hd.pdf (2015)
8. Debiemme-Chouvy, C.: 17 Les surfaces antimicrobiennes: un atout dans la lutte contre le développement des biofilms. In: Interactions Matériaux-Microorganismes: Bétons et métaux plus résistants à la biodétérioration, pp. 371–394. EDP Sciences, Les Ulis (2017). https://doi.org/10.1051/978-2-7598-1877-8.c019
9. Yong, L.X., Calautit, J.K.: A Comprehensive Review on the Integration of Antimicrobial Technologies onto Various Surfaces of the Built Environment. Sustainability. **15**(4), 3394 (2023). https://doi.org/10.3390/su15043394
10. Sharpe, P.A., Schmidt, M.G.: Control and mitigation of healthcare-acquired infections: Designing clinical trials to evaluate new materials and technologies. HERD. **5**(1), 94–115 (2011). https://doi.org/10.1177/193758671100500109
11. Jessica, J.: Développement d'une technologie de surface bactéricide à base d'aluminium. CORE Research Archive (2023). https://core.ac.uk/download/pdf/553633222.pdf
12. Bouchart, F.: Biomatériaux céramiques et lutte contre les infections nosocomiales en chirurgie ostéo-articulaire. Université Polytechnique des Hauts-de-France (2023). https://uphf.hal.science/tel-04291352/
13. Marfoglia, A.: Développement, caractérisation et évaluation biologique d'une bioencre antimicrobienne pour des applications en bioimpression 3D. Thesis, Theses.fr (2024). https://theses.fr/2024TLSES064
14. Freitas, J., et al.: Consensus-based guidelines for best practices in the selection and use of examination gloves in healthcare settings. Nurs. Rep. **15**(1), 9 (2025) https://pmc.ncbi.nlm.nih.gov/articles/PMC11767351/
15. Morris, H.V.: Textile based medical devices and their role in sustainable health systems. Keele University Repository. https://keele-repository.worktribe.com/output/956321 (2024)
16. Alhmoud, S.: Sustainability of development and application of nanomaterials in healthcare within hospital settings. https://www.researchgate.net/publication/381843159 ResearchGate (2024)
17. Medical Design Briefs. Nanotexturing creates bacteria-killing spikes on stainless steel surfaces. https://www.medicaldesignbriefs.com/component/content/article/28355-nanotexturing creates-bacteria-killing-spikes-on-stainless-steel-surfaces (2017)
18. https://www.unifiedalloys.com/blog/medical-stainless-steel
19. Montanstahl. Role of stainless steel in the medical and healthcare industries: Protecting human health. https://www.montanstahl.com/blog/the-role-of-stainless-steel-in-protecting-human-health/ (2024)
20. Stuffey. Benefits of stainless steel surfaces in hospitals. https://blog.pattersonpope.com/blog/benefits-of-stainless-steel-surfaces-in-hospitals (2020)
21. Advanced Plastiform. Using PVC in medical supplies and applications. https://advancedplastiform.com/using-pvc-in-medical-supplies-and-applications/ (2023)
22. Schweitzer, M., Gilpin, L., Frampton, S.: Healing spaces: Elements of environmental design that make an impact on health. J. Altern. Complement. Med. **10**(Suppl 1), S71–S83 (2004) https://pubmed.ncbi.nlm.nih.gov/15630824/
23. Salgado, C.D., Sepkowitz, K.A., John, J.F., Cantey, J.R., Attaway, H.H.: Copper surfaces reduce the rate of healthcare-acquired infections in intensive care units. Infect. Control Hosp. Epidemiol. **34**(5), 479–486 (2013). https://doi.org/10.1086/670207
24. Jose, A., Gizdavic-Nikolaidis, M., Swift, S.: Antimicrobial coatings: Reviewing options for healthcare applications. Appl. Microbiol. **3**(1), 145–174 (2023). https://doi.org/10.3390/applmicrobiol30100128007/3/1/12

25. Kramer, A., Schwebke, I., Kampf, G.: How long do nosocomial pathogens persist on inanimate surfaces? A systematic review. BMC Infect. Dis. **6**(1), 130 (2006). https://doi.org/10.1186/1471-2334-6-130
26. Jerman, M., & Žnidaršič, A. : Comparison of primary laundering effects on hospital textiles by various laundering procedures. ResearchGate (2006)

Chemical Extraction of Cellulosic Fibers from Forsskaolea tenacissima: Characterization, and Application in Dye Adsorption

Bouthaina Missaoui[1(✉)], Mahjoub Jabli[1,2], and Saber Ben Abdessalem[1]

[1] Textile Materials and Processes Research Unit, Tunisia National Engineering School of Monastir, University of Monastir, Monastir, Tunisia
Missaouibouthaina33@gmail.com

[2] Department of Chemistry, College of Science, Majmaah University, Al-Majmaah 11952, Saudi Arabia

Abstract. Cellulose is an interesting biopolymer due to its employment in many motivating applications such as composites, pharmacy, textiles, decolorization, etc. The current work describes the extraction of fibrillated cellulose from Forsskaolea tenacissima leaves using alkalization and bleaching processes. The techniques of SEM, XRD, FT-IR, and TGA-DTA were used to analyze the untreated substrates and the extracted fibrillated cellulose. FT-IR data indicated that, after alkalization and bleaching processes, lignin and hemicellulose initially present in untreated leaves, were removed. The crystallinity index [CrI (%)] values calculated for extracted fibrillated cellulose and Forsskaolea tenacissima leaves were equal to 42.85% and 23.2%, respectively. The high CrI value calculated for the extracted fibers demonstrated the removal of the amorphous components in untreated leaves. TGA analysis showed that alkalization and bleaching processes affected the thermal events. At equilibrium, the adsorption capacities of bleached fibers were 220 mg/g and 185 mg/g for methylene blue (MB) and crystal violet (CV), respectively. The adsorption process was exothermic and non-spontaneous. Overall, the adsorption results confirmed that the extracted fibers could be used as effective adsorbents of cationic dyes from water

Keywords: Forsskaolea tenacissima · Fibrillated cellulose · Bleaching · Methylene blue · Crystal violet

1 Introduction

This study explores the extraction of cellulose from Forsskaolea tenacissima leaves, a plant native to arid regions, using alkalization and bleaching techniques. Cellulose, a renewable and abundant biopolymer, is highly valued for its applications in composites, textiles, and environmental remediation. The research addresses the urgent need for sustainable and cost-effective adsorbents to remove hazardous cationic dyes, such as methylene blue (MB) and crystal violet (CV), from wastewater. Traditional dye removal methods, including coagulation and chemical oxidation, often suffer from drawbacks like high energy consumption and sludge formation. In contrast, adsorption stands out as an eco-friendly and efficient alternative.

M. Hamdaoui et al. (Eds.): ITDM'3 2025, PTT, pp. 26–30, 2026.
https://doi.org/10.1007/978-981-95-7704-0_3

2 Methods

The cellulose extraction process involved treating the leaves with a 5 wt.% NaOH solution at 80 °C for 2 h to remove lignin and wax, followed by bleaching with a hydrogen peroxide and acetic acid mixture at 90 °C to obtain pure cellulose (Fig. 1).

Fig. 1. Photographs demonstrating Forsskaolea tenacissima plant and extracted fibrillated cellulose

The extracted cellulose was characterized using scanning electron microscopy (SEM), X-ray diffraction (XRD), Fourier-transform infrared spectroscopy (FT-IR), and thermogravimetric analysis (TGA-DTA). Batch adsorption experiments were conducted to evaluate the removal efficiency of MB and CV under varying conditions: pH (3–9), contact time (0–120 min), initial dye concentration (0–600 mg/L), and temperature (23–53 °C). Dye concentrations were measured using UV-Vis spectroscopy at wavelengths of 660 nm for MB and 590 nm for CV.

3 Results and Discussion

3.1 Characterization

FT-IR analysis (Fig. 2) confirmed the successful removal of lignin and hemicellulose, as indicated by the disappearance of the peak at 1733 cm^{-1}. The persistence of characteristic cellulose peaks, such as the hydroxyl group stretch at 3292 cm^{-1}, confirmed the preservation of the cellulose structure.

SEM images (Fig. 3) revealed that untreated leaves had rough, compact fibers covered with non-cellulosic materials, whereas the extracted fibers were homogeneous, well-separated, and exhibited slight deformation due to chemical treatments.

XRD patterns (Fig. 4) identified cellulose I crystallinity, with prominent peaks at 17°, 23°, and 35°. The crystallinity index (CrI) increased from 23.2% in raw leaves to 42.85% in extracted cellulose, demonstrating the effective removal of amorphous components like hemicellulose and lignin.

TGA-DTA curves (Fig. 5) showed that raw leaves underwent four-stage degradation, with the final stage at 490 °C corresponding to crystalline cellulose decomposition. In contrast, bleached cellulose exhibited a single major decomposition event at 351.5 °C and a higher total weight loss (98%), attributed to the elimination of lignin, a hydrophobic component.

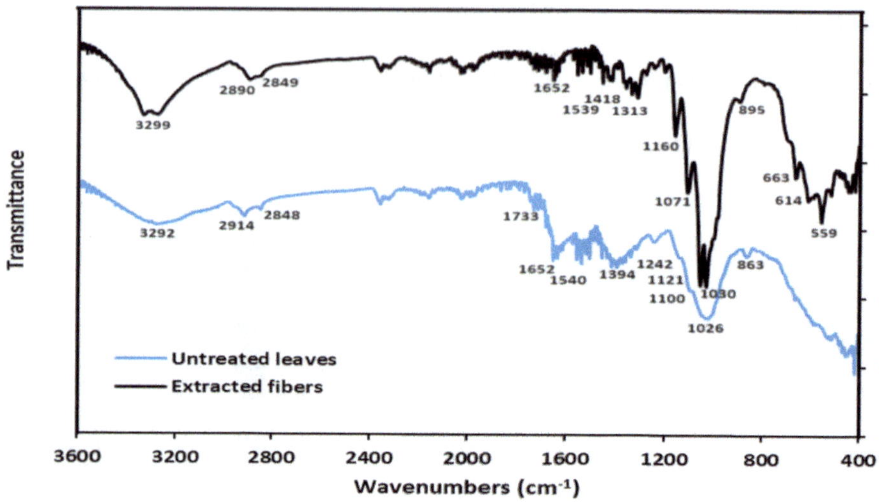

Fig. 2. FT-IR spectroscopy of Forsskaolea tenacissima leaves and extracted fibrillated cellulose

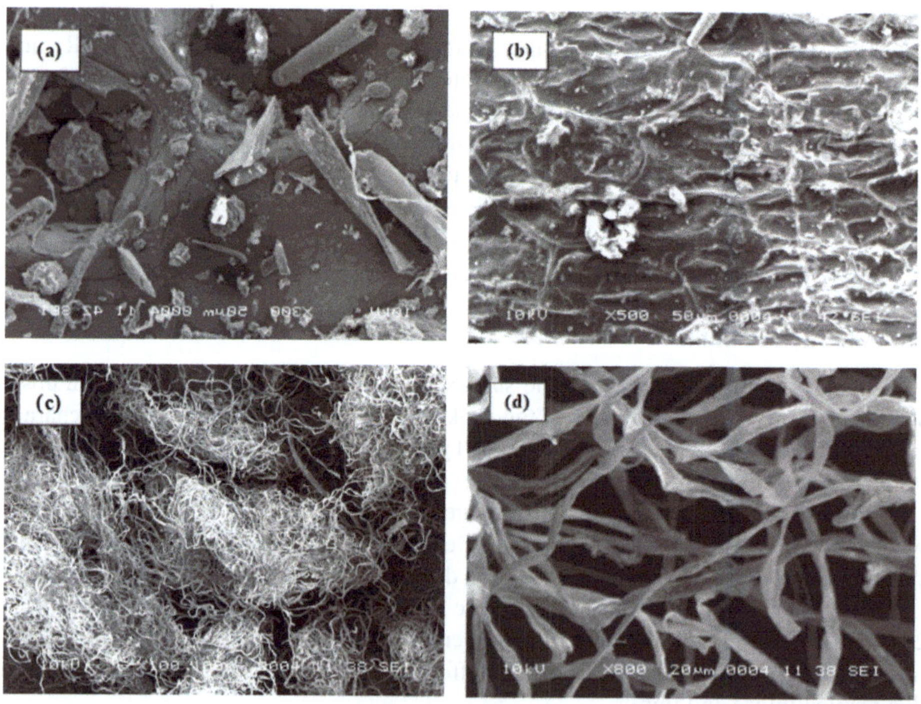

Fig. 3. SEM pictures of: (a, b) untreated Forsskaolea tenacissima leaves, and (c, d) extracted fibrillated cellulose

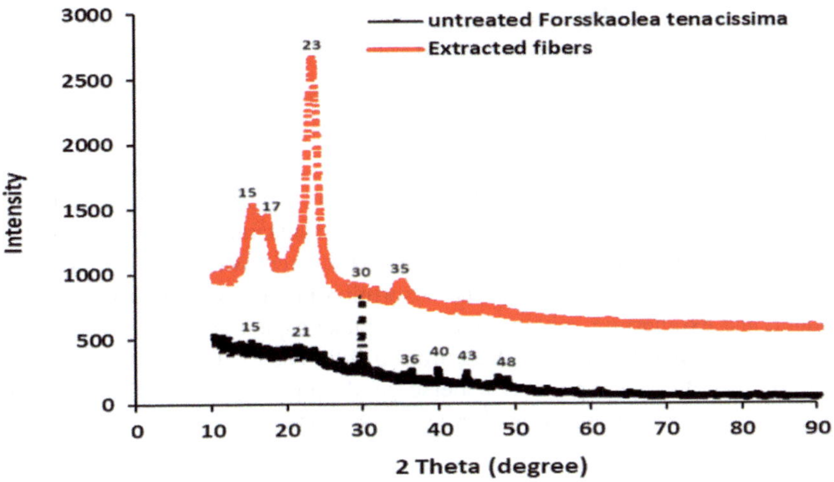

Fig. 4. XRD patterns of untreated Forsskaolea tenacissima leaves, and extracted fibrillated cellulose

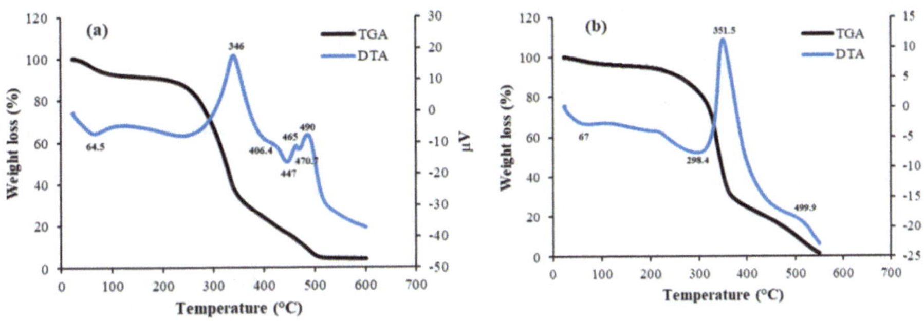

Fig. 5. TGA/DTA curves of: (a) untreated Forsskaolea tenacissima leaves, and (b) extracted fibrillated cellulose

3.2 Adsorption Performance

Effect of pH: Maximum adsorption capacities for MB and CV were achieved at pH 6. In acidic conditions, the positively charged dye ions repelled the adsorbent surface, while at higher pH, the negatively charged cellulose surface facilitated electrostatic attraction. Kinetic Studies: Equilibrium was reached within 40 min. The adsorption data fitted well with the pseudo-second-order model ($R^2 > 0.99$) and intraparticle diffusion models, suggesting chemisorption and multilayer adsorption mechanisms.

Isotherm Analysis: The Freundlich isotherm ($R^2 \geq 0.98$) provided the best fit, indicating heterogeneous adsorption surfaces. The maximum adsorption capacities were 220 mg/g for MB and 185 mg/g for CV at 23 °C, outperforming many reported adsorbents, such as acid-treated black cumin seed powder (73.5 mg/g) and corncob cellulose-based hydrogel (55.3 mg/g). Thermodynamic Parameters: Negative enthalpy values ($\Delta H° = -6.69$ kJ/mol for MB, 6.56 kJ/mol for CV) confirmed the exothermic nature of adsorption. Positive Gibbs free energy values ($\Delta G^\circ = 1.98$–3.27 kJ/mol) indicated a non-spontaneous process requiring external energy.

4 Conclusion

The study successfully extracted high-purity cellulose from Forsskaolea tenacissima leaves, demonstrating its potential as an effective and sustainable adsorbent for cationic dyes. Comprehensive characterization confirmed the removal of non-cellulosic components and enhanced crystallinity. The adsorption process was highly efficient, pH-dependent, and exothermic, with notable dye uptake capacities. Future research could explore chemical modifications of the extracted cellulose to broaden its application for removing other pollutants. This work underscores the potential of agricultural byproducts in developing eco-friendly solutions for water purification.

References

1. Atoui, S., Belaadi, A., Xian Chai, B., Abdullah, M.M.S., Al Khawlani, A., Ghernaout, D.: Extracting and characterizing novel cellulose fibers from Chamaerops humilis rachis for textiles' sustainable and cleaner production as reinforcement for potential applications. Int. J. Biol. Macromol. **276**(Part 2), 134029 (2024)
2. Ferfari, O., Belaadi, A., Bedjaoui, A., Alshahrani, H., Khan, M.K.A.: Characterization of a new cellulose fiber extracted from Syagrus Romanzoffiana rachis as a potential reinforcement in biocomposites materials. Mater. Today Commun. **36**, 106576 (2023)
3. Hadou, A., Belaadi, A., Alshahrani, H., Khan, M.K.A.: Extraction and characterization of novel cellulose fibers from Dracaena draco plant. Mater. Chem. Phys. **313**, 128790 (2024)
4. Gheribi, H., et al.: Statistical study of the mechanical behavior of the new fiber from the strelitzia juncea plant fibers: Application in ecological yarns. J. Nat. Fibers. **21**(1), 21–40 (2024)
5. Teyar, S., et al.: Analyzing the Strelitzia juncea cellulosic fibers mechanical properties' experimental data using various statistical methods. J. Nat. Fibers. **21**(1), 1–20 (2024)
6. Lalaymia, I., Belaadi, A., Bedjaoui, A., Alshahrani, H., Khan, M.K.A.: Extraction and characterization of fiber from the flower stalk of the Agave plant for alternative reinforcing biocomposite materials. Biomass Convers. Biorefinery. **14**, 32271–32287 (2023)
7. Faruk, O., Bledzki, A.K., Fink, H.-P., Sain, M.: Biocomposites reinforced with natural fibers: 2000–2010. Prog. Polym. Sci. **37**, 1552–1596 (2012)
8. Gurukarthik Babu, B., Prince Winston, D., Senthamarai Kannan, P., Saravanakumar, S., Sanjay, M.: Study on characterization and physicochemical properties of new natural fiber from Phaseolus vulgaris. J. Nat. Fibers. **16**, 1035–1042 (2019)

Microcapsules Properties and Their Fixation on Gauze for Medical Applications: Influence of Solution and Process Parameters

Khmais Zdiri[1,2](✉), Hend Felhi[1], Ayda Baffoun[1], and Tarek Baati[3]

[1] Unité de Recherche Matériaux et Procédés Textiles (MPTex), Université de Monastir, ENIM, UR17ES33, 5019 Monastir, Tunisie
khmaiszdiri@gmail.com

[2] Centre de Recherche en Sciences et Technologie de Textile à Monastir (CRSTTM), Université de Monastir, 5000 Monastir, Tunisie

[3] Laboratoire des Substances Naturelles, Institut National de Recherche et d'Analyse Physico-chimique (INRAP), BiotechPole Sidi Thabet, Ariana, Tunisia

Abstract. The microencapsulation of essential oils represents a promising approach to addressing challenges in the healing of chronic wounds. This study aims to elaborate high-performance biopolymer microcapsules and to fix them to gauze dressings for biomedical applications. The influence of various formulation and process parameters on the physical and functional properties of the microcapsules was investigated. These parameters included clay concentration, crosslinking agent concentration (calcium chloride ($CaCl_2$)), the molecular weight of sodium alginate, as well as stirring speed and duration. A kinetic study was conducted to evaluate the encapsulation efficiency and release profile of the essential oil within the microcapsules. Spectrophotometric analysis revealed that optimal encapsulation efficiency was achieved with a moderate clay concentration, medium molecular weight alginate, low $CaCl_2$ concentration, and gentle stirring for a short duration.

The microcapsules were applied to gauze dressings using the exhaustion method with different types of binders. Synthetic binders, such as polyurethane and polyacrylic, as well as natural binders, including sodium alginate and chitosan, were tested. Morphological analysis of the fixed microcapsules was conducted to assess their distribution and integrity on the gauze dressings. The kinetic study demonstrated that the binder type plays a key role in the essential oil release profile. Notably, natural binders facilitated a rapid release, whereas synthetic binders resulted in a more sustained release. However, this study is limited by the absence of clinical trials, which would involve applying the dressings to patients to assess potential allergic reactions, irritations, or other adverse effects. These investigations will be the focus of a future study.

Keywords: Microencapsulation · Essential oils · Alginate · Clay · Biomedical Applications

M. Hamdaoui et al. (Eds.): ITDM'3 2025, PTT, pp. 31–46, 2026.
https://doi.org/10.1007/978-981-95-7704-0_4

1 Introduction

Wound healing is a complex process that can be impaired in the presence of deep, extensive wounds or those associated with vascular, metabolic, or neuropathic disorders (e.g., chronic vascular or diabetic wounds). In such cases, delayed healing increases the risk of infections, potentially leading to severe complications or even amputations. Chronic wounds thus represent a major medical challenge. While various solutions exist to support healing, those that directly promote tissue regeneration often rely on natural tissue grafts, which are complex and costly to obtain. Additionally, some advanced bioactive dressings that combine cells, matrices, and growth factors remain difficult to access due to regulatory constraints [1, 2].

This study aims to develop microcapsules based on naturally derived biopolymers that are readily available and cost-effective. These microcapsules will incorporate components designed to enhance wound healing, stimulate cell migration, and promote neovascularization while mimicking the functions of the extracellular matrix. Enriched with active ingredients that support tissue regeneration and enhance the immune response against infections, these microcapsules will be used to functionalize gauze dressings.

To optimize the performance of microcapsules, it is essential to investigate the parameters influencing the microencapsulation process, including formulation, mixing conditions, and processing methods. This constitutes one of the key objectives of this study. In this work, Tunisian clay is used as a reinforcing agent. Its application, still relatively unexplored in the field of biocomposites, not only enhances the mechanical properties of the microcapsules but also modulates the release profile of the encapsulated active agents [3].

In this study, alginate-based microcapsules incorporating clay and essential oil were synthesized via ionic gelation. The essential oil was selected as the active ingredient, and the resulting microcapsules were immobilized onto gauze dressings using the exhaustion method. Widely recognized for their ability to meet the stringent requirements of the medical field, gauze dressings offer outstanding performance, including high efficiency, excellent air permeability, superior comfort, remarkable tear resistance, and effective control of cross-contamination risks. They also play a crucial role in stabilizing the developed microcapsules, enabling their use as high-quality bandages for the treatment of complex burns. These functionalized dressings are specifically designed for medical applications, with a particular focus on complex wound management [4, 5].

2 Experimental

2.1 Materials

As part of our study, we performed a comparative analysis of Low Molecular Weight (LMW) and Medium Molecular Weight (MMW) sodium alginate (Table 1). This strategy allowed us to gain an in-depth understanding of the influence of molecular weight variations on the microencapsulation processes.

The clay used in our study is a Tunisian clay sourced from the Khanguet Rihana region, located in central Tunisia, less than 25 km south of the Regueb area. The properties of the clay used in this study are presented in the Table 2.

Table 1. Physicochemical properties of medium and low molecular weight sodium alginate.

Property	Sodium Alginate (Low Molecular Weight)	Sodium Alginate (Medium Molecular Weight)
Chemical Formula	$(C_6H_7O_6Na)n$	
pH	5–8	5.5–8
Appearance	Color: white to beige Form: powder	Color: white Form: powder
Viscosity	5–40 mPas	350–550 mPas
Brand	SIGMA-ALDRICH	Natriumalginate Biochemica
Solubility	Soluble in water	Soluble in water
Loss on Drying	< 15.5%	< 15%

Table 2. Physicochemical Characteristics of the Tunisian Clay [6].

Clay Family	Chemical Formula	Physicochemical Properties
Smectite	$(Na,Ca)_{0.33}(Al,Mg)_2(Si_4O_{10})(OH)_2 \cdot nH_2O$	Internal surface area: 700–800 m^2/g Dehydroxylation: occurs between 400 and 600 °C Density: 2–2.7

Alkyl ammoniums are compounds frequently used to modify the properties of clay. In the context of the reaction between clay and alginate, we selected TetraButylAmmonium Bromide (TBAB) (supplied by SIGMA ALDRICH) due to its long alkyl chain, which can enhance compatibility with alginate as well as with other organic polymers.

We used Tea Tree essential oil as the active ingredient. Tea Tree essential oil is characterized by its high concentration of terpinen-4-ol (up to 50%) and other monoterpene compounds. These molecules act directly on pathogens by disrupting their cell membranes. Moreover, Tea Tree essential oil is well tolerated by the skin, and its active compounds do not cause irritation, making disinfection painless—unlike alcohol-based disinfectants, which can cause a stinging sensation when applied to open wounds [7].

Tween 80 is a non-ionic surfactant widely used in oil-in-water (O/W) systems for its emulsifying and stabilizing properties, with a hydrophilic-lipophilic balance (HLB) value of 15.0. In our study, it was used at a concentration of 2%.

Gauze compresses were used as the substrate in our study. Their characteristics are presented in the Table 3:

To evaluate the adhesion of microcapsules to the textile substrate, we tested four types of binders based on polysaccharide polymers, namely sodium alginate and chitosan, as well as two synthetic polymers: polyurethane and polyacrylic.

Table 3. Characteristics of the textile substrate (gauze compress).

Fabric	Composition	Dimension
Compress	100% cotton	7.5*7.5 cm

2.2 Purification and Organo-Modification of Tunisian Clay

The purified and organo-modified clays are prepared according to the protocol used by Zdiri, K. et al. [8].

2.3 Preparation of Microcapsules by Droplet Gelation

Clay, alginate, and essential oil based microcapsules are prepared by slowly adding a solution containing the essential oil, distilled water, and Tween 80 to a mixture of clay and sodium alginate, which is kept under rapid stirring. The suspension is stirred until homogenized. Cross-linking of the microcapsules is then carried out by dropwise introduction of the solution, using a syringe, into a bath containing calcium chloride.

To study the influence of certain solution and process parameters on microcapsule formation, we varied factors such as clay concentration, calcium chloride concentration, sodium alginate molecular weight, and stirring speed and duration, as shown in the Table 4.

Table 4. Designation and composition of prepared samples.

Sample	Alginate Type	Alginate (% w/v)	Clay (% w/v)	Oil (% w/v)	Tween80 (% w/v)	$CaCl_2$ (% w/v)	Agitation (rpm)	Agitation (min)
MICALG1.1	MMW	2	1	2	2	2	720	30
MICALG1.2							720	45
MICALG1.3							720	60
MICALG1.4							960	30
MICALG1.5							960	45
MICALG1.6							960	60
MICALG1.7							1200	30
MICALG1.8							1200	45
MICALG1.9							1200	60
MICALG2	MMW	2	3	2	2	2	720	30
MICALG3	MMW	2	5	2	2	2	720	30
MICALG4	MMW	2	1	2	2	4	720	30
MICALG5	MMW	2	3	2	2	4	720	30
MICALG6	MMW	2	5	2	2	4	720	30

(continued)

Table 4. *(continued)*

Sample	Alginate Type	Alginate (% w/v)	Clay (% w/v)	Oil (% w/v)	Tween80 (% w/v)	$CaCl_2$ (% w/v)	Agitation (rpm)	Agitation (min)
MICALG7	LMW	2	1	2	2	2	720	30
MICALG8	LMW	2	3	2	2	2	720	30
MICALG9	LMW	2	5	2	2	2	720	30

2.4 Determination of Encapsulation Rate

We determined the encapsulation rate of the essential oil using a spectrophotometer. Based on the equation below, we were able to calculate the amount of oil encapsulated within the microcapsules.

$$TE\% = (A_0 - A_1)^* 100/A_0 \ or \ TE\% = (C_0 - C_1)^* 100/C_0 \quad (1)$$

A_0: the absorbance of the initial solution; C_0: the initial concentration of the stock solution.

A_1: the absorbance of the crosslinking solution; C_1: the concentration of the crosslinking solution.

2.5 Fixation of Microcapsules to Gauze Pads

Table 5. Microcapsule Application Protocol on Gauze Compresses.

Distilled Water (mL)	Microcapsule Amount (g)	Synthetic Binders		Natural Binders	
		Polyurethane	Polyacrylic	Alginate	Chitosan
100	0.2				
Concentration (%)		10		2	
Stirring Speed (rpm)		30		30	
Stirring Duration (min)		15	15	30	
Drying Temperature (°C)		120	140	25	
Drying Time (min)		15	4	1440	

We opted for the exhaustion method to apply microcapsules onto gauze compresses. This technique involves immersing the compresses in a solution containing distilled water, synthetic or natural binding agents, and microcapsules. Through gentle agitation and the application of heat, the microcapsules become firmly attached to the textile substrate. The data related to the parameters used to evaluate the essential oil release process are summarized in the Table 5.

2.6 Morphological Properties

A scanning electron microscope (SEM) analysis was performed using an FEI Quanta 250 microscope equipped with a Thermo Fisher EDX spectrometer. This instrument is installed at the National Engineering School of Monastir (ENIM). It was used to analyze the surface morphology of gauze compresses onto which microcapsules were fixed. The specimens were mounted on a sample holder using adhesive carbon tape and metallized with gold. SEM images were analyzed using Sigma Scan Pro image analysis software.

2.7 Spectrophotometric Analysis

We used the spectrophotometer to conduct a kinetic study aimed at determining the encapsulation rate of the oil within alginate–clay-based microcapsules, as well as the release rate of the essential oil. Absorbance measurements were taken every 27 min throughout the crosslinking reaction.

3 Results and Discussion

3.1 Characterization of the Elaborated Microcapsules

3.1.1 Microcapsule Diameter

The diameter of the elaborated microcapsules was evaluated using the optical microscope "Projectina." This analysis also enabled the calculation of the coefficient of variation (CV), which serves as a key indicator for assessing the dimensional uniformity of the microcapsules (see Fig. 1). Indeed, size consistency is a critical parameter that directly influences the behavior, performance, and efficiency of microcapsules in various industrial or biomedical applications.

The CV expresses the dispersion of data relative to the mean:

A low CV ($< 5\%$) indicates a relatively homogeneous size distribution, which is generally desirable.

Conversely, a high CV reflects significant heterogeneity, which may adversely affect the reproducibility and functional properties of the microcapsules.

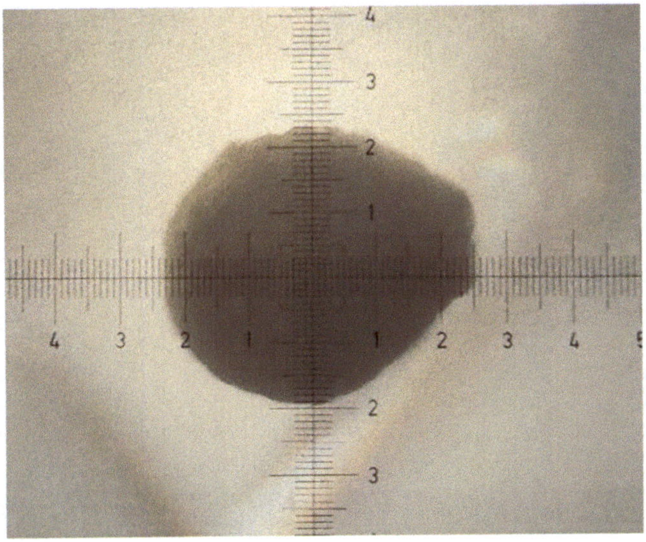

Fig. 1. Microcapsule observed under the "Projectina" microscope (magnified view).

To obtain a representative estimate of the size distribution, ten independent measurements were performed. The results are summarized in Table 6:

Table 6. Microcapsule Diameter Measurements.

Diameter (μm)	100	85	110	105	95	90	100	102	90	95
Mean	97.2									
Standard Deviation	7.64									
CV (%)	7.86									

The analysis of these data reveals a relatively dispersed diameter distribution, with a mean value of 97.2 μm and a coefficient of variation of 7.86%, which exceeds the commonly accepted threshold of 5%. These results indicate a certain degree of variability in the morphology of the microcapsules, which are generally spherical but exhibit irregular shapes.

This lack of uniformity is mainly attributed to the manufacturing method used—namely, the drop-by-drop gelation technique. Although easy to implement, this method presents limitations in terms of precise control over the size and shape of the microcapsules, which may pose a constraint depending on the requirements of the intended application.

3.1.2 Morphological Analysis

The scanning electron microscopy (SEM) images of the elaborated microcapsules are presented in Fig. 2.

Fig. 2. SEM images of the elaborated microcapsules.

A general observation of the microcapsules reveals that they predominantly exhibit a spherical morphology.

However, a more detailed analysis of the SEM images highlights a certain degree of heterogeneity in both the size and shape of the microcapsules. Some capsules appear slightly deformed or show significant dimensional deviations. These observations could be attributed to an instability in the flow rate during the droplet gelation fabrication process. Indeed, an irregular flow rate can result in droplets of varying sizes prior to solidification, which directly affects the size distribution and morphological uniformity of the resulting microcapsules.

Moreover, factors such as the viscosity of the solution, the droplet fall height, and the gelation conditions (e.g., crosslinking agent concentration, and the speed and duration of agitation) could also influence the process stability and contribute to the observed defects. Therefore, optimization of these parameters is necessary to improve the uniformity of the microcapsules, ensure better reproducibility of the process, and, consequently, guarantee consistent performance for the intended applications.

3.2 Effect of Process and Solution Parameters on Microencapsulation

3.2.1 Impact of Stirring Speed and Duration

The influence of stirring speed and duration on the encapsulation efficiency of microcapsules is illustrated in the Fig. 3.

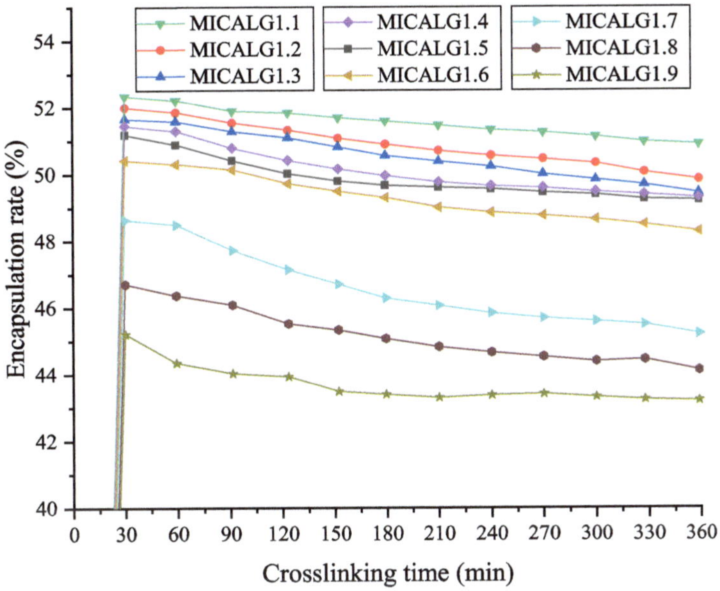

Fig. 3. Microencapsulation Kinetics as a Function of Stirring Speed and Stirring Time.

Following the interaction between calcium chloride and the mixture of sodium alginate, clay, and essential oil, the formation of microcapsules was observed after 30 min. This phenomenon is primarily attributed to the kinetics of the crosslinking reaction. At the onset of the reaction, Ca^{2+} ions interact with the uronic acid groups present in the alginate, progressively strengthening the structure of the forming microcapsules [9].

After 30 min, a general slowdown of the crosslinking reaction is observed, leading to a gradual decrease in the encapsulation efficiency of the essential oil. This can be explained by several factors: the slower diffusion of Ca^{2+} ions through an increasingly dense matrix, and the progressive saturation of available crosslinking sites [10].

Regarding the influence of stirring speed, the results indicate that essential oil encapsulation is less efficient at a high stirring speed (1200 rpm). Conversely, reducing the stirring speed significantly improves the oil fixation rate. Indeed, a high stirring speed can cause coalescence of the oil droplets, leading to phase separation and thus hindering the formation of a stable emulsion necessary for effective encapsulation. Moreover, excessive agitation can disrupt the structuring of the matrix around the oil droplets, compromising the stability of the formed microcapsules [11].

Additionally, a higher encapsulation rate is achieved with a shorter stirring duration (30 min). Increasing the stirring time leads to a decrease in encapsulation efficiency, likely due to enhanced coalescence of the oil droplets, which reduces their interaction surface with the encapsulating agents. Furthermore, prolonged mixing can cause degradation of the formulation materials, weakening their ability to form a stable matrix around the essential oil and thereby negatively impacting the overall encapsulation efficiency [12].

For the remainder of the study, microcapsules were prepared using a stirring speed of 720 rpm for a duration of 30 min.

3.2.2 Impact of Alginate Molecular Weight

The efficiency of the oil encapsulation process also depends on the molecular weight of the polymer used. A kinetic study of the encapsulation process was conducted by varying the molecular weight of sodium alginate, as illustrated in the Fig. 4.

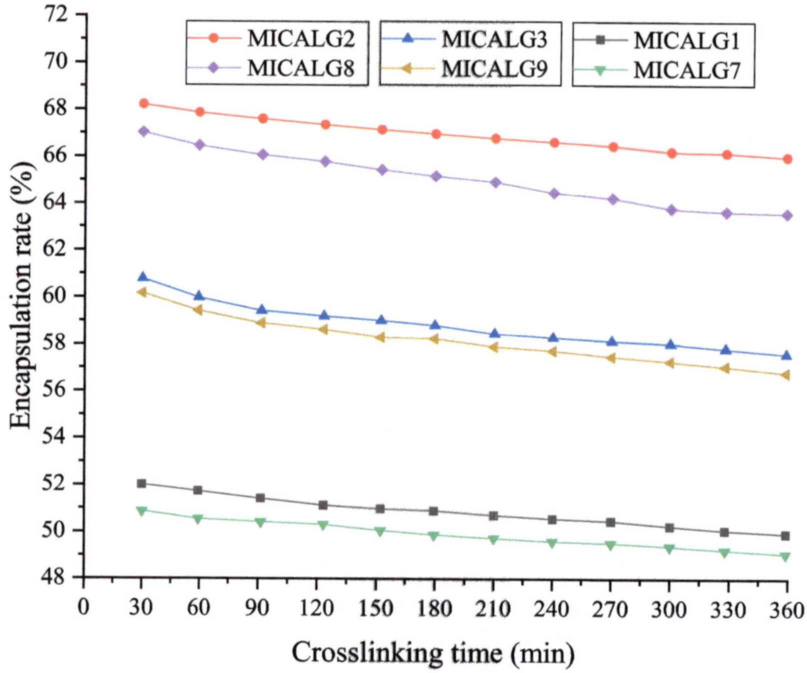

Fig. 4. Encapsulation kinetics for different alginate molecular weights.

The results show that sodium alginate with a medium molecular weight delivers remarkable performance and proves to be the most suitable for essential oil encapsulation.

The variation in the molecular weight of sodium alginate in alginate-clay-based microcapsules significantly influences the oil encapsulation rate due to its impact on membrane permeability. Low molecular weight alginates tend to form more porous membranes, facilitating the leakage of oil from the microcapsules and thereby reducing the encapsulation efficiency. In contrast, medium molecular weight alginates create denser membranes, enhancing oil retention within the microcapsules and improving the overall encapsulation efficiency [13].

3.2.3 Impact of Clay Concentration

The study of the encapsulation kinetics as a function of clay concentration is illustrated in the Fig. 5.

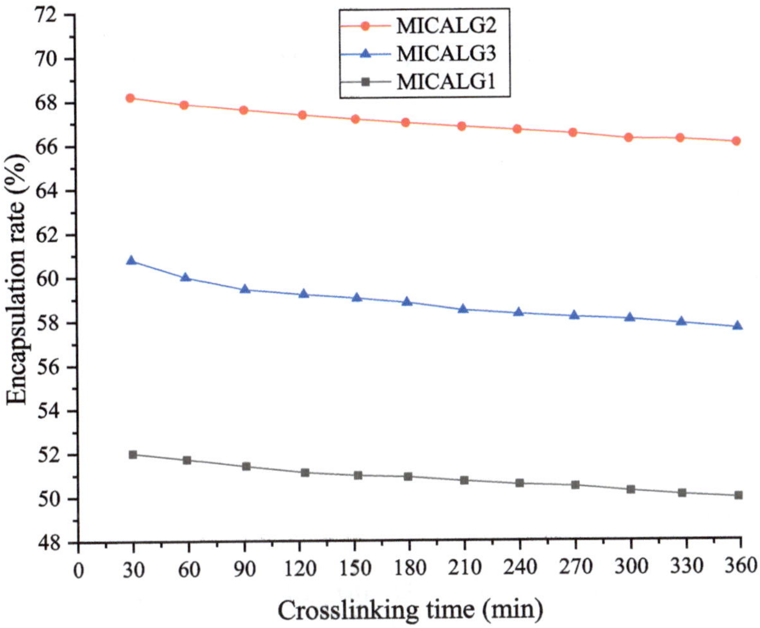

Fig. 5. Encapsulation kinetics at different clay concentrations.

The MICAALG2 sample, formulated with 2% alginate, 3% clay, and 2% $CaCl_2$, exhibited the highest essential oil encapsulation capacity. This result can be attributed to an optimal amount of clay particles available to effectively adsorb and trap the oil, thus promoting strong interactions between the oil and the clay, and enabling the formation of a well-balanced matrix with optimal barrier properties [14].

Conversely, a higher clay concentration (5%) led to a decrease in the encapsulation rate, although it remained higher than that observed at a 1% clay concentration. This decline in performance can be explained by the excessive aggregation of clay particles, the increased viscosity of the medium, which hampers oil dispersion, and the reduced oil-clay interactions.

3.2.4 Impact of $CaCl_2$ Concentration

The effect of varying the concentration of the crosslinking agent ($CaCl_2$) on the essential oil encapsulation rate is shown in the Fig. 6.

By varying the $CaCl_2$ concentration, the MICALG2 sample exhibited the highest encapsulation rate. Furthermore, for samples with lower clay concentrations, a significant decrease in oil fixation rates was observed when the $CaCl_2$ concentrations were also low. A reduction in encapsulation efficiency was similarly noted for samples containing 5% clay and 4% $CaCl_2$.

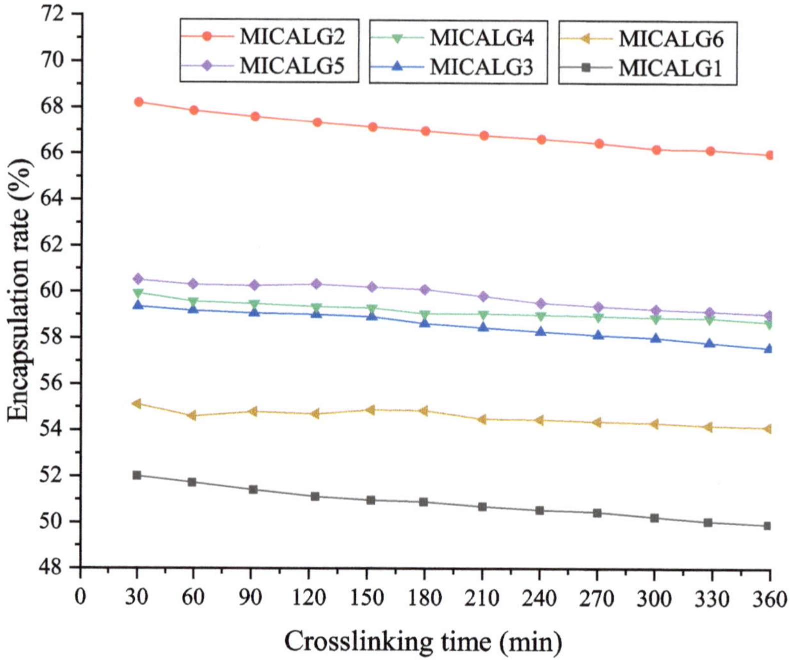

Fig. 6. Encapsulation kinetics at different $CaCl_2$ concentrations.

This decline can be attributed to the acceleration of the gelation process caused by the increased $CaCl_2$ concentration, leading to the formation of thicker, more rigid, and compact gel structures. Such structures hinder the uniform distribution of the oil within the matrix, thereby reducing encapsulation efficiency. Additionally, a higher $CaCl_2$ concentration promotes the aggregation of clay and alginate particles, further complicating the homogeneous dispersion of the oil [15].

Based on these results, the optimal conditions for essential oil encapsulation are determined to be: 3% clay concentration, 2% $CaCl_2$ concentration, a stirring speed of 720 rpm, and a stirring time of 30 min. These parameters, which have shown to deliver the best performance, will be adopted for the continuation of this work.

3.3 Application of Microcapsules on a Textile Support

In this section, the exhaustion method was employed to fix the microcapsules onto a cotton gauze compress. The evaluation of the fixation of sodium alginate– and clay–based microcapsules onto the cotton compress was carried out through:

- SEM analysis to assess the adhesion of microcapsules on the compress surface.
- Diffusion kinetics studies using spectrophotometry to investigate the release mechanism of the essential oil from the microcapsules fixed onto the compress.

3.3.1 Morphological Analysis

SEM images of the essential oil–loaded microcapsules fixed onto textile supports using polyurethane and polyacrylic as binding agents are presented in Fig. 7.

No SEM images were recorded for the compresses where microcapsules were fixed using alginate or chitosan, as only a small quantity of microcapsules was observed on these samples. Furthermore, the adhesion of the microcapsules was insufficient with these natural binders, as they could easily be detached by light manual rubbing of the compresses.

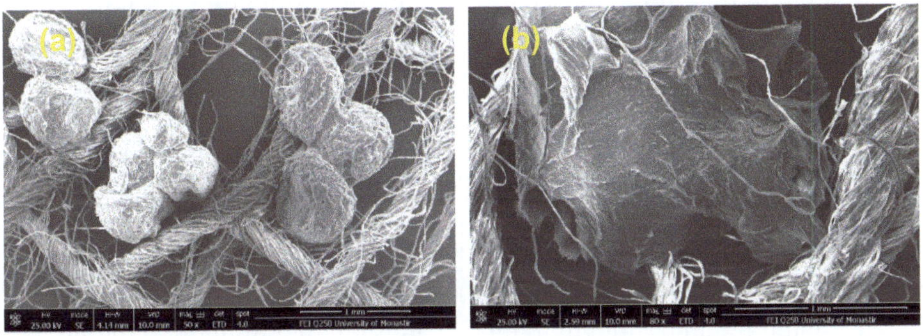

Fig. 7. Microcapsules fixed onto a textile substrate using a polyurethane binder (a) and a polyacrylic binder (b).

Comparative analysis of the SEM images reveals that polyurethane, due to its elastic properties and excellent adhesion capacity, appears to form a thicker coating around the microcapsules. This observation could be attributed to the ability of polyurethane to stretch and closely envelop the microcapsules, resulting in a more pronounced coating.

Conversely, while polyacrylic also exhibits good adhesion, its greater flexibility compared to polyurethane may lead to a thinner and less prominent coating. This difference could stem from a distinct interaction between the microcapsules and polyacrylic, resulting in the formation of a less substantial coating.

3.3.2 Release Kinetics

The objective of the study on the release kinetics of essential oil from microcapsules applied to these compresses was to assess how different fixation mechanisms and parameters influence the release rate of the essential oil.

We analyzed the release rate of the essential oil over time in three different pH environments (pH = 3, pH = 5, and pH = 7), while also evaluating the impact of various binding agents used for microcapsule fixation.

The binding agents included natural polymers such as sodium alginate and chitosan, as well as synthetic polymers like polyurethane and polyacrylic acid.

The results regarding the diffusion of essential oil from the microcapsules fixed onto the compresses, according to the type of binding agent used, are presented in the Fig. 8.

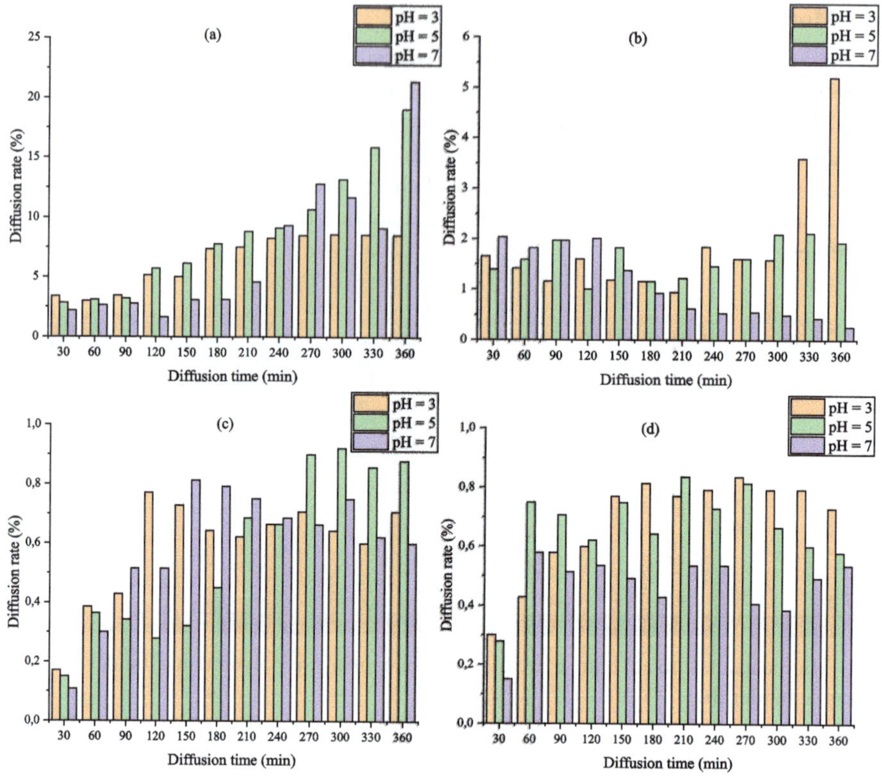

Fig. 8. Diffusion rate of essential oil from microcapsules fixed on a compress using different binding agents: alginate (a), chitosan (b), polyurethane (c), and polyacrylic (d).

The use of sodium alginate and chitosan as binding agents resulted in the highest diffusion rates, despite their poor fixation on the gauze, as evidenced by the morphological analysis.

When using alginate, at pH = 5, the membrane becomes more porous due to interactions between alginate and H^+ ions, promoting faster oil release. With chitosan as the binding agent, the essential oil diffusion was lower compared to alginate but remained higher than with synthetic binders (polyurethane and polyacrylic). Release was faster at lower pH levels, owing to the degradation of chitosan and the microcapsule membrane.

In contrast, the use of synthetic binding agents such as polyurethane and polyacrylic resulted in a lower diffusion rate compared to natural agents, despite their superior fixation performance, as previously demonstrated by morphological analysis. However, polyurethane exhibited significantly higher diffusion at pH = 7, which may be attributed to network swelling at this pH, thereby facilitating the release of molecules [16]. Conversely, polyacrylic promoted greater release at pH = 3, likely due to the ionization of its functional groups under acidic conditions, which enhances molecular diffusion [17].

4 Conclusion and Outlook

The microencapsulation of essential oils emerges as a promising strategy to address the challenges associated with chronic wound healing. The objective of our final year project was to develop high-performance biopolymer microcapsules by studying the various parameters influencing their design. To achieve this goal, we analyzed the impact of several factors on the microencapsulation process.

Sodium alginate was selected as the encapsulating agent due to its gelling properties, and clay was chosen as a reinforcing agent. The droplet gelation technique was employed for the microencapsulation process. In order to optimize the procedure, we investigated the encapsulation kinetics by evaluating the effects of various parameters, including stirring speed and duration, clay concentration, crosslinking agent concentration (calcium chloride), and the molecular weight of sodium alginate. Spectrophotometric monitoring of the encapsulation rate revealed that the best performance was achieved with a medium clay concentration (3%), a medium molecular weight alginate, a low $CaCl_2$ concentration (2%), a moderate stirring speed (720 rpm), and a duration of 30 min.

In the final phase of our study, we applied the microcapsules onto gauze compresses using the exhaustion method. Various binders, both synthetic (polyurethane, polyacrylic) and natural (sodium alginate, chitosan), were tested to fix the microcapsules onto the fabric. Morphological analysis of the fixed microcapsules, performed using SEM, provided a visual assessment of their distribution and integrity on the gauze. The study of oil release kinetics showed distinct release profiles depending on the type of binder: a rapid release was observed with natural binders, contrasting with the more gradual release achieved with synthetic binders.

Finally, although this study enabled the synthesis and characterization of the microcapsules as well as the analysis of the release mechanisms of essential oils from functionalized gauze compresses, it now appears relevant to focus on reducing the microcapsule size in order to enhance encapsulation efficiency while maintaining controlled release. To this end, exploring new methods of microcapsule preparation and designing alternative release mechanisms are envisaged. These future developments will also require clinical trials, including the application of compresses to a panel of patients, with particular attention to monitoring allergic reactions, irritation, or other potential adverse effects.

References

1. Frykberg, R.G., Banks, J.: Challenges in the treatment of chronic wounds. Adv. Wound Care. **4**(9), 560–582 (2015). https://doi.org/10.1089/wound.2015.0635
2. Nunan, R., Harding, K.G., Martin, P.: Clinical challenges of chronic wounds: Searching for an optimal animal model to recapitulate their complexity. Dis. Model. Mech. **7**(11), 1205–1213 (2014). https://doi.org/10.1242/dmm.016782
3. Zdiri, K., Elamri, A., Hamdaoui, M., Khenoussi, N., Harzallah, O., Brendle, J.M.: Impact of Tunisian clay nanofillers on structure and properties of post-consumer polypropylene-based nanocomposites. J. Thermoplast. Compos. Mater. **32**(9), 1159–1175 (2019). https://doi.org/10.1177/0892705718792377

4. Liakos, I. et al.: Fibrous wound dressings encapsulating essential oils as natural antimicrobial agents. J. Mater. Chem. B. **3**(8), 1583–1589 (2015)
5. Wang, H. et al.: Antibacterial polysaccharide-based hydrogel dressing containing plant essential oil for burn wound healing. Burns Trauma. **9**, tkab041 (2021)
6. Jlassi, K., Ben Rhaiem, M., Amara, B.H.A.: Characterization and physicochemical properties of Tunisian clay minerals: Application to an industrial scale. Appl. Clay Sci. **28**(1–4), 155–165 (2005)
7. Hammer, K.A., Carson, C.F., Riley, T.V.: Antimicrobial activity of essential oils and other plant extracts. J. Appl. Microbiol. **86**(6), 985–990 (1999). https://doi.org/10.1046/j.1365-2672.1999.00780.x
8. Zdiri, K., Elamri, A., Hamdaoui, M., Harzallah, O., Khenoussi, N., Brendlé, J.: Valorization of post-consumer PP by (un)modified Tunisian clay nanoparticles incorporation. Waste Biomass Valor. **11**(5), 2285–2296 (2020). https://doi.org/10.1007/s12649-018-0427-2
9. Shanli, O., Olukman, M.: Encapsulated eucalyptus oil in ionically cross-linked alginate microcapsules and its controlled release. Waste Biomass Valor. **6**(5), 839–847 (2015). https://doi.org/10.1007/s12649-015-9391-1
10. Noppakundilograt, S., Piboon, P., Graisuwan, W., Kiatkamjornwong, S.: Preparation of alginate complex capsules containing eucalyptus essential oil and its controlled release. Carbohydr. Polym. **122**, 1–9 (2015). https://doi.org/10.1016/j.carbpol.2015.01.027
11. Huang, Y., Sun, B., Muhoza, B.: Influence of emulsification mode, stirring speed and volume on ibuprofen-loaded PLGA microparticles. Czech Slovak Pharm. **2021**(1), 22–28 (2022). https://doi.org/10.14748/csf.v2021i1.126748
12. Ananingsih, V.K., Soedarini, B., Andriani, C., Konstantia, B.A.A., Santoso, B.D.: Optimization of encapsulated agents and stirring speed on the physicochemical characteristics of vacuum dried nutmeg seed oleoresin (Myristica fragrans). J. Food Process. Preserv. **46**(12), e17151 (2022). https://doi.org/10.1111/jfpp.17151
13. Gåserød, O., Smidsrød, O., Skjåk-Braek, G.: Microcapsules of alginate–chitosan. II. A study of capsule stability and permeability. Biomaterials. **19**(20), 1821–1829 (1998). https://doi.org/10.1016/S0142-9612(98)00073-8
14. Huang, Y., Sun, B., Muhoza, B.: Hybrid microcapsules for encapsulation and controlled release of rosemary essential oil. Polymers. **15**(4), 823 (2023). https://doi.org/10.3390/polym15040823
15. Abang, S., Chan, E.S., Poncelet, D.: Effects of process variables on the encapsulation of oil in Ca-alginate capsules using an inverse gelation technique. J. Microencapsul. **29**(5), 417–428 (2012). https://doi.org/10.3109/02652048.2012.655331
16. Kim, J., Martin, J.G., Jambhrunkar, S., Kentish, S.E., O'Connor, A.J.: Design and development of pH-responsive polyurethane membranes for intravaginal release of nanomedicines. Eur. J. Pharm. Biopharm. **132**, 143–151 (2018). https://doi.org/10.1016/j.ejpb.2018.09.013
17. Chen, J., Liu, H., Zhao, C., Qin, M.: Polyacrylic acid functionalized nanographene as a nanocarrier for loading and controlled release of doxorubicin hydrochloride. J. Nanomater. **2013**, 345738 (2013). https://doi.org/10.1155/2013/345738

Optimizing the Mechanical and Thermal Properties of Nettle Fiber-Reinforced Composite Materials Using Manufacturing Parameters: Literature Review

Nesrine Bhouri(✉) and Ayda Baffoun

Textile Materials and Processes Research Unit, UR17ES33, University of Monastir, Monastir, Tunisia
bhourinesrine@yahoo.fr

Abstract. The development of nettle fiber-reinforced polymer composites presents a sustainable alternative to synthetic materials, offering advantages such as biodegradability, recyclability, and reduced environmental impact. The fabrication method, matrix type, and fiber characteristics significantly influence the composite's mechanical and thermal properties. Techniques like hand layup, melt mixing, extrusion, and injection molding are commonly employed, with fiber orientation and length playing a critical role in load transfer and tensile performance. Thermoplastics (e.g., PLA, PP) and thermosets (e.g., epoxy, polyester) are used as matrices, with PLA gaining attention for its renewability and biodegradability. However, natural fibers like nettle are hydrophilic, leading to poor fiber-matrix adhesion. To address this, chemical (e.g., alkaline, silane) and physical (e.g., plasma, UV) treatments are used to improve surface compatibility. Mechanical studies show optimal performance at 50 wt% fiber loading, beyond which interfacial adhesion weakens. Thermal analysis reveals that nettle fiber composites maintain stability up to 300 °C, with enhanced storage modulus at optimal fiber content. These findings highlight the importance of fiber treatment, matrix selection, and processing methods in designing high performance natural fiber composites for applications in packaging, construction, and automotive sectors.

Keywords: Nettle fiber · composite material · TGA · Mechanical properties · Physical treatment · Chemical treatment

1 Introduction

Composite materials are engineered to achieve enhanced properties that individual components alone cannot provide. Natural fibers are gaining attention as sustainable alternatives to synthetic reinforcements (glass, ceramic, or metal fibers). However, due to their differing chemical compositions, a key challenge lies in the poor interfacial adhesion between natural fibers and polymer matrices. To enhance bonding, natural fibers are often treated chemically to introduce functional groups, improving their interaction with the matrix and their mechanical properties. Common matrix materials used with nettle

M. Hamdaoui et al. (Eds.): ITDM'3 2025, PTT, pp. 47–55, 2026.
https://doi.org/10.1007/978-981-95-7704-0_5

fibers include polypropylene, polyester, epoxy, and poly(lactic) acid (PLA), offering varied advantages for bio-composite development.

2 Fabrication Methods

2.1 Technique de Fabrication

The processing method has a direct impact on the final properties of fiber-reinforced composites. The choice of technique depends largely on factors such as fiber type and length, as well as the polymer matrix used. Common fabrication methods include extrusion, extrusion/injection molding, melt mixing, compression molding, pultrusion, casting, and electrospinning [1]. The hand layup technique is widely used due to its simplicity, cost-effectiveness and versatility. It has been frequently used in the fabrication of nettle-reinforced composites [2]. Other methods, such as compression molding, hot pressing, and injection molding have been less commonly explored for this application. The injection molding involves numerous parameters, and high-speed processing has been shown to promote preferential fiber alignment, resulting in anisotropic materials with enhanced mechanical properties such as tensile strength and elastic modulus [3]. Melt mixing is a simple method that entails the direct dispersion of fibers within the polymer matrix, promoting homogeneous distribution. Extrusion, typically performed using a twin-screw extruder at controlled temperature and rotation speed settings, exhibit a predetermined cross-sectional shape. This technique offers significant advantages, including reduced fiber agglomeration due to the mechanical shear forces generated during processing.

2.2 Matrix Impact

In natural fiber composites, the matrix plays a dual role: it binds the fibers and transfers load to them through bonding or friction, improving the overall strength and stiffness of the material [4]. It also determines the composite's shape, surface appearance, and environmental durability. Thermoplastics (e.g., PP, PLA, PBAT) and thermoset polymers (e.g., epoxy, polyester) are commonly used as matrices. However, choices are limited by the thermal degradation of natural fibers at temperatures above 200 °C [5]. Thermoplastics offer low cost, design flexibility, and easy processing, while thermosets provide high strength, durability, and stability. Bio-based thermoplastics like PLA, PBAT, and PHA are increasingly used as eco-friendly alternatives to petroleum-based resins due to their sustainability and renewability.

2.3 Fiber Impact

Fiber length: Long fibers enhance the mechanical performance of composites by improving stress transfer between the matrix and the fibers and promoting orthotropic behavior. However, short fibers reduce stress transfer efficiency and lead to more isotropic composites. Thus, effective reinforcement requires fiber lengths significantly greater than the critical length [6].

Fiber arrangement: The mechanical properties of composite materials are primarily influenced by fiber orientation within the polymer matrix. Fibers can be arranged in unidirectional (UD), bidirectional (2D), or three-directional (3D) configurations. Optimal tensile strength is achieved when fibers are aligned with the direction of the applied load. However, such precise alignment is difficult to realize in practice. Partial alignment can be obtained through injection molding or manual placement of long fibers [7].

Fiber ratio: The mechanical characteristics of natural fiber reinforcement composites are strongly related to the fiber content within the matrix. While incorporating natural fibers can enhance composite performance, excessive fiber loading may decrease mechanical properties due to fiber agglomeration and poor dispersion [8].

3 Treatment of Nettle Fibers for Composite Production

Natural fiber-reinforced composites are gaining attention for their biodegradability, recyclability, and environmental benefits. However, their hydrophilic nature causes poor interfacial adhesion with the matrix, reducing mechanical performance. To overcome this, chemical and physical surface modification are essential to enhance fiber-matrix compatibility.

3.1 Chemical Treatments

Various techniques, such as sodium hydroxide treatment, bleaching, acetic acid treatment, silane treatment, polymer coating, and graft copolymerization, are commonly employed to mitigate the hydrophilic nature of cellulose plant fibers. Viju et al. [9] showed that sodium hydroxide treatment significantly increased the cellulose content in nettle fibers by removing non-cellulosic components such as hemicellulose, lignin, and extractives. This increase in cellulose improves the fibers' tensile strength, tenacity, elongation at break, and crystallinity, as cellulose plays a key role in providing mechanical strength to the fibers. Table 1 presents the chemical composition of nettle fiber following various chemical treatments. The data show a significant increase in cellulose content after treatment, particularly when sodium chlorite is applied, typically following sodium hydroxide treatment. This combination effectively cleanses the fiber and enhances its cellulose content, indicating improved compatibility with the polymer matrix.

Table 1. Chemical composition of nettle fiber after various chemical treatments

Material	Cellulose (%)	Hemicellulose (%)	Lignin (%)	References
Raw nettle	72.12 ± 0.8	10.96 ± 0.6	7.56 ± 0.6	[10]
NaOH-treated fiber	78.85 ± 1.1	8.14 ± 0.9	5.65 ± 0.5	[11]
H2O2-treated fiber	81.62 ± 0.9	7.12 ± 0.7	4.12 ± 0.4	[12]
NaClO2-treated fiber	83.1 ± 0.7	7.54 ± 0.5	3.41 ± 0.9	[9]
SiH4-treated fiber	82.67 ± 0.9	7.12 ± 0.7	4.69 ± 0.8	[13]

Figure 1 displays SEM images depicting the treated nettle fibers. The analysis reveals that untreated fibers (Fig. 1(a)) have a smooth surface. In contrast, treated fibers display varying degrees of surface protrusions compared to the untreated ones. This suggests the removal of an external layer, exposing the fibrils to chemical agents. These surface protrusions enhance mechanical interlocking between the fiber and the matrix during tensile loading, thereby contributing to improved tensile properties [14].

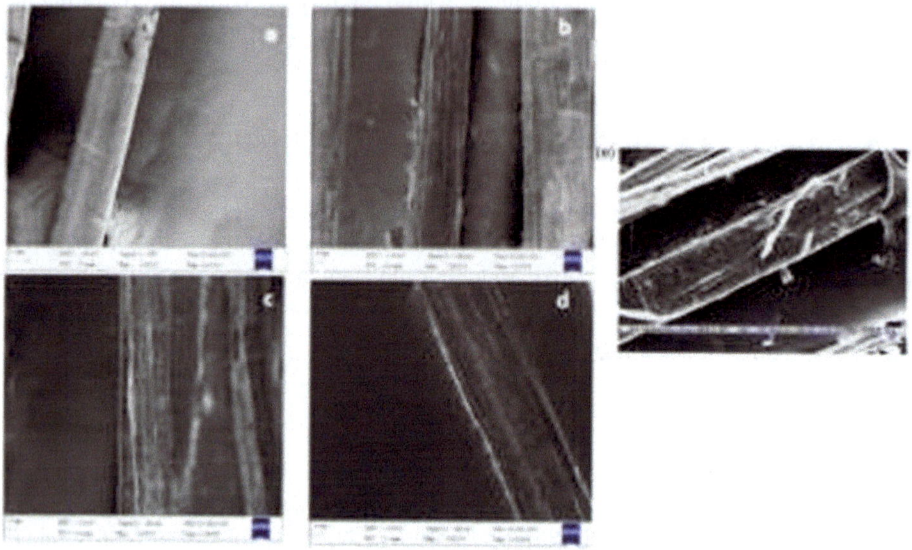

Fig. 1. SEM images of treated nettle fibers: (a) Raw nettle; (b) sodium hydroxide treated fiber; (c) Hydrogen peroxide treated fiber; (d) sodium chlorite treated fiber; (e) Silane treated fiber [15]

3.2 Physical Surface Treatments

Physical treatments applied to natural fibers aim to enhance interfacial bonding with polymer matrices without altering the fibers' chemical structure. Although being more expensive than chemical treatments, they are environmentally friendly and classified as green surface modification methods. Common approaches include plasma, corona, ultraviolet (UV), and thermal treatments [16, 17].

Plasma treatment, generated through radiofrequency gas discharge, involves energetic particles that cause physical and chemical changes on the surface materials. These interactions increase surface energy and improve the compatibility between fibers and resin. Corona treatment can also change the surface energy of the fibers and enhance the surface roughness of natural fibers. However, corona treatment is not applicable either on 3D products or fibrous materials like woven fabrics. Because of these limitations, other methods like ozone treatment or ultraviolet radiation are needed for modifying surfaces on three-dimensional structures. UV treatment is a clean method that generates photochemical reactions in natural fibers, improving their wettability with the matrix.

4 Characterization of Composite Materials

4.1 Mechanical Performance

To obtain a composite with good mechanical performance, the fibers and the matrix must have, as much as possible, some criteria as described in Fig. 2.

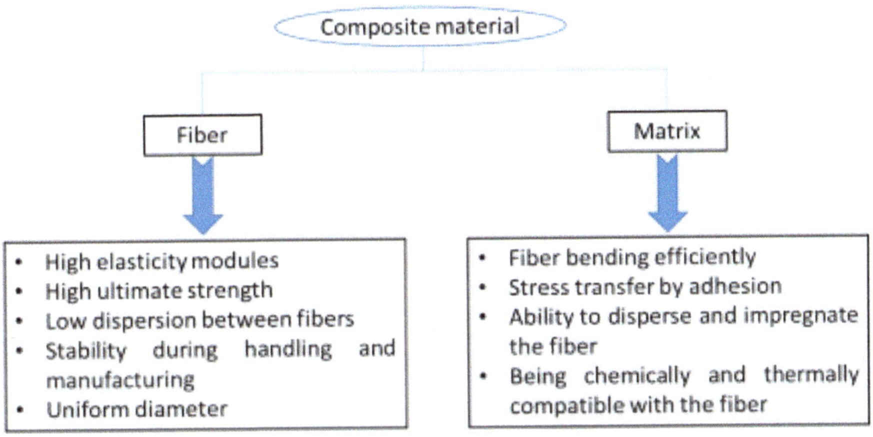

Fig. 2. Criteria for optimal composite material

Pokhriyal et al. [18] have prepared a nettle fabric reinforced polyester composite material using a hand layup technique. It has been found that the composite panel made with 15 wt% of fiber loading showed optimum characteristics among all composite samples (Fig. 3). The decrease of the tensile strength at 20 wt% may be attributed to the poor wetting of fibers with the matrix material. Indeed, the high fiber content of the composite material led to poor interfacial adhesion and inefficient stress transfer, leading to a decrease in mechanical strength.

Paukszta et al. [19] found that polypropylene (PP) composite reinforced with 10 wt% of nettle fibers exhibits lower tensile strength, elongation at break, impact strength, and hardness as compared to unfilled PP. However, the addition of nettle fiber resulted in a slight increase in Young's modulus. The mechanical properties of PP composite reinforced with nettle fibers did not show significant differences from composites filled with other lignocellulosic materials. The primary cause of the low mechanical properties was attributed to the poor interface between the non-polar matrix and the polar natural fiber. The crucial aspect for enhancing interfacial interaction and achieving better properties was highlighted through the surface treatment of the fiber or by the application of PP grafted with maleic anhydride acid serving as a non- adhesion promoter [17].

Nettle fiber-reinforced PLA composites have been extensively studied (Table 2). Kumar et al. [20] reported that increasing nettle fiber content improved mechanical characteristics such as tensile strength, elongation at break, Young's modulus, and flexural strength. However, beyond 50 wt%, these properties declined due to poor interfacial adhesion between the fibers and PLA matrix, caused by insufficient fiber dispersion.

Authors [21] also observed the formation of micro-spaces in the composites during tensile loading, which disrupted stress propagation from the matrix to the fibers. While these micro-spaces contributed to increasing stiffness and tensile strength at moderate fiber loadings, they became detrimental at higher contents.

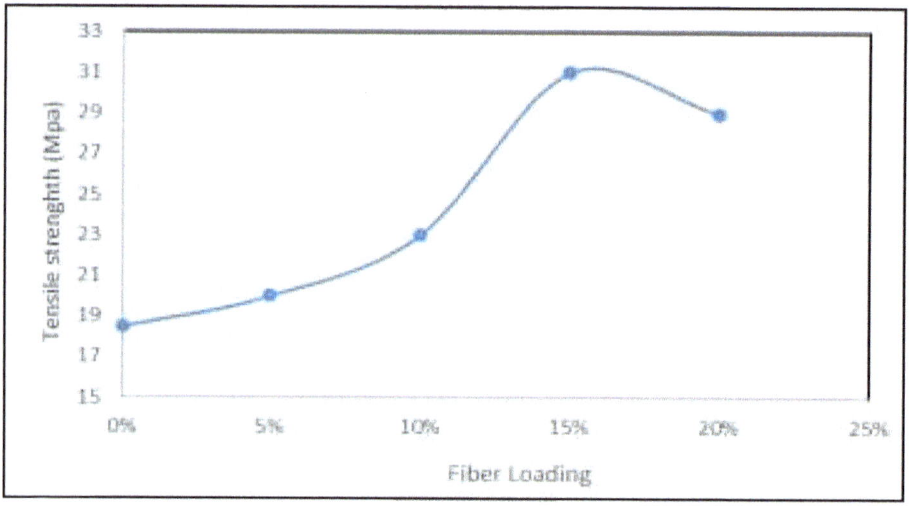

Fig. 3. Influence of fiber content on tensile strength of nettle fabrics composites [18]

Table 2. Mechanical properties of nettle/PLA composite material

Nettle composite	Fiber proportion (wt%)	Young modulus (GPa)	Tensile strength (MPa)	Elongation at break (%)
	0	2.21	2.21	0.83
	10	2.37	15.54	142
Nettle	25	3.11	36.70	1.80
/PLA	50	3.34	50.82	2.45
	75	2.33	27.04	1.80
	90	2.19	15.92	1.11

4.2 Thermal Analysis

The thermal stability of composite materials is mainly influenced by the matrix, often improved by increasing polymer crystallinity or incorporating intermediate substances to raise the glass transition temperature. However, the introduction of fibrous materials such as nettle alters these thermal properties. Although natural fibers offer advantages, they are sensitive to high temperatures (processing temperatures above 200–240 °C) can degrade them. To overcome this, high-performance thermoplastics with higher processing limits (over 300 °C) are often used. Heat treatment can also enhance the mechanical performance of thermoplastic resin composites. As an illustration, maintaining a ramie/PLA composite at 115 °C for one hour, followed by cooling, increases tensile strength and stress, while reducing elongation at break, due to the higher PLA crystallinity obtained during the heat curing time [22].

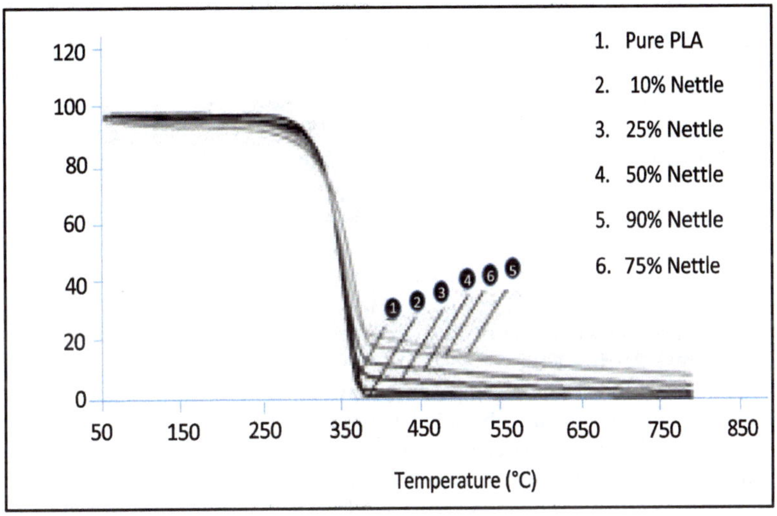

Fig. 4. TGA curves of Nettle/PLA composite material [20]

The mass of nettle fiber remains relatively constant up to 300 °C [23]. This characteristic results in a nettle fiber composite that exhibits thermal stability at 260–340 °C and in some cases even higher temperatures up to 380 °C [18]. According to Kumar et al. (2017) [20], adding more nettle fiber improves thermal stability, as confirmed by TGA analysis (Fig. 4). They found that a 50% fiber content doubled the storage modulus from 2.42 GPa (neat PLA) to 4.024 GPa at 35 °C. Additionally, a higher fiber ratio reduced the temperature sensitivity of the storage modulus, likely due to restricted polymer chain mobility. Equal fiber to matrix proportions yielded optimal energy absorption, consistent with findings in other bio-composites like PLA/Kenaf/APP and Linen/HDPE.

References

1. Rabbi, M.S., Islam, T., Islam, G.M.S.: Injection-molded natural fiber-reinforced polymer composites–a review. Int. J. Mech. Mater. Eng. **16**, Article 21 (2021)
2. Prasad, L., Kumain, A., Patel, R.V., Yadav, A., Winczek, J.: Physical and mechanical behavior of hemp and nettle fiber-reinforced polyester resin-based hybrid composites. J. Nat. Fibers. **19**, 2632–2647 (2020)

3. Sun, Z.-Y., Han, H.-S., Dai, G.-C.: Mechanical properties of injection-molded natural fiber-reinforced polypropylene composites: Formulation and compounding processes. J. Reinf. Plast. Compos. **29**, 637–650 (2009)
4. Zhan, J. et al.: Review on the performances, foaming and injection molding simulation of natural fiber composites. Polym. Compos. **42**, 1305–1324 (2020)
5. Summerscales, J., Dissanayake, N., Virk, A., Hall, W.: A review of bast fibres and their composites. Part 2 – Composites. Compos. Part A Appl. Sci. Manuf. **41**, 1336–1344 (2010)
6. Pickering, K.L., Efendy, M.G.A., Le, T.M.: A review of recent developments in natural fibre composites and their mechanical performance. Compos. Part A Appl. Sci. Manuf. **83**, 98–112 (2016)
7. Ben Amor, I., Rekik, H., Kaddami, H., Raihane, M., Arous, M., Kallel, A.: Effect of palm tree fiber orientation on electrical properties of palm tree fiber-reinforced polyester composites. J. Compos. Mater. **44**, 1553–1568 (2009)
8. Ku, H., Wang, H., Pattarachaiyakoop, N., Trada, M.: A review on the tensile properties of natural fiber reinforced polymer composites. Compos. Part B Eng. **42**, 856–873 (2011)
9. Viju, S., Thilagavathi, G.: Characterization of surface modified nettle fibers for composite reinforcement. J. Nat. Fibers. **19**, 1819–1827 (2020)
10. Dreyer, J., Müssig, J., Koschke, N., Ibenthal, W.D., Harig, H.: Comparison of enzymatically separated hemp and nettle fibre to chemically separated and steam exploded hemp fibre. J. Ind. Hemp. **7**, 43–59 (2002)
11. Viju, S., Thilagavathi, G.: Effect of alkali treatment of nettle fibers on oil absorbency. J. Nat. Fibers. **18**, 2092–2101 (2020)
12. Guo, A., Sun, Z., Satyavolu, J.: Impact of chemical treatment on the physiochemical and mechanical properties of kenaf fibers. Ind. Crops Prod. **141**, 111726 (2019)
13. Suarsana, I., et al.: Flexural strength of hybrid composite resin epoxy reinforced stinging nettle fiber with silane chemical treatment. AIMS Mater. Sci. **8**, 185–199 (2021)
14. Kalusuraman, G., Siva, I., Aslan, M., Anandkumar, S., Satyanarayana, K.G.: Evaluation of surface pre-treatments on the surface characteristics modifications in sponge gourd (Luffa Cylindrica) fibers. Mater. Res. Express. **6**, 095106 (2019)
15. Dabi, G.G., Wakjira, Y.T., Feysa, H.E., Abebe, W.M.: Development and characterization of laminated fiber reinforced bio-composite From nettle and poly lactic acid fiber. J. Ind. Text. **52**, 1–19 (2022)
16. Rahman, A.N.M.M., Ahmed, F., Alimuzzaman, S., Khan, R.A.: Enhancement of thermo-mechanical properties of okra fiber by photografting technique. J. Inst. Eng. India Ser. E. **102**, 45–59 (2021)
17. Liu, X., Hao, S.-J., Cui, Y.-H., Chen, H.-Y.: Improvement on the interfacial compatibility of jute fiber-reinforced polypropylene composites by different surface treatments. J. Ind. Text. **49**, 906–922 (2018)
18. Pokhriyal, M., Prasad, L., Rakesh, P.K., Raturi, H.P.: Influence of fiber loading on physical and mechanical properties of Himalayan nettle fabric reinforced polyester composite. Mater. Today Proc. **5**, 16973–16982 (2018)
19. Paukszta, D., Mańkowski, J., Kołodziej, J., Szostak, M.: Polypropylene (PP) composites reinforced with stinging nettle (Utrica dioicaL.) fiber. J. Nat. Fibers. **10**, 147–158 (2013)
20. Kumar, N., Das, D.: Fibrous biocomposites from nettle (Girardinia diversifolia) and poly(lactic acid) fibers for automotive dashboard panel application. Compos. Part B Eng. **130**, 54–63 (2017)

21. Mudoi, M.P., Sinha, S., Parthasarthy, V.: Polymer composite material with nettle fiber reinforcement: A review. Bioresour. Technol. Rep. **16**, 100860 (2021)
22. Chen, X., Yan, N.: A brief overview of renewable plastics. Mater. Today Sustain. **7–8**, 100031 (2020)
23. Raj, M., Fatima, S., Tandon, N.: An experimental and theoretical study on environment-friendly sound absorber sourced from nettle fibers. J. Build. Eng. **31**, 101395 (2020)

Ballistic Impact of Filled Composites: Finite Element Modeling

Amira Hassouna[1,2](✉) and Salah Mezlini[1]

[1] Laboratoire de Génie Mécanique, Ecole Nationale d'Ingénieurs de Monastir, Université de Monastir, Monastir, Tunisia
Hassounaamira0@gmail.com

[2] Institue Préparatoire Aux Etudes d'Ingénieurs de Gafsa, Université de Gafsa, Gafsa, Tunisia

Abstract. Composite materials reinforced with nanofillers, have received particular attention in the defense and aerospace fields due to their superior strength-to weight ratio and energy absorption abilities. The present study investigates the effect of adding nanofillers (carbon nanotubes and graphene nanoplatelets) on the ballistic performance of body armor made of Kevlar composite materials. Therefore, a finite element model is developed using Abaqus explicit. The Hashin criterion is used to describe the behavior of composite materials. Residual velocity and target damage are analyzed for different percentages of nanofillers. It was found that there is a good agreement between the results of our numerical model and the results of the literature for different impact velocities. Moreover, it is revealed that the volume percentage of nanofillers has a considerable influence on the target damage but has a negligible effect on the residual velocity. The minimum target damage is observed for a nanofiller volume percentage of 0.5%.

Keywords: Finite Element Model · Composite · Nanofillers · Body Armor

1 Introduction

Composite materials are widely used in various fields, thanks to their outstanding properties, which make them ideal for protective structures, particularly ballistic armor. The soft body armor is generally made from Kevlar fiber composite (Kumar et al., 2010) or ultra higher molecular weight polyethylene material (Krishnan et al., 2010; Pundhir et al., 2019; Yang et al., 2022). The performance of these traditional fiber-reinforced composites can be further enhanced by the incorporation of nano filler such as Carbon Nanotube (CNTs) or graphene nanoplatelets. In this regard, El Moumen et al. (Moumen et al., 2018) carried out experimental and numerical ballistic impact tests at low impact velocities on a carbon fiber-reinforced epoxy filled with carbon nanotubes. The results showed that the energy absorption capacity of vests containing carbon nanotubes (CNT) was greater than those lacking CNT. In addition, they found that the addition of CNTs significantly reduced damage to the composites. Vignesh et al. (Vignesh et al., 2021) tested the ballistic impact of two types of composites using finite element modeling, the first is a kevlar-29 composite laminates and the second formed of Kevlar layers and

M. Hamdaoui et al. (Eds.): ITDM'3 2025, PTT, pp. 56–63, 2026.
https://doi.org/10.1007/978-981-95-7704-0_6

graphene laminates. They deduced that the addition of graphene reduced the total deformation of the body armor and the equivalent stress in the target. Therefore, the strength of the body armor is improved by the graphene reinforcement between the Kevlar layers. Further more, they revealed that the graphene is more bendable than Kevlar, which translates into fewer breaks in the weaving and knitting processes. Sonmez et Ghasemnejad (Sonmez & Ghasemnejad, 2019) developed numerical models simulating the ballistic impact of graphene, aluminum and carbon fiber reinforced polymer (CFRP) plates using a spherical projectile. The comparative study showed that the CFRP has the best performance in terms of specific absorption energy for an impact energy of 20 J and the aluminum target has the best specific absorption energy for an impact energy of 80 J. However, the ballistic limit of graphene, which is 3375 m/s, is 30 times higher than the ballistic limits of aluminum and CFRP, which are just over 100 m/s. Pandya et Naik (Pandya & Naik, 2015) performed an experimental and analytical study to describe the ballistic impact behavior of pure epoxy and epoxy containing multi-walled carbon nanotube (MWCNT). They found that adding MWCNT to pure epoxy increases V50 by up to 5% and absorbed energy by up to 10%.

Despite research into the ballistic impact of composites filled with carbon nano tubes or graphene, the study of the impact behavior of Kevlar composites reinforced with both carbon nanotubes and graphene nanoplatelets against Piercing projectile using numerical modeling remains a gap. Therefore, this study aims to analyze the effect of adding nanofillers (carbon nanotubes and graphene nanoplatelets) to Kev lar fiber composites on the residual velocity and the target damage. A finite element model is developed using explicit Abaqus code to simulate the ballistic impact of composites. Firstly, the results of the numerical model are validated with those of the literature. Then, the effect of the impact velocity on the residual velocity is investigated. Finally, the influence of varying nanofiller volume percentage on the residual velocity and the target damage is examined.

2 Finite Element Modeling

Figure 1 shows the geometry of the numerical model, boundary conditions and meshing. The target is a square plate, 120 mm side length and 8 mm thick.

The projectile has an ogival nose with a diameter of 10.9 mm and a mass of 30 g. The boundary conditions applied to the target are an embedding on its outer edges. For the projectile, a variable velocity (V) is applied along the Z axis. The impact velocities selected during this study are: 250 m/s, 300 m/s, 400 m/s, 500 m/s, 600 m/s and 700 m/s (Rodríguez Millán et al., 2016). In this model, two types of mesh are used. The composite plate is meshed by SC8R (8-node Continuum Shell with reduced integration) elements with a total number of 44376 and 52983 nodes. The number of composite layers is 16 oriented at 0 and 90° ([0/90°] 3S). For the projectile, the mesh type used is C3D8R (Eight-node brick element with reduced integration) with an element number of 19440 and a node number of 22022. The choice of element size is based on the literature. The 'GENERAL CONTACT' algorithm available in Abaqus/Explicit is applied to define the contact between the projectile and the target with a coefficient of friction equal to 0.1 (Rodríguez Millán et al., 2016).

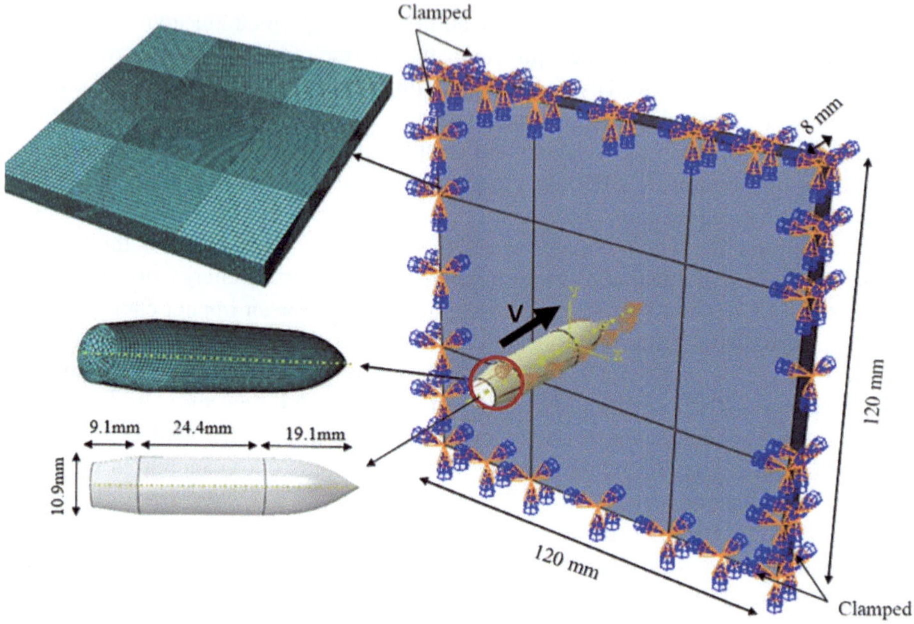

Fig. 1. Finite element model

In this study, the composite plate is made of Kevlar fiber and epoxy matrix mixed with nanofillers (carbon nanotubes and graphene nanoplatelets). This material is characterized by a Lamina-type behavior law and a damage model based on the Hashin criterion (I. ABAQUS, 2011).

In the literature, the mechanical properties of composite depend on several parameters, particularly the volume percentage of nanofillers (carbon nanotubes and graphene nanoplatelets) added to the composite (Ma et al., 2023). In this study, several numerical tests are carried out by modifying the volume percentage of nano fillers to analyze its effect on the residual velocity and the target damage. Table 1 illustrates properties of the composite material as a function of nanofiller volume percentage.

Table 1. Chemical composition of nettle fiber after various chemical treatments

Configuration	C1	C2	C3	C4
nanofiller volume percentage	0%	0.5%	1%	2%
E1 = E2 (GPa)	15.2	16.8	20.1	17
⋮	0.2	0.2	0.2	0.2
G12 = G13 = G23 (GPa)	1.56	1.65	1.65	1.5
Xt = Yt (MPa)	380.6	384.3	513.7	411
Xc = Yc (MPa)	94.4	97.2	106.1	102.9
Sc (MPa)	30.2	38	32.5	32.4

3 Results and Discussion

3.1 Validation of the Finite Element Model

The validation of the numerical model consists in assessing its ability to predict results from literature. In this case, the same geometries, boundary conditions and material properties used by Rodríguez Millán et al. (2016) are used. The target is made of composite material with a Kevlar fiber and an epoxy matrix. Figure 2 shows the evolution of the residual velocity as a function of the impact velocity for our numerical results and those reported in the literature.

Various impact velocities ranging from 250 m/s to 750 m/s are tested. It can be seen that our numerical results are in good agreement with those of the literature with an average relative error of approximately 4%. Moreover, it is observed that the residual velocity increases as the impact velocity increases. For an impact velocity of 700 m/s, the residual velocity is 637.54 m/s, representing a decrease of about 10%. However, for an impact velocity of 250 m/s, the residual velocity decreases by around 21%. In fact, increasing the impact velocity reduces the contact time between the projectile and the target, so the projectile takes less time to penetrate the thickness of the target. The same observation was made by Kumar et al. (Kumar et al., 2010) when impacting kevlar/epoxy using a cylindrical projectile with a flat tip.

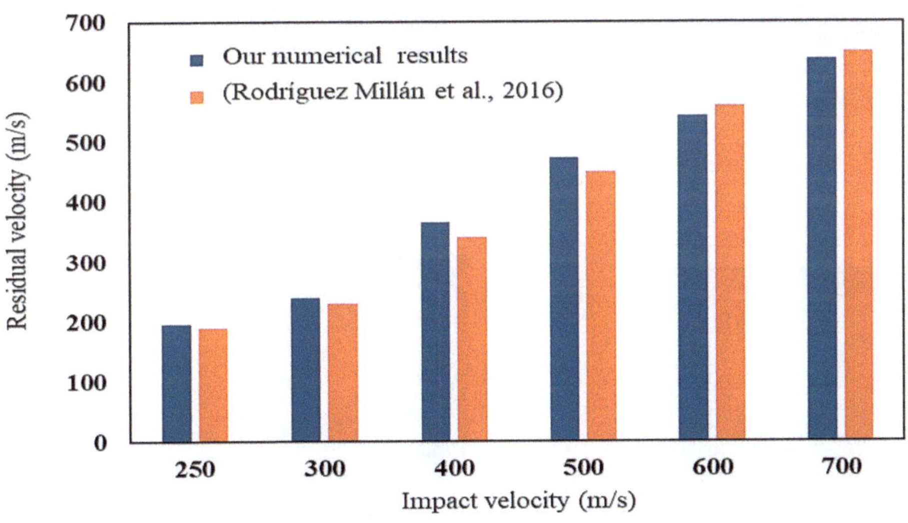

Fig. 2. Variation of the residual velocity as a function of the impact velocity

3.2 Effect of Nanofillers Volume Percentage

Target Damage

Figures 3 and 4 illustrate the Hashin criterion damage on the front and rear faces of the composite plate for the four configurations mentioned above (Table 1). Results show that the compressive matrix damage under the Hashin criterion varies from one configuration to another for the front and rear faces. In fact, the damage caused to the target in the configuration C1 (without nanofillers) is greater than the damage caused to the target loaded with nanofillers. For configuration C1, the damaged diameter at the front of the target is around 40 mm, but it is around 23 mm for a plate loaded with 0.5% nanofillers and 26 mm for a plate loaded with 1% nanofillers. For configuration C4, where the volume percentage of nanofillers is 2%, the damage diameter is 33 mm. The minimum damage around the contact zone is observed for a nanofiller volume percentage of 0.5%. In this context, Bhowmik et al. (Moumen et al., 2018) confirmed that the presence of carbon nanotubes in woven carbon fiber composites slowed crack propagation and reduced the number of damaged elements. These tests were carried out on open-hole tensile (OHT) and short-beam shear (SBST) specimens.

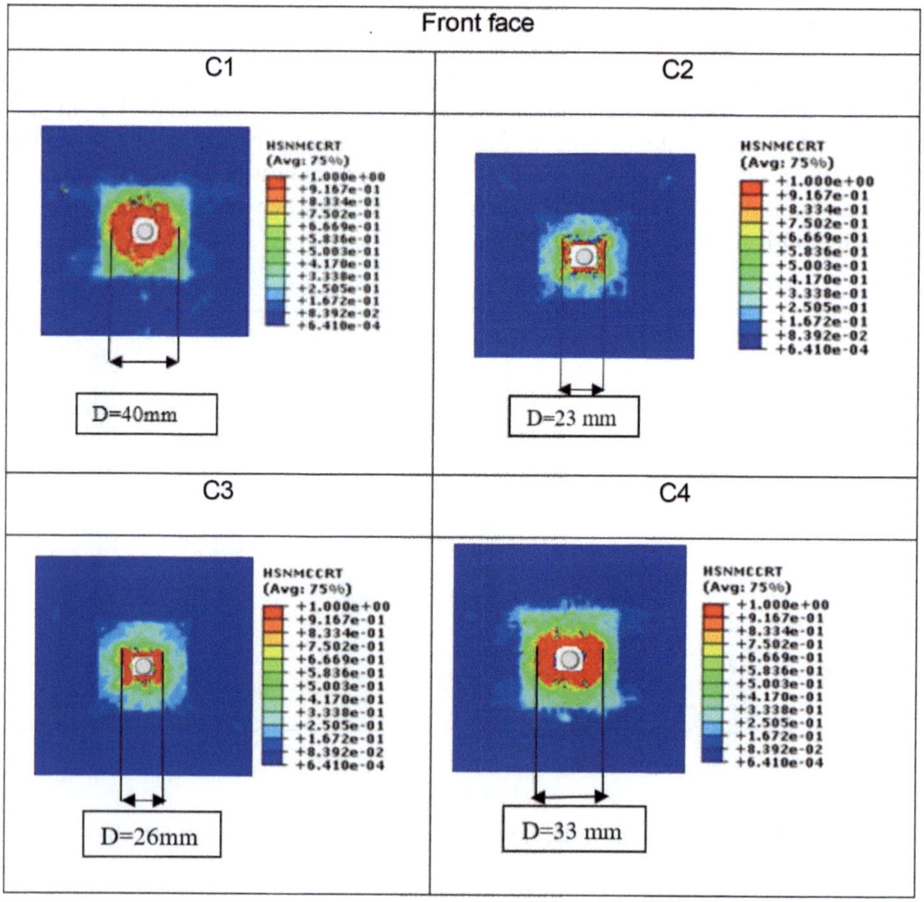

Fig. 3. Target damage in the front face

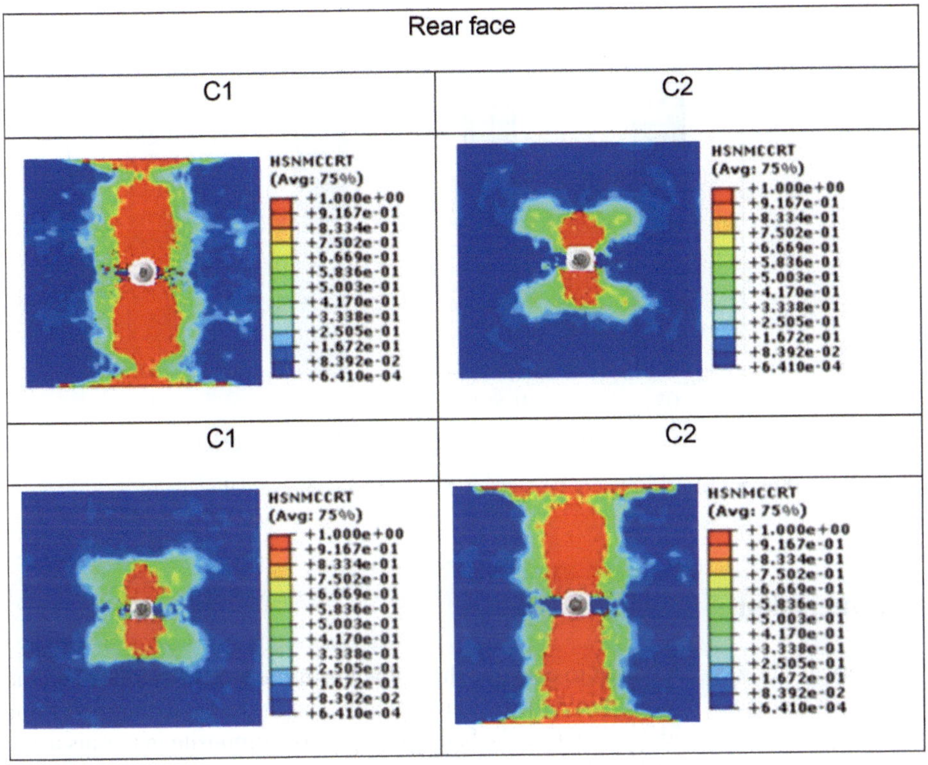

Fig. 4. Target damage in the rear face

Residual Velocity

Figure 5 shows the variation of the residual velocity as a function of the impact velocity for different volume percentages of nanofiller. The analysis of this figure revealed that the nanofillers volume percentage has a negligible influence on the residual velocity. This can be explained by the small variation of material parameters with volume percentage, namely Young's modulus and yield strength in tension and compression (see Table 1). In this context, Moumen et al. (Moumen et al., 2018) showed that the impact force and the absorbed energy were little influenced by the addition of carbon nanotubes when the volume percentage varied from 0 to 2%. However, increasing the volume percentage of carbon nanotubes to 4% reduces both impact force and absorbed energy. These tests were carried out on a carbon fiber composite using a hemispherical impactor.

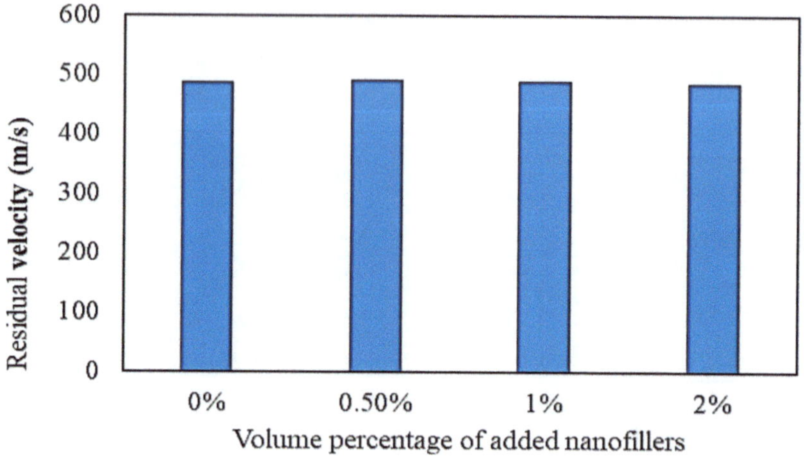

Fig. 5. Effect of nanofillers volume percentage on residual velocity

4 Conclusion

A numerical study is carried out to examine the influence of the impact velocity and the volume percentage of nanofillers on the residual velocity and the target damage. For this purpose, a numerical model simulating the ballistic impact of composite materials using Piercing projectiles is developed.

The results of this study are as follows:

- Good consistency between our numerical model and literature results for different impact velocities.
- The absorbed velocity decreases with increasing the impact velocity.
- The volume percentage of nanofillers has a significant influence on the target damage but has little effect on the residual velocity. Indeed, the target damage decreases with increasing the volume percentage of nanofillers. The minimum target damage is observed at a volume percentage of 0.5%.

References

ABAQUS: Abaqus Theory Manual, v.6.11. USA: Hibbitt, Karlsson and Sorensen, "ABAQUS." (2011)

Krishnan, K., Sockalingam, S., Bansal, S., Rajan, S.D.: Numerical simulation of ceramic composite armor subjected to ballistic impact. Compos. B Eng. **41**(8), 583–593 (2010)

Kumar, S., Gupta, D.S., Singh, I., Sharma, A.: Behavior of kevlar/epoxy composite plates under ballistic impact. J. Reinf. Plast. Compos. **29**(13), 2048–2064 (2010)

Ma, D., Matheus, C., Campos, S., Giglio, M., Manes, A.: A novel numerical method for stochastic study of fiber-reinforced composites with nanoparticles under impact loading. Int. J. Impact Eng **180**(February), 104662 (2023)

Moumen, A. El, Tarfaoui, M., Hassoon, O.: Experimental study and numerical modelling of low velocity impact on laminated composite reinforced with thin film made of carbon nanotubes. Appl. Compos. Mater. 309–320 (2018). https://doi.org/10.1007/s10443-017-9622-8

Pandya, K.S., Naik, N.K.: International journal of impact engineering analytical and experimental studies on ballistic impact behavior of carbon nanotube dispersed resin. Int. J. Impact Eng **76**, 49–59 (2015). https://doi.org/10.1016/j.ijimpeng.2014.09.003

Pundhir, N., Goyal, D., Singh, P., Pathak, H., Zafar, S.: Numerical simulation of composite armor subjected to ballistic impact. Mater. Today Proc. **18**, 696–703 (2019). https://doi.org/10.1016/j.matpr.2019.06.470

Rodríguez Millán, M., Moreno, C.E., Marco, M., Santiuste, C., Miguélez, H.: Numerical analysis of the ballistic behaviour of Kevlar® composite under impact of double-nosed stepped cylindrical projectiles. J. Reinf. Plast. Compos. **35**(2), 124–137 (2016). https://doi.org/10.1177/0731684415608004

Sonmez, M.B., Ghasemnejad, H.: Transverse impact response analysis of graphene panels: impact limits. J. Compos. Mater. **9**(2), 124–144 (2019). https://doi.org/10.4236/ojcm.2019.92006

Vignesh, S., Surendran, R., Sekar, T., Rajeswari, B.: Ballistic impact analysis of graphene nanosheets reinforced kevlar-29. Mater. Today Proc. **45**(xxxx), 788–793 (2021). https://doi.org/10.1016/j.matpr.2020.02.808

Yang, Y., Ling, T., Xue, S., Liu, Y., Zhou, J.: Meso-scale numerical modeling of uni-directional composite panel under ballistic impact. Appl. Compos. Mater. **29**(1), 205–218 (2022). https://doi.org/10.1007/s10443-021-09930-x

Porous Alginate-Clay Based Biocomposite as a Potential Platform for Drug Delivery

Sirine Guedouar[1(✉)], Khmais Zdiri[1,2], Ayda Baffoun[1], and Tarek Baati[3]

[1] Unité de Recherche Matériaux et Procédés Textiles (MPTex), UR17ES33, Université de Monastir, ENIM, 5019 Monastir, Tunisie
sirine.guedouar11@gmail.com

[2] Centre de Recherche en Sciences et Technologie de Textile à Monastir (CRSTTM), Université de Monastir, 5000 Monastir, Tunisie

[3] Laboratoire des Substances Naturelles, Institut National de Recherche et d'Analyse Physico-chimique (INRAP), BiotechPole Sidi Thabet, Ariana, Tunisia

Abstract. This study aims to produce a highly porous stable alginate/clay -based biocomposide with a regular spherical shape as a potential candidate for bioactives molecules delivery. Alginate/clay-based microspheres, with the ability to tune their structures and porosities, are well suited to serve as carriers for sustainable drug delivery such as caffeine. The impact of formulation parameters, encompassing clay content, $CaCl_2$ concentration (used as the crosslinking agent), and pH, was studied to enhance porosity, mechanical strength and stability of this biocomposite. The results show that 2.5% of alginate leads to the formation of regular spherical shape of the biocomposite. In addition, the variation of clay concentration up to 7% and neutral pH increases the encapsulation efficiency of caffeine. These microspheres could achieve higher caffeine drug-loading content close to 70% leading to a controlled drug release up to 24 h under incubation at 37 C in PBS. Taken together, these novel microspheres generation open the way for their application in the biomedical field, particularly in topical applications requiring localized and efficient drug delivery systems.

Keywords: Alginate · Clay · Microspheres · Encapsulation and Sustainable Release

1 Introduction

In recent years, biopolymers have attracted significant interest in biomedical research due to their biodegradability, biocompatibility, and environmental friendliness. Among them, alginate a natural polysaccharide extracted from brown seaweed has shown great potential in the development of drug-loaded microspheres. Its ability to form hydrogels through ionic crosslinking with divalent cations, particularly calcium, makes it a suitable candidate for encapsulation applications.

However, pure alginate microspheres often suffer from limitations such as poor mechanical strength and low porosity [1, 2], which can compromise encapsulation efficiency and control over drug release. To overcome these challenges, clay was incorporated into the alginate matrix as a reinforcement agent. The clay not only enhances

M. Hamdaoui et al. (Eds.): ITDM'3 2025, PTT, pp. 64–81, 2026.
https://doi.org/10.1007/978-981-95-7704-0_7

the mechanical integrity of the microspheres but also contributes to increased porosity, thereby improving their drug-loading and release performance.

The main goal of this study was to produce microspheres that are morphologically uniform and perfectly spherical. To achieve this, we developed hybrid alginate/clay microspheres using two different encapsulation strategies. In the first protocol, the active ingredient (caffeine) was added to the alginate/clay solution before the gelation step. In the second, caffeine was introduced into the coagulation bath containing calcium chloride during the crosslinking process.

Formulation parameters including clay content, $CaCl_2$ concentration, and pH were varied and analyzed to determine which protocol offered the best control over microsphere morphology and encapsulation efficiency. After identifying the optimal formulation, the release of caffeine was investigated under two simulated physiological conditions: PBS at pH 7, representing skin application, and an acidic medium at pH 1.2, simulating gastric conditions. These release studies were essential to evaluate the potential of the developed microspheres for targeted biomedical applications.

2 Materials and Methods

2.1 Materials

The clay used in this study was collected from the Khanguet Rihana region, located less than 25 km south of Regueb, Tunisia. To improve its compatibility with organic matrices, the raw clay underwent several purification and modification steps [3, 4]. These treatments aimed to transform the native clay into an organophilic clay [5], making it suitable for incorporation into biopolymeric systems such as alginate-based microspheres. Sodium alginate was also used as the main biopolymer matrix for the development of the microspheres. The alginate employed was Natriumalginate Biochemica, a white powder, purchased from PanReac AppliChem (ITW Reagents). Table 1 shows the properties of used algiante.

Table 1. The properties of sodium alginate

	Sodium alginate
Chemical formula	$(C_6H_7O_6Na)_n$
Molecular weight	10 000–60 000 g/mol
pH	5.5–8
Viscosity	350–550 mPas
Solubility	Soluble in water
Perte au séchage	< 15%

This sodium alginate, due to its excellent water solubility and capacity to form hydrogels in the presence of calcium ions, was selected for its suitability in forming spherical hydrogel-based microspheres. Also, we used caffeine as an active ingredient.

2.2 Development of an Alginate-Clay Based Biocomposite

Microspheres are produced by interfacial polycondensation [6]. The microspheres are gelled in a bath containing a solution of calcium chloride $CaCl_2$ using two different protocols.

First protocol: The aim of this protocol is to produce microspheres already loaded with the active ingredient AI. The microcapsules are produced by adding dropwise, using a syringe with a 0.3 mm diameter needle, a solution of sodium alginate- Clay- AI to a bath containing a solution of calcium chloride $CaCl_2$.

Second protocol: The aim of this protocol is to elaborate microspheres and then load them with active ingredients AI. Gelled microspheres are produced by adding dropwise, using a syringe with a 0.3 mm diameter needle, a sodium alginate-clay solution to a bath containing a calcium chloride solution and the active ingredient AI.

To study the influence of certain parameters on the microcapsule elaboration process, we varied factors such as clay concentration, calcium chloride concentration and pH according to the following table.

2.3 Determining Encapsulation Rate

To measure encapsulation rate, several steps are required:

The supernatant from each coagulation bath is separated from the capsules formed once the required gelation time has elapsed. 100 µl of each supernatant is taken and diluted twice with a mobile phase solution of 60% PBS (phosphate-buffered saline) and 40% ACN (acetonitrile). Each diluted sample is analyzed by HPLC to determine the desired absorption rate. This process enables precise, quantitative assessment of the encapsulation rate of the samples analyzed [7].

2.4 Determination of AI Release Rate

Caffeine release was carried out in two media:

first medium: by immersing caffeine-loaded microspheres in phosphate buffer solution (PBS; 0.05 M, pH 7.4) at 37 °C under continuous two-dimensional agitation (250 rpm, MultitonOrbitale incubator, Infors HT, France), at different incubation times (5mn,15mn,30 mn, 1H,3H,24H,48H, 72H and 7 days), and each diluted sample is analyzed by HPLC to determine the desired AI release rate.

Second medium: by immersing caffeine-loaded capsules in a simulated stomach medium, a solution of ultra- pure water at acid pH (pH = 1.2 adjusted with hydrochloric acid) maintained under continuous stirring at 37 °C.Samples of 100 µL supernatant were taken at different incubation times (5mn, 15mn, 30mn, 1H, 3H) at 37 °C and each diluted sample is analyzed by HPLC to determine caffeine release kinetics.

2.5 Morphological Characterization of Microspheres

This characterization is performed using an optical microscope. The aim of this characterization is to study the morphology of the microspheres and the dispersion quality of the clay [8].

2.6 Structural Characterization of Microspheres

Infrared spectroscopy (FTIR: Fourier Transform Infrared Spectroscopy) used to determine the structure of the microspheres [9]. The FTIR spectra were recorded using a Spectrum Two FT-IR Spectrometer (PerkinElmer, USA).

3 Results

3.1 Characteristic Infrared Spectrum of Clay

To analyze the spectrums, we examined the main differences absorption peaks in the different regions [10].

Figure 1 shows the infrared spectrum of the clay samples before and after various purification treatments. It shows significant changes, reflecting the removal of impurities and the chemical modification of the clay.

A comparison between the infrared spectrum of raw clay and purified clay shows that in the region 1300–1600 cm^{-1}: This range is associated with vibrations of functional groups such as carbonyl (C=O) and carbonate bonds. A marked decrease in peak intensity in this region after HCl treatment indicates the effective reduction of carbonates. Around 3400 cm^{-1}: This broad band is attributed to O–H stretching vibrations. The noticeable decrease in intensity for the sample treated with DCB (dithionite–citrate–bicarbonate) demonstrates the dissolution of hydroxides, confirming the efficiency of this treatment in removing iron and aluminum oxides from the clay surface.

At 870 cm^{-1}: For the H_2O_2-treated sample, this band appears significantly weaker compared to the DCB-treated sample. This attenuation is attributed to the removal of organic matter by hydrogen peroxide, which may have contributed to the original intensity of this band.

Figure 2 shows the infrared spectrum of organophilic clay, we notice that in the region 2800–3000 cm^{-1}: The appearance of a new peak in this region corresponds to the presence of surfactant molecules. These bands indicate C–H stretching vibrations from aliphatic chains, confirming the introduction of an organic surfactant. This chemical modification transforms the clay into an organophilic material, characterized by increased interlayer spacing that facilitates the intercalation of polymer chains.

These spectral changes clearly demonstrate the progressive purification and functionalization of the clay, making it suitable for incorporation into biopolymer matrices.

3.2 Morphological Characterization of Prepared Microspheres

Our aim in this work is to obtain spherical microspheres with a regular shape. We therefore carried out preliminary tests low alginate concentrations (1%, 1.5%), Figs. 3 and 4 show that the capsules we obtained were not spherical, nor were they regular in shape before drying, even after drying.

Based on previous results, the alginate content was increased to 2.5%. Figure 5 shows that spherical microspheres were successfully obtained.

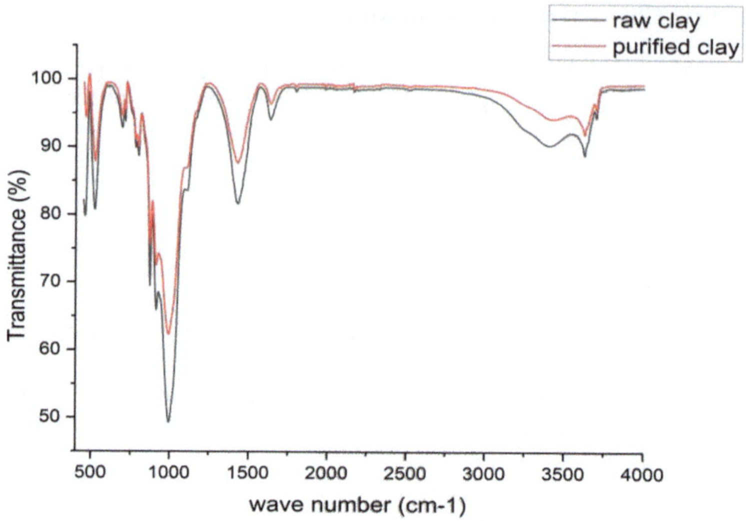

Fig. 1. Infrared spectra of raw clay and purified clay

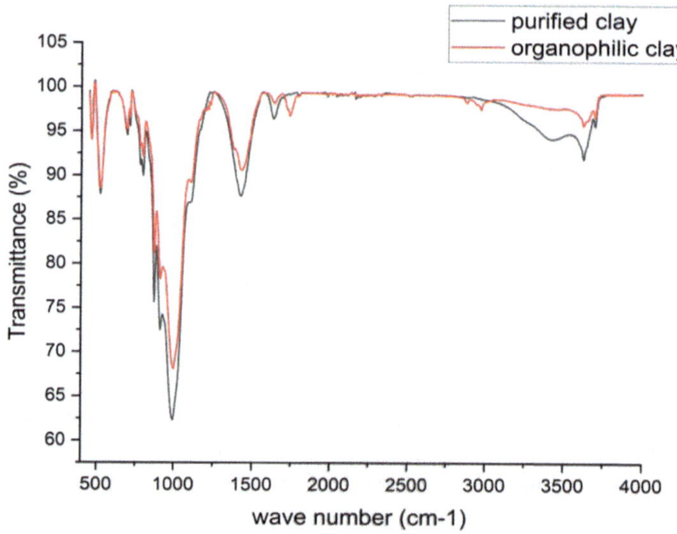

Fig. 2. Infrared spectra of purified and organophilic clay

As we move from lower to higher concentrations, the shape of the beads becomes more spherical. Figure 6 show that they are less deformed and more regular, even after drying.

A higher concentration of alginate solution produces a more viscous gel, resulting in stronger, even more spherical drops.

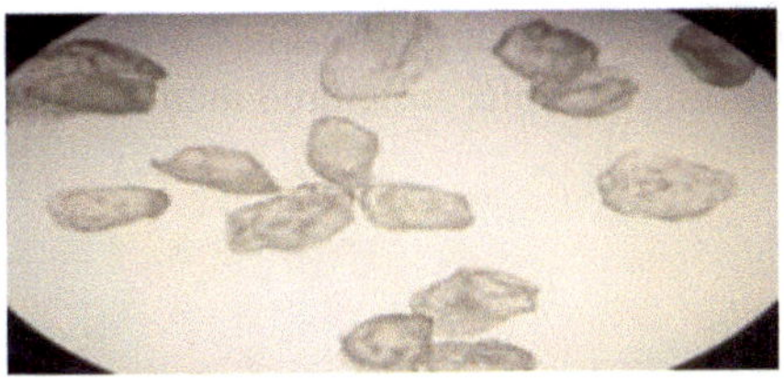

Fig. 3. Microspheres before drying

Fig. 4. Microspheres after drying

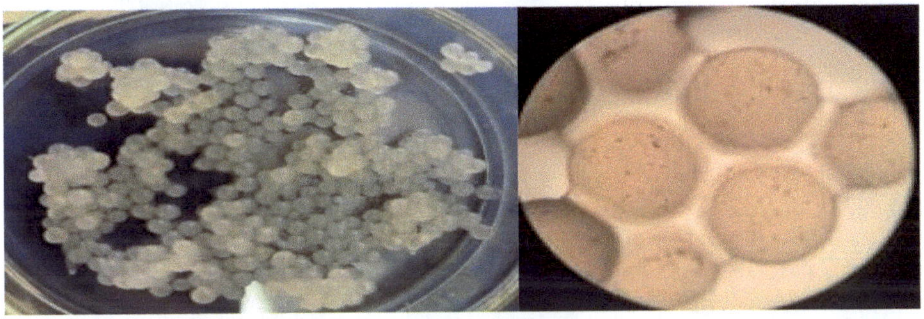

Fig. 5. Microspheres before drying

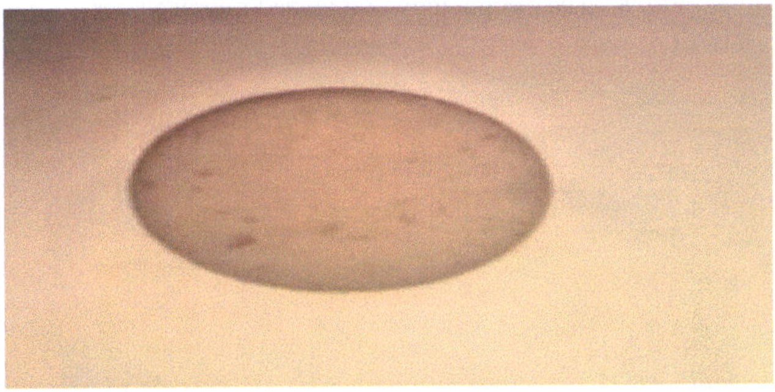

Fig. 6. Microsphere after drying

Indeed, once they reach the surface of the cross-linking bath, if droplet viscosity and surface tension forces are unable to overcome the surface tension exerted by the cross-linking solution, the resulting beads may be deformed.

On the other hand, a high concentration of cross-linking solution enables the beads to gel rapidly while retaining their initial shape. The latter depends on the alginate concentration.

Figures 7 and 8 show that the characterization using an optical microscope has the advantage of giving an initial idea of the size and shape of the microspheres, the dispersion of the clay particles and the surface of the microspheres.

Figure 9 shows the surface of the microspheres, alginate cross-links through the binding of Ca^{2+} ions to carboxylate groups, forming a stable, structured membrane.

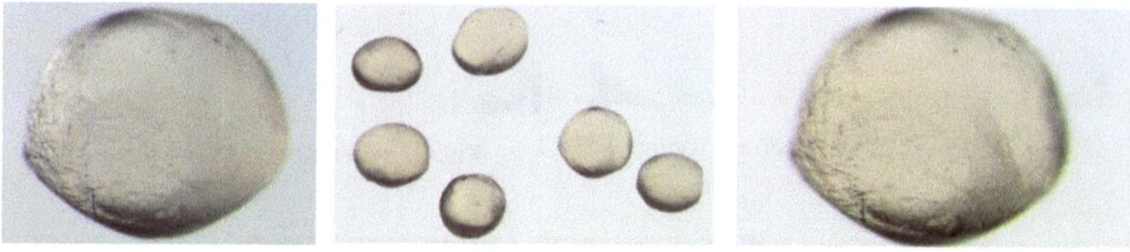

Fig. 7. The morphology of an empty microspheres

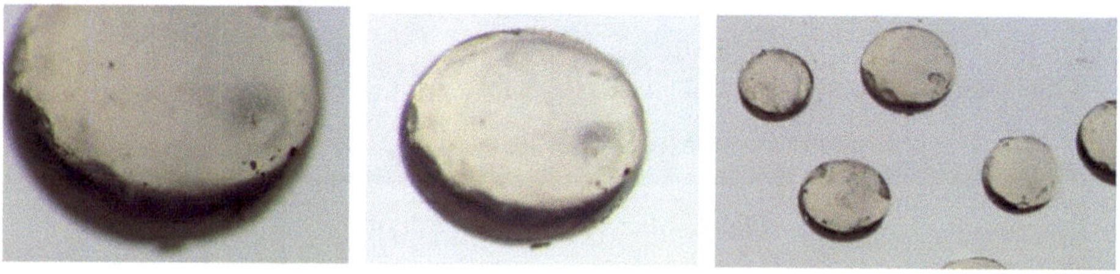

Fig. 8. The morphology of a loaded microspheres

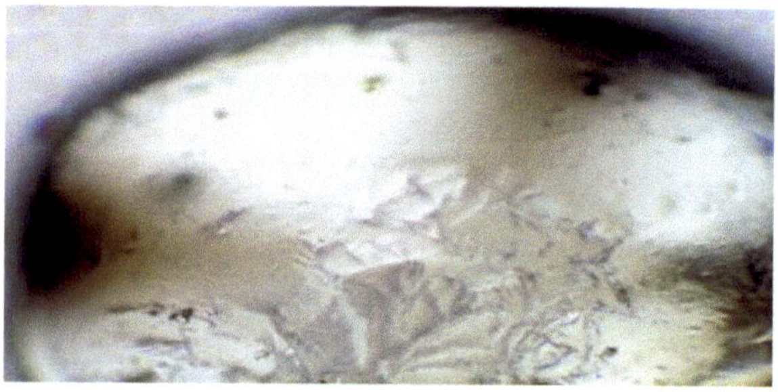

Fig. 9. Cross-linking of alginate

3.3 Structural Characterization of Microspheres

The IR spectrum of full and empty microspheres is shown in the Fig. 10. The infrared absorption peaks indicate the different chemical bonds present in each sample:

- Around 1700 cm^{-1}: A strong peak can be seen in both spectra, typical of the C=O stretching vibrations (carbonyl groups) present in caffeine and alginate.

– Between 1600–1500 cm^{-1}: These bands are often associated C=C stretching vibrations or N-H deformation vibrations (present in caffeine).

Comparing the two curves, that of the empty capsules shows less intense absorption in the above-mentioned regions, confirming the absence of caffeine. And that of the charged capsules shows more intense peaks at 1700 cm^{-1} and 1500 cm^{-1} than that of the empty capsules, explained by the presence of caffeine, which has specific vibrations in these regions [11].

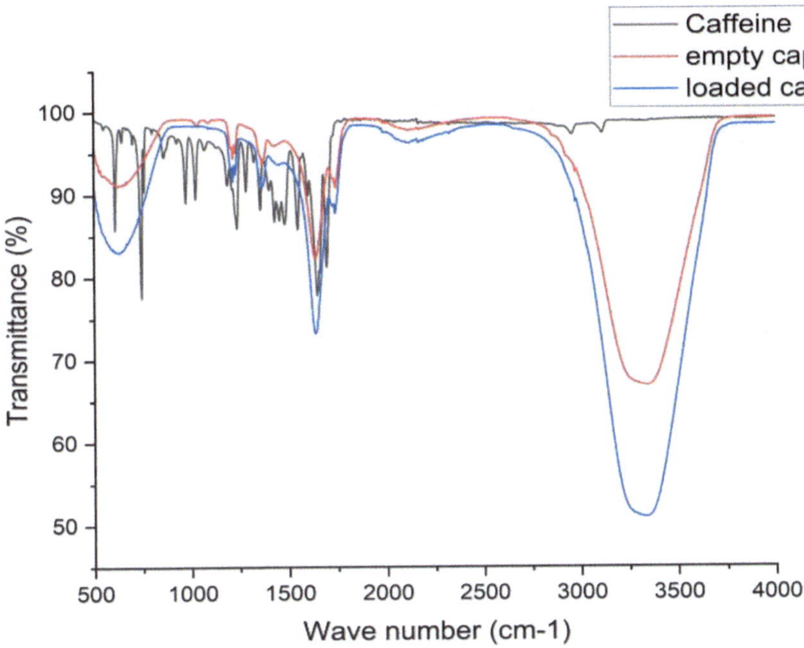

Fig. 10. Infrared spectrum of caffeine, an empty and a loaded capsule

3.4 Determination of Encapsulation Rate According First and Second Protocol

The encapsulation rate was measured according to both encapsulation protocols, and the results obtained are presented in the following two tables.

Table 2 shows that according to the first protocol, the encapsulation rate ranges between 57,45% and 74.60%, besides Table 3 shows that the encapsulation rate ranges between 64.27% and 79.36%.

The second protocol, based on an adsorption mechanism where caffeine diffuses through the microsphere wall relies on several physicochemical phenomena. As a result, various parameters can influence the encapsulation efficiency:

Caffeine concentration in the $CaCl_2$ bath: A higher concentration gradient between the external medium ($CaCl_2$ bath) and the capsule core promotes diffusion into the capsules. Once equilibrium is reached between the two phases, diffusion ceases.

Table 2. Encapsulation rate according to the first protocol

N°	Caffeine (mg/ml)	Alginate	Clay	$CaCl_2$	pH	Encapsulation rate (%)	Diameter (mm)
1	10	2,50%	1%	2%	5	70,37	1.76
2					7	71,58	1.47
3				3%	5	57,45	1.66
4					7	62,23	1.86
5				4%	5	72,70	1.53
6					7	70,07	1.67
7			3%	2%	5	70,69	1.7
8					7	71,61	2.23
9				3%	5	64,25	1.6
10					7	63,37	1.42
11				4%	5	72,45	1.7
12					7	73,13	1.65
13			5%	2%	5	69,60	1.3
14					7	72,50	1.42
15				3%	5	63,38	1.7
16					7	63,92	1.86
17				4%	5	70,99	1.58
18					7	69,42	2.02
19			7%	2%	5	68,17	1.62
20					7	70,98	1.54
21				3%	5	65,96	1.53
22					7	69,79	1.6
23				4%	5	71,85	1.62
24					7	74,60	1.75

Porosity and internal structure of the capsules (alginate + clay): Increased porosity facilitates caffeine diffusion. In contrast, highly dense or overly crosslinked capsules (excessive Ca^{2+}) act as barriers to diffusion.

Degree of crosslinking of the alginate gel: A highly crosslinked gel (due to high $CaCl_2$ concentration or prolonged gelation time) results in tighter pore networks, thereby limiting caffeine penetration.

Capsule size variability: Although the capsules are generally spherical, size heterogeneity has been observed, which may lead to differences in encapsulation efficiency.

Due to the complexity and sensitivity of these parameters, the first protocol was selected for the continuation of our study.

Table 3. Encapsulation rate according to the second protocol

N°	Caffeine (mg/ml)	Alginate	Clay	$CaCl_2$	pH	Encapsulation rate (%)	Diameter (mm)
1	10	2,50%	1%	2%	5	79,36	1.64
2					7	76,16	1.72
3				3%	5	69,07	1.92
4					7	77,90	1.86
5				4%	5	69,53	1.72
6					7	70,10	1.70
7			3%	2%	5	67,93	2.23
8					7	65,62	1.88
9				3%	5	64,76	1.95
10					7	65,10	1.69
11				4%	5	68,23	1.53
12					7	68,49	1.66
13			5%	2%	5	70,59	1.7
14					7	64,21	1.75
15				3%	5	67,17	1.85
16					7	66,34	1.67
17				4%	5	69,67	1.76
18					7	71,45	1.69
19			7%	2%	5	67,25	2.03
20					7	76,42	1.53
21				3%	5	62,01	1.58
22					7	68,87	1.4
23				4%	5	70,85	1.78
24					7	70,16	1.57

3.5 Effect of Various Factors on the Encapsulation Rate

3.5.1 Effect of pH on Encapsulation Rate

Figure 11 shows the effect of pH variation on the encapsulation rate for $CaCl_2$ concentrations of 2%, 3% and 4%.

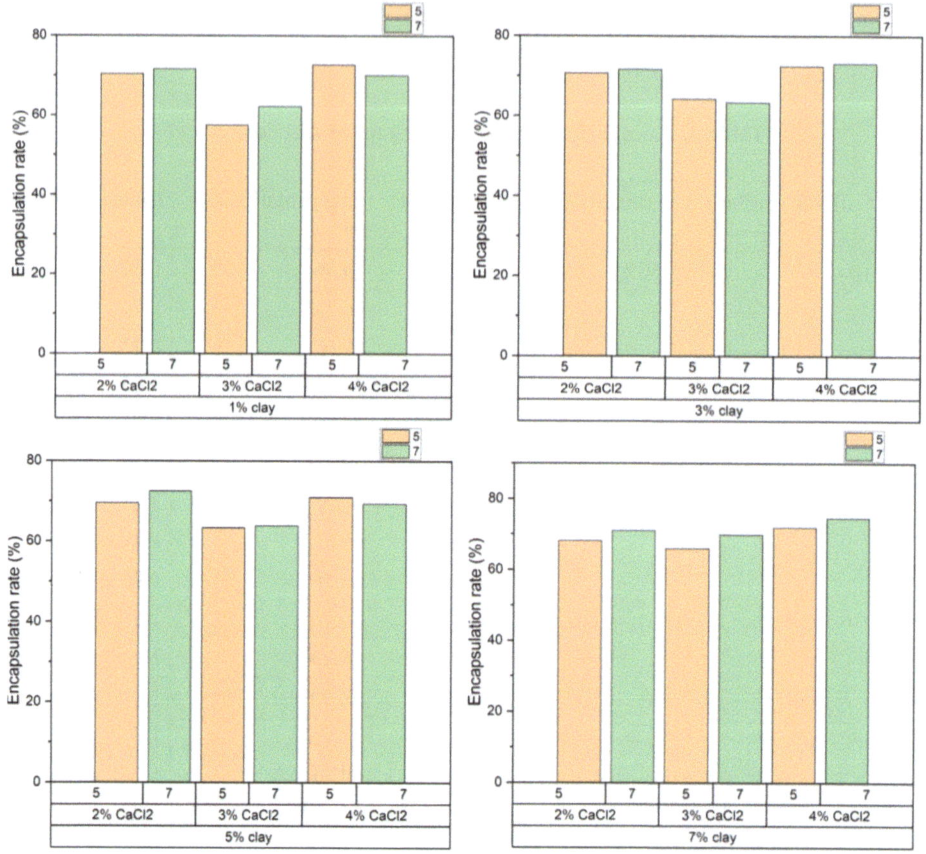

Fig. 11. Variation in encapsulation rate as a function of pH variation

At pH 5: average encapsulation rate is 68.15%. Besides at pH 7: The average encapsulation rate is 69.44%.

The encapsulation rate is slightly higher at pH 7 than at pH 5. This difference may indicate that encapsulation is more effective under neutral conditions (pH 7) than under slightly acidic conditions (pH 5).

Before studying the effect of pH on the rate of encapsulation, we studied the effect of pH on each component of the microsphere.

Effect of pH on Alginate: At pH 5 (Acid): Alginate is an anionic polysaccharide that can be sensitive to pH changes. At acidic pH, acidity can protonate some of alginate's carboxylate groups reducing their ability to interact with calcium ions (Ca^{2+}) in the $CaCl_2$ bath. This can lead to less effective cross-linking, and hence a less dense, more permeable gel structure, reducing the encapsulation rate.

At pH 7 (Neutral): alginate's carboxylate groups remain deprotonated and can interact more effectively with calcium ions, forming a denser, less permeable gel network. This can lead to better encapsulation and a higher encapsulation rate.

Effect of pH clay: At pH 5 (Acid): Clay contains minerals that can also be affected by pH. In an acidic environment, some clay minerals can partially dissolve, altering the microsphere's structure and ability to retain substances. The resulting more permeable structure may result in a lower encapsulation rate.

At pH 7 (Neutral): the clay retains its natural structure, contributing to the stability and density of the capsule, which improves encapsulation rates.

Gelation in a $CaCl_2$ bath: Calcium ions (Ca^{2+}) play a crucial role in alginate gelation, forming "bridges" between polysaccharide chains. The pH of the alginate/clay solution can affect the efficiency of this gelation.

At neutral pH, cross-linking is more efficient, forming a stronger, less permeable gel matrix and increasing encapsulation rates.

At acidic pH, this interaction is less effective, forming a less dense and more permeable gel, which reduces the encapsulation rate.

So we can deduce that at pH 5 (Acid) we have protonation of alginate carboxylate groups, partial dissolution clay minerals, and less efficient gelation in the $CaCl_2$ bath, resulting in a more permeable capsule structure and lower encapsulation rate. And at pH 7 (Neutral) we have deprotonated carboxylate groups, stable clay structure, and efficient gelation in $CaCl_2$ bath, forming a denser, less permeable gel matrix with a higher encapsulation rate.

3.5.2 Effect of Clay Concentration on the Encapsulation Rate

Figures 12 and 13 show the effect of varying clay concentration on encapsulation rate at two different pH values (pH = 5 and pH = 7). The analysis was carried out on capsules gelled in a $CaCl_2$ solution of varying concentration (2%, 3% and 4%).

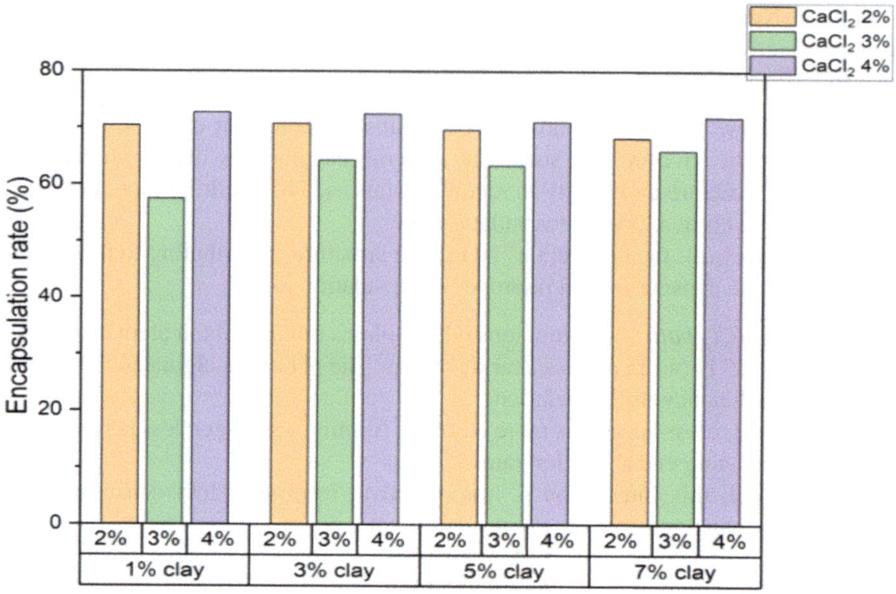

Fig. 12. Variation of encapsulation rate in function of the variation of clay concentration at pH = 5

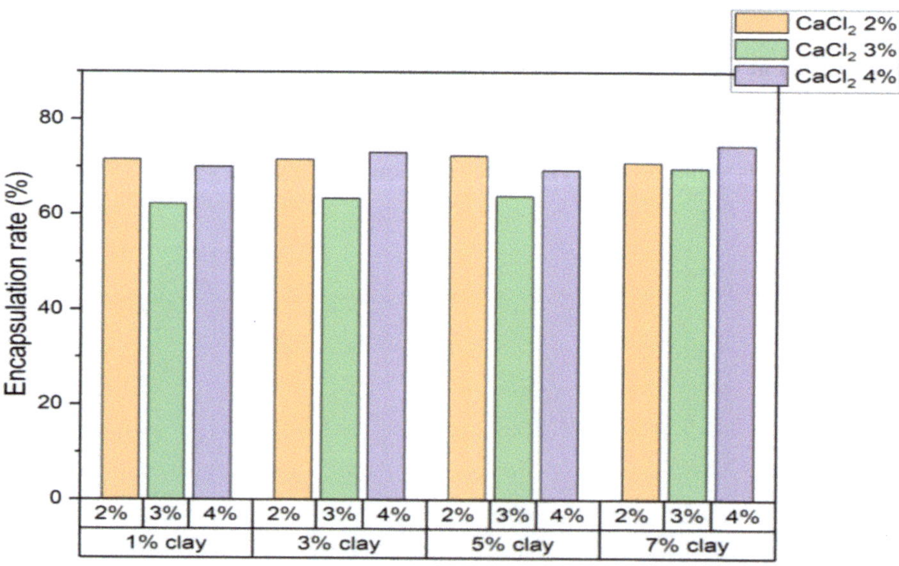

Fig. 13. Variation of encapsulation rate in function of the variation of clay concentration at pH = 7

It can be seen that at pH = 5 and pH = 7, the maximum encapsulation rate is obtained when the clay content is around 7%.

Capsules which contain 7% clay can form more porous structures, which promotes better encapsulation of substances through increased surface area for exchange.

3.5.3 Effect of $CaCl_2$ Levels on the Encapsulation Rate

Figures 14 and 15 show the effect of varying $CaCl_2$ concentration on encapsulation rate at two different pH values (pH = 5 and pH = 7). The analysis is carried out on capsules containing the four variations in clay content (1%, 3%, 5% and 7%).

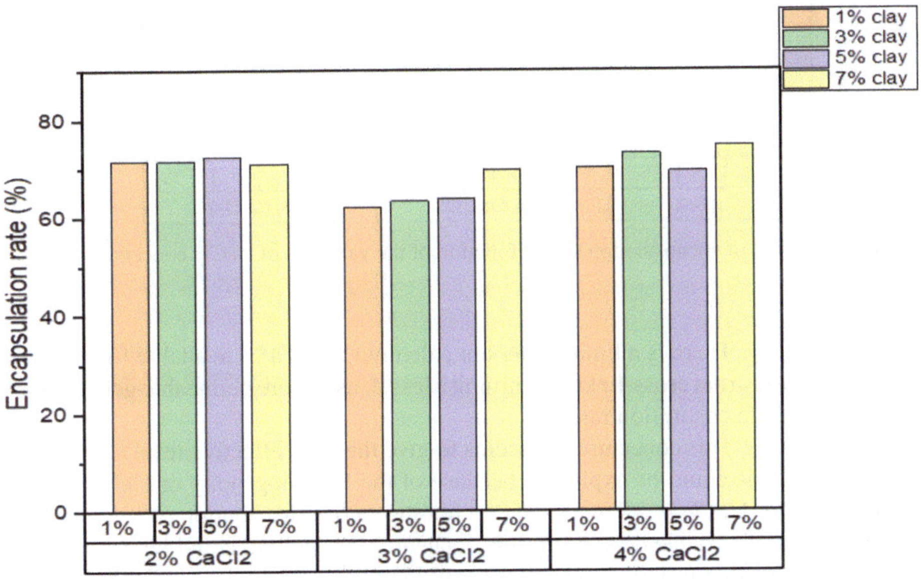

Fig. 14. Variation of encapsulation rate in function of the variation of $CaCl_2$ concentration at pH = 7

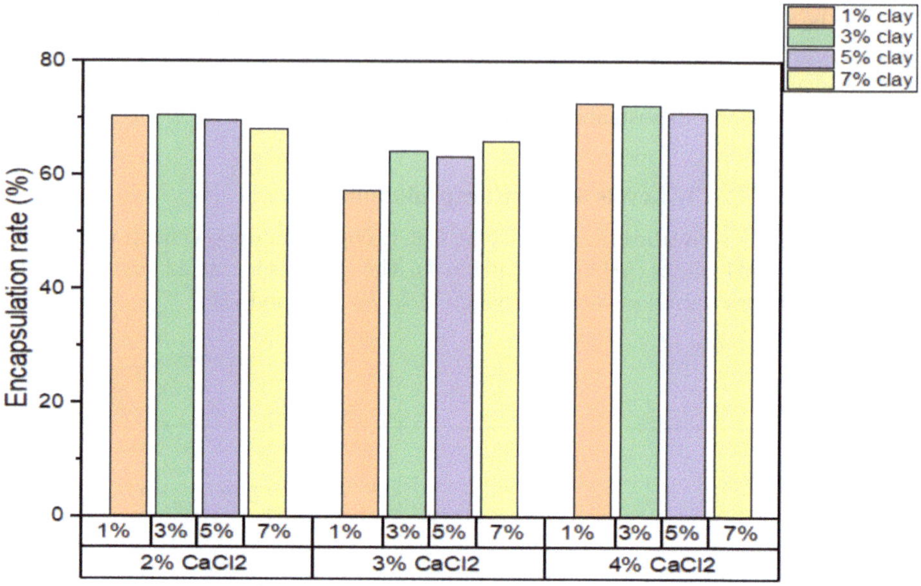

Fig. 15. Variation of encapsulation rate in function of the variation of $CaCl_2$ concentration at pH = 5

At 2% $CaCl_2$: There is a limited amount calcium ions (Ca^{2+}) available for gelation. This can lead to partial cross-linking, forming a less dense, more permeable gel, resulting in an intermediate absorption rate.

At 3% $CaCl_2$: This concentration seems to give the least effective results for encapsulation rate. This could be explained because of the Inhomogeneity of Gel, when the concentration of $CaCl_2$ is 3% we haven't a homogeneous distribution of calcium ions, creating zones of variable density. At this concentration, gelling could be partially disordered, creating denser looser gel zones, resulting in an uneven, less efficient structure for encapsulation.

At 4% $CaCl_2$: At this concentration, there are sufficient Ca^{2+} ions for optimal alginate cross-linking. This forms a denser, less permeable gel network, maximizing the encapsulation rate.

3.6 Release Kinetics of Active Ingredient (Caffeine)

3.6.1 Release Kinetics of Caffeine in a Simulated Physiological Medium (PBS at pH = 7)

The microspheres enable AI to be transferred between the internal and external environments by controlling the physical and chemical properties of the membrane.

The amount of caffeine released into PBS at pH = 7; increases progressively to 56% after 3 h incubation, then stabilizes at 65% until 7d incubation, showing a plateau. Figure 16 shows that the active ingredient is gradually released from the microcapsules into the alkaline medium.

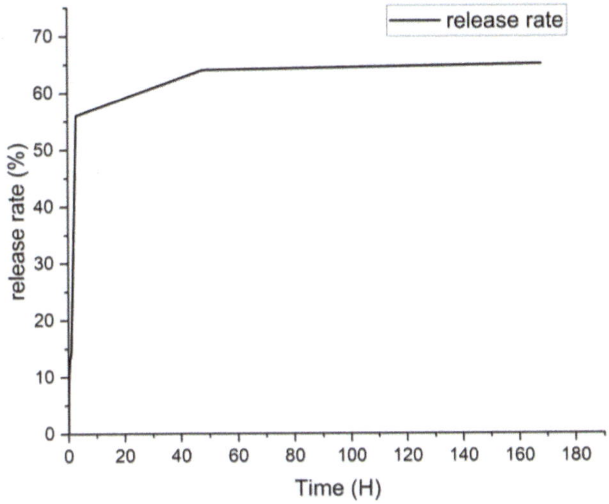

Fig. 16. Release kinetics in a pH = 7 medium

After an incubation period of 7 days, we observed the presence of small particles from the alginate and clay microspheres. Figure 17 indicates that the microspheres swelled and eventually burst in PBS medium at pH 7.

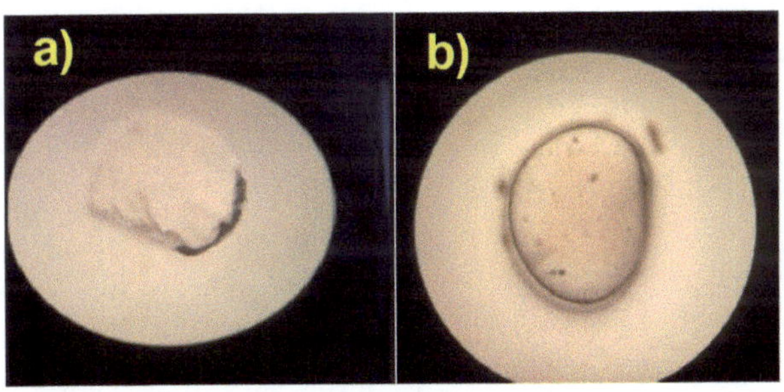

Fig. 17. a) bursted microsphere, b) swelled microsphere

3.6.2 Release Kinetics of the Caffeine in a Simulated Physiological Medium (pH = 1.2)

Figure 18 shows that the amount of caffeine released into the stomach medium, simulated at pH = 1.2, increases rapidly and exponentially, reaching 100% after 3 h incubation.

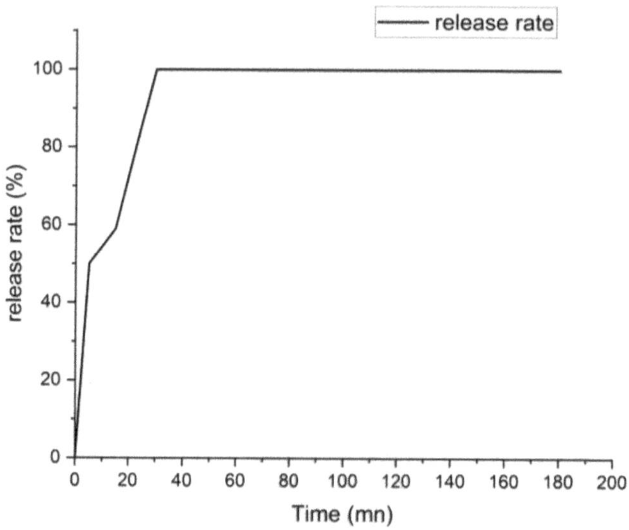

Fig. 18. Release kinetics in a pH = 1.2 medium

After a 3-h incubation period, caffeine is diffused through the microcapsules. Figure 19 shows that the microspheres did not swell or erode, but their size has decreased remarkably.

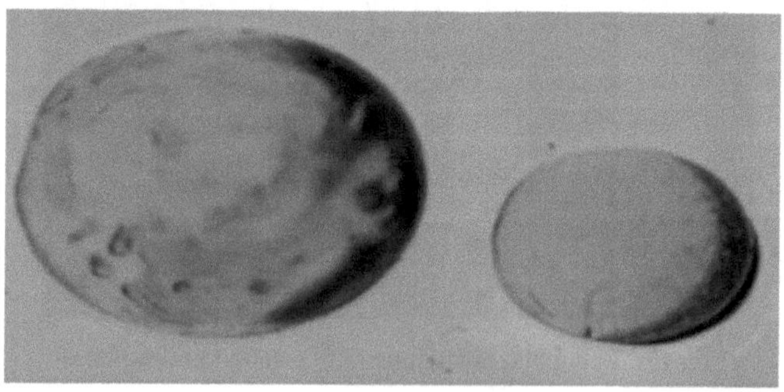

Fig. 19. The reduction of the microsphere's size

4 Conclusion

This study demonstrated the relevance of developing alginate–clay microspheres for biomedical applications, with a focus on obtaining perfectly spherical and morphologically uniform structures across all batches. The incorporation of clay significantly improved the mechanical strength and porosity of the microspheres, which are critical factors for effective encapsulation and controlled release.

By comparing two different encapsulation protocols, it was shown that the method involving the incorporation of caffeine into the alginate–clay mixture before gelation provided the most consistent results in terms of shape regularity and encapsulation efficiency. The optimization of key formulation parameters namely clay content, $CaCl_2$ concentration, and pH was essential to achieving these results.

Furthermore, the release behavior of caffeine was successfully evaluated in two simulated physiological environments: PBS (pH 7), mimicking the skin, and an acidic medium (pH 1.2), simulating gastric conditions. These findings underline the potential of the developed microspheres for future applications in localized and controlled drug delivery systems, especially in the context of wound healing or oral administration.

References

1. Paul, A., Cantor, A., Shum-Tim, D., Prakash, S.: Superior cell delivery features of genipin crosslinked polymeric microcapsules: Preparation, in vitro characterization and pro-angiogenic applications using human adipose stem cells. Mol. Biotech. **48**, 116–127 (2011)
2. Tam, S., et al.: Biocompatibility and physicochemical characteristics of alginate– polycation microcapsules. Acta Biomater. **7**, 1683–1692 (2011)
3. Bergaya, F., Theng, B.K.G., Lagaly, G.: Handbook of Clay Science, vol. 1, pp. 1–1224. Elsevier, Amsterdam (2006)
4. Tsiantos, C., Gionis, V., Chryssikos, G.D.: Smectite in bentonite: Near infrared systematics and estimation of layer charge. Appl. Clay Sci. **160**, 81–87 (2018)
5. Lertwimolnun, W.: Réalisation de nanocomposites polypropylène/argile par extrusion bivis. Thèse de doctorat, École Nationale Supérieure des Mines de Paris (2006)
6. Faure A. L'encapsulation chimique et physico-chimique d'actifs. Polymer Expert. Parution: 03-03-2023/Modifié le: 21-07-2023
7. Colomb F. Académie de Normandie–Sciences physiques: HPLC: Principe et appareillage (2020)
8. Techno-Science.net: Microscope optique. https://www.techno-science.net/glossaire- definition/Microscope-optique.html, consulté le 24 avril 2025
9. Cours IR. Chapitre III: La spectroscopie infrarouge. Document pédagogique, Master Sc-Anal M2, Université de Sfax (2025)
10. Torchi, S., Chennoufi, O.: Effet de l'activation chimique de la montmorillonite sur la photocatalyse des colorants. Mémoire de Master, Université Mohamed Khider de Biskra (2020)
11. Mehalli, A., Belaicha, F.Z.: Étude de l'activité de la caféine extraite de marc de café comme inhibiteur de corrosion acide d'un acier. Mémoire de Master, Université Saad Dahleb de Blida (2018)

Technical and Smart Textiles

Development and Challenges of Conductive Nanofiber—Based Textile Sensors

Parian Mohamadi[1], Elham Mohsenzadeh[2,3](✉), Cedirc Cochrane[2], and Vladan Koncar[2]

[1] Department of Environmental and Occupational Health, Université de Montréal, 2375 Chem. de La Côte-Sainte-Catherine, Montréal, QC H3T 1A8, Canada

[2] Univ. Lille, ENSAIT, ULR 2461 - GEMTEX - Génie et Matériaux Textiles, F-59000 Lille, France

elham.mohsenzadeh@junia.com

[3] Junia, F-59000 Lille, France

Abstract. Flexible textile sensors are increasingly used in wearable electronics, structural monitoring, and biomedical applications due to their adaptability and responsiveness to external stimuli. These sensors rely on conductive materials embedded in polymeric matrices to achieve electrical functionality while maintaining flexibility. However, optimizing their electrical conductivity without compromising mechanical properties remains a key challenge.

This study focuses on fabricating electrospun thermoplastic polyurethane (TPU) nanofibers enhanced with carbon-based conductive fillers to detect air filter clogging. Different concentrations of conductive particles were incorporated into the polymer solution, and the resulting nanofibrous membranes were evaluated for structural, mechanical, and electrical properties. The impact of filler dispersion and percolation on conductivity was analyzed, highlighting the trade- offs between increased electrical performance and potential material aggregation. Alternative methods, such as surface modifications, were explored to improve charge transport pathways.

In addition to applications in smart textiles and biomedical sensing, these conductive membranes show promise for real-time clogging detection in air filtration systems. By monitoring changes in electrical resistance, they can provide early warnings of filter clogging, optimize maintenance schedules, and improve efficiency. The findings offer valuable insights into tailoring electrospun conductive nanofibers for both wearable technology and industrial sensing applications.

Keywords: Smart textiles · Conductive nanoparticles · Integration method

1 Introduction

These days, flexible textile sensors, such as those for pressure, strain, humidity, and temperature, have attracted considerable attention in personalized health monitoring, human motion detection, structural health monitoring, brilliant garments, and robotics [1]. The important factors for evaluating these sensors are their sensitivity, stretchability, and

M. Hamdaoui et al. (Eds.): ITDM'3 2025, PTT, pp. 85–98, 2026.
https://doi.org/10.1007/978-981-95-7704-0_8

repeatability. The materials, structures, and methods involved affect the performance of flexible textile sensors. They can be made from natural or synthetic fibers such as cellulose acetate, cotton, wool, polyamide, polyester, and polyurethane, and manufactured through various processes such as melt spinning, weaving, knitting, electrospinning, printing, and screening [2]. Additionally, by using flexible polymers like polyurethane and embedding conductive materials, namely, carbon nanotubes, flexible textile sensors can be fabricated with the capability of integration into electronic devices or attachment to human skin [3]. Through chemical modifications, polymerization, synthesis, or the addition of conductive materials such as metallic particles, polymers can become conductive and responsive to various stimuli, including mechanical, chemical, electrical, and thermal [4].

Additionally, electronic devices can be integrated into textile structures using weaving, embroidery, and printing due to their flexibility and deformability. The sensing mechanism is influenced by the structure of the component, the fabrication method, and the materials applied, also, it is determined by four principal factors: i) behavior of the sensing material, ii) formation of cracks in the sensing layer, iii) formation of the conduction network, and iv) geometrical transformation [5].

Kim et al. fabricated a fiber strain sensor by embedding AgNPs using an in-situ formation process. This system exhibited an initial electrical resistance of 0.9 Ω/cm, a good strain sensing range (220%), high sensitivity (5.8×104), and was able to withstand 5000 cycles of stretching and releasing [6]. Ouyang et al. added MWCNTs and silver particles to a polyurethane solution before electrospinning, with different concentrations for biosensor applications. They reported that the interaction between MWCNTs and AgNPs improves the uniform distribution of AgNPs along the nanofibers, and the assembly of the two particles helps electron transfer and enhances the electrical properties [7].

In various research papers, it is claimed that incorporating conductive particles into the nanofiber structure causes low conductivity due to the insulating polymer wrapping layer around the particles [8, 9]. Therefore, increasing particle concentration is recommended, but it may also increase the viscosity of the solution and make electrospinning more difficult. In contrast, a high concentration of nanoparticles above the percolation threshold causes agglomeration in composites, resulting in a decrease in their mechanical properties. Darbandi et al. incorporated MWCNTs (2–12 wt.%) into polyurethane nanofibers and found that the conductivity increased for the sample containing up to 6% of carbon particles, while the tensile strength and elongation at break were reduced [10]. Another study reported that a high concentration of conductive fillers is required to achieve suitable conductivity, but this could have negative effects on the polymeric solution and the agglomeration of particles, which can decrease the electrical conductivity [11]. Almuhamed et al. reported the volume resistivity of MWCNTs/PAN nanofibers due to the lack of a conductive pathway on the surface of the webs [12]. Conclusively, to create sufficient conductive pathways on the surface of nanofibers, coating, immersion, and electrospray methods can be used. However, these methods could block the porosity of nanofibers and may not be suitable for applications such as air filtration. Moreover, the length, diameter, and surface area of the particles can affect the final properties [13].

TPU polymeric solution before electrospinning. After characterizing the solutions, the electrospinning parameters were optimized, and conductive membranes were produced. Finally, the morphological, mechanical, and electrical properties of the developed samples were investigated to assess the benefits of integrating conductive particles into the nanofiber structure. To evaluate their suitability for detecting air filter clogging applications, the final samples were tested in a ventilation tunnel to measure pressure drop and electrical resistance changes during strain and deformation. These tests aimed to assess the feasibility of using conductive nanofiber membranes for real- time clogging detection in air filters. By enabling early detection of filter clogging, these membranes could be used for predictive maintenance, reducing energy consumption and improving system efficiency.

2 Materials

TPU (Laripur 5760, MW: 100,000–200,000 Da) was purchased from Coim Group, Italy. N, N-Dimethylformamide (DMF, 98% purity) and tetrahydrofuran (THF, 98% purity) were provided by Sigma-Aldrich. To improve the conductivity and mechanical properties of polyurethane nanofibers, MWCNTs were bought from Nanocyl, Belgium, and carbon black from Ketjenblack, USA, was used. The properties of MWCNTs were reported to be 9.5 nm in diameter and 1.5 μm in length.

3 Method

3.1 Solution Preparation

The first two solutions of 20 wt.% and 11 wt.% of TPU (5760) were prepared in a mixture of DMF: THF (1:1 v/v) and stirred for 5 h at room temperature [14]. To prepare the solution with particles, different percentages of MWCNTs (0.5, 1, 1.5, and 2 wt.%) and CB (2, 3, 5, and 7 wt.%) were mixed in DMF: THF (1:1 v/v) with an ULTRA-TURRAX at 2000 rpm for 15 min and then for 60 min in an ultrasonic bath (37 kHz frequency and 70% power). Finally, TPU granules (20 and 11 wt.%) were added, and the solutions were stirred for 5 h. It is worth mentioning that DMF is a good solvent for dissolving and dispersing MWCNTs. After solution preparation, the solutions were electrospun with electrospinning parameters of 0.5–1.5 mL/h feed rate, 15–30 cm distance, and 12–20 kV voltage using a dynamic collector with a speed of 200 rpm, as shown in Fig. 1 [1].

3.2 Optimizing Ultrasonic Times by UV

To avoid a decrease in particle length, UV-Vis spectroscopy was used to determine the appropriate duration of ultrasonication for solution preparation. After each 15-min ultrasonication interval, the UV spectra of the solutions were recorded, and the absorbance curve was analyzed to observe the dispersion of the particles. The duration of ultrasonication was increased until the absorbance curve stabilized, indicating that the particles were well dispersed and no longer agglomerated. According to the literature, the absorption peak of MWCNTs is typically around 386 nm [15]. As presented in Table 1, the optimal

Fig. 1. The electrospinning setup.

ultrasonication time for dispersing the MWCNTs was determined to be 60 min, as after this point, the UV absorbance of the solutions did not show any further change. As the dispersion of CB was easier and less agglomeration was observed for those particles, this test was not done on the solutions including CB.

Table 1. The UV absorbance of the polymeric solution in different durations of ultrasonication

Time of ultrasonication (min)	Absorbance
15	0.06
30	0.08
45	0.16
60	0.35
75	0.35

3.3 Viscosity of Solutions

In electrospinning, the polymer entanglement, the molecular weight, and the viscosity of the solution must be sufficient to have the spinning capability, but not so high that they prevent stretching the jet under the electrical field. However, the viscosity of the electrified solution has a significant effect on the formation of beads, which is one of the most important defects that occur during the electrospinning of polymeric solutions. Research has shown that these defects happen more severely at low viscosities. The concentration of solutions influences viscosity and surface tension. Increasing the concentration raises the mass of the polymer in the jet, and in this condition, the viscosity goes up and creates a stronger bond between the polymer chains, making extension more difficult and causing an increase in the diameters [14]. In this study, a viscometer (Rheomat RM 100) was used to measure the viscosity of solutions containing different concentrations of particles to investigate the effect of carbon particles on the polymeric solutions. Table 2 shows the viscosity of TPU solutions 20 wt.% alone and with MWCNTs, whereas each measurement was repeated three times.

Table 2. The viscosity of solutions including different MWCNT percentages at a shear rate of 25 (1/s)

Solutions (wt.%)	Viscosity (mPa/s)
TPU (S860) 20	405 ± 11.1
TPU/0.5 MWCNTs	765 ± 23.4
TPU/1 MWCNTs	531 ± 29.3
TPU/1.5 MWCNTs	724 ± 22.6
TPU/2 MWCNTs	835 ± 36.1

As reported in earlier studies, adding CNTs increases the viscosity and conductivity of solutions, which is also deduced from Table 2. Therefore, it is recommended to reduce the polymeric concentration to balance the viscosity and decrease its effect on diameters. Eivazi Zadeh et al. decreased the PU concentration from 11 to 10 wt.% for 0.5 and 1 wt.% of CNTs to achieve similar diameters [13]. Darbandi et al. stated that the viscosity of solutions and diameters of nanofibers depend on MWCNTs' content and dispersion [10].

Table 3 presents the viscosity of TPU solutions with 11 and 20 wt.% concentrations containing various percentages of CB and MWCNTs. The comparison of the data indicates that CB causes a lower increase in the viscosity of solutions compared to MWCNTs and results in less particle agglomeration. The suitable viscosity range for the electrospinning machine (Bioinicia) is 170–400 mPa.s, Hence, the 11 wt.% polymeric solution and CB as the conductive particle were chosen to continue the research and produce uniform membranes.

3.4 Conductivity of Solutions

The significant effect of solution conductivity on nanofibers' diameters and final properties has resulted in much research. In this study, the conductivity of thermoplastic polyurethane solutions, including different concentrations of MWCNTs and CB, was measured using a Conductimeter (inoLab Cond 7110-WTW). The measurements were repeated three times for each solution.

As shown in Tables 4 and 5, the conductivity of the solutions increased significantly by increasing the percentage of particles increased. Although MWCNT particles

Table 3. The viscosity of solutions including conductive particles.

Solutions (wt.%)	Viscosity (mPa/s)
TPU 11	176 ± 9.8
TPU 11/2 CB	363 ± 15.1
TPU 11/3 CB	400 ± 13.2
TPU 11/5 CB	458 ± 16.7
TPU 11/7 CB	501 ± 20.9
TPU 20	336 ± 13.9
TPU 20/3 CB	804 ± 27.7
TPU 20/3 MWCNTs	1883 ± 34.1

Table 4. The conductivity of TPU (S860) solutions containing different MWCNT contents

Polymeric solutions (wt.%)	Conductivity (mS)
TPU 11	0.004 ± 0.001
TPU/0.5 MWCNTs	0.01 ± 0.02
TPU/1 MWCNTs	0.016 ± 0.002
TPU/1.5 MWCNTs	0.025 ± 0.004
TPU/2 MWCNTs	0.033 ± 0.004

are more conductive, using CB reduces the agglomeration of particles and allows for higher solution concentrations [16]. Almuhamed et al. reported that adding CNTs to PAN solutions increases the conductivity of the solution and lessens the diameter of the nanofibers [12]. The UV test results determined the optimal ultrasonication time for carbon particles, effectively reducing agglomeration and ensuring a homogeneous solution. Additionally, viscosity measurements showed that lowering the polymer concentration improved solution homogeneity and enhanced its suitability for electrospinning. Among the conductive fillers tested, carbon black particles proved to be the most suitable for this application, as they did not significantly alter the viscosity of the solution. This characteristic is beneficial for maintaining a stable electrospinning mechanism, leading to the consistent formation of conductive nanofiber membranes.

4 Characterizations

4.1 Scanning Electron Microscopy (SEM)

The morphology of the nanofibers was examined using SEM (Phenom ProX, ThermoFisher Scientific, US) and Jeol microscope. A 1 cm2 sample was cut from the membranes and mounted on an SEM sample holder using conductive carbon tape. Then,

Table 5. The conductivity of TPU solutions containing different percentages of CB

Polymeric solutions (wt.%)	Conductivity (mS)
TPU 11	0.0017 ± 0.0002
TPU 11/3 CB	0.0026 ± 0.0004
TPU 11/5 CB	0.0032 ± 0.0004
TPU 11/7 CB	0.0053 ± 0.0004

they were coated with a thin layer of gold for 1 min before experimenting. The average diameter of prepared nanofibers was determined using Digimizer software by measuring the mean of 50 fibers.

4.2 Transmission Electron Microscopy (TEM)

To confirm the presence of particles in the structure of nanofibers, transmission electron microscopy (TEM) images were acquired using a Tecnai G2 f20 S-Twin electron microscope (FEI, Eindhoven, the Netherlands) operating at 200 kV at the Université de Picardie Jules Verne. TEM samples were prepared by direct electrospinning the nanofibers onto holey carbon-coated copper grids (Ted Pella Inc., USA).

4.3 Mechanical Properties

In this study, the mechanical properties of polyurethane nanofibers produced using different electrospinning parameters and having various diameters were investigated using an MTS Criterion machine. Typically, ISO 527-3, 1, and ASTM D882 (Standard Test Method for Tensile Properties of Thin Plastic Sheeting) standards are applied for the mechanical testing of nanofibers, therefore, the test was carried by these standards and repeated three times for each sample.

4.4 Electrical Properties

To measure the electrical linear resistance at different distances, surface resistance, and volume resistance of nanofibers at different locations, a multimeter (Keithley, France), a four-probe sheet resistance meter (OSSILA, France), and a Resistivity Test Fixture Box (Keithley, France) were used in sequence. The resistance and capacitance of samples were evaluated using an impedance card (Analog Discovery Impedance Digilent - 50 Hz frequency).

5 Results

5.1 SEM Images and Diameter Measurements

After finding the optimal polymeric concentration and electrospinning parameters, MWCNTs were added to the solution at 0.5, 1, 1.5, and 2 wt.% and electrospun. Figure 2b shows an SEM image of the analyzed samples. As expected, the addition of MWCNTs decreased the diameters of the nanofibers from 940 ± 110 nm to 693 ± 120 nm due to the increased conductivity of the solutions. However, the resulting structures were not satisfactory, particularly for the samples with 2 wt.% particles, as they formed a partially film-like structure. Thus, the parameters needed to be optimized to account for changes in the viscosity and conductivity of the polymeric solutions. Darbandi et al. and Tondnevis et al. reported that an increase in viscosity can lead to droplet formation and bead defects. Additionally, to ensure a good conductive pathway, the nanofiber structure should be uniform with thinner fibers to ensure that the particles are on the surface of the samples and have sufficient contact between them [10, 16]. Then, CB was incorporated into TPU, as illustrated in Fig. 2a. The SEM image indicated the presence of uniform nanofibers, suggesting that CB was better dispersed in the polymeric solutions compared to MWCNTs, and their effects on the solution's viscosity and nanofibers morphology were less significant, regarding

Table 5. Moreover, polymer LP 5760 was suitable for electrospinning owing to its molecular weight and chain entanglement. As a result, the structure of samples containing CB was free of beads, uniform, smooth, and homogeneous. The diameters were measured to be 630 ± 70 nm and 1.2 ± 0.25 μm for samples a and b, respectively.

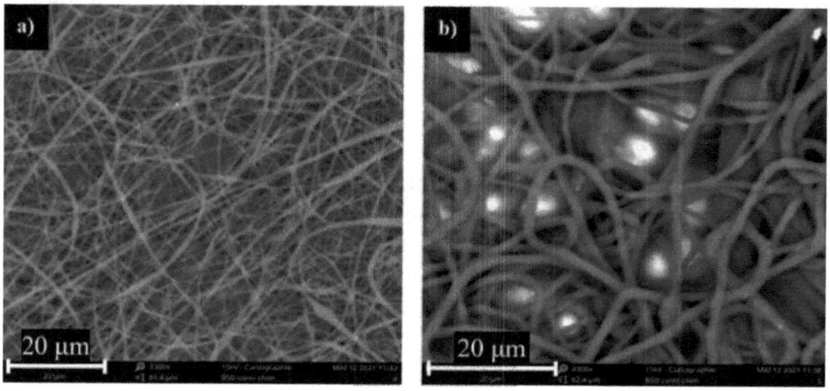

Fig. 2. The SEM images of TPU nanofibers containing a) 2 wt.% CB with electrospinning parameters of 12 kV, 15 cm, and 1.5 ml/h, and b) 2 wt.% MWCNTs, with 22 kV, 20 cm, 0.2 ml/h.

Figure 3 shows that the electrospinning parameters have been properly adjusted, and the nanofibers are bead-free, uniform, and smooth, even with a high concentration of CB. The calculated diameters were 800 ± 79 nm, 780 ± 105 nm, 731 ± 116 nm, and 670 ± 101 nm for TPU/2 wt% CB, TPU/3 wt.% CB, TPU/5 wt.% CB, and TPU/7 wt.% CB, respectively, with a voltage of 12 kV, distance of 15 cm, and feed rate of 1.5 ml/h. The diameters decreased slightly by adding CB, which could be due to the higher conductivity of polymeric solutions. An SEM image was taken of the TPU membrane containing 3 wt.% of CB using a Jeol IT200LA to see the presence of particles on the surface of the nanofilaments (Fig. 3e). To compare the surfaces of the samples, the surface of the nanofibers containing particles demonstrated more roughness, while the surface of the neat TPU was flat and uniform, which confirms the presence of CB on the surface of the samples.

5.2 TEM Images

Figures 3 and 4 show TEM and SEM images taken for TPU/3 wt.% CB to confirm the presence of carbon particles in the structure of nanofibers. It can be observed that the round-shaped CB is distributed evenly in the nanofiber polymer matrix without any agglomeration. To further validate the findings, it should be noted that Haider et al. applied TEM to detect the presence of Nano-Hydroxyapatite and 1 wt.% of carbon black in PAN nanofibers and reported the agglomeration of both particles on the surface of samples [17]. According to Datsyuk et al., carbon nanotubes are observed to have an aligned shape in the structure of nanofibers, while carbon black particles exhibit a round shape [18].

5.3 Mechanical Properties

Using thermoplastic polyurethane as a synthetic polymer is recommended to boost the poor mechanical properties of nanofibrous structures. In addition, this polymer could be convenient in composites and textile sensor applications due to its high elasticity and flexibility as about 70% strain was obtained in this research [19]. Previous research by Wang et al. showed that adjusting the speed of the electrospinning drum collector influenced fiber alignment and diameter, with higher speeds resulting in finer, more stretched fibers. Inspired by these findings, we employed a rotating collector in our electrospinning setup to control fiber morphology and alignment, ensuring uniform fiber formation and improved mechanical properties. This setup helped prevent excessive fiber entanglement and promoted better dispersion of conductive fillers such as carbon black (CB). Unlike MWCNTs, which increased solution viscosity and obstructed electrospinning, CB maintained good processability, forming conductive pathways without compromising fiber structure. The use of a rotating collector was essential to optimizing fiber alignment and mechanical performance, enabling the production of TPU-based conductive nanofibers suitable for applications in strain sensing and air filtration. By ensuring controlled fiber morphology, our approach aimed to enhance both conductivity and mechanical stability, aligning with the requirements for real-time clogging detection and predictive maintenance in air filtration systems [20].

Darbandi et al. demonstrated that incorporating MWCNTs into PU nanofibers significantly enhances mechanical properties, particularly tensile strength and Young's modulus, up to an optimal concentration of 6 wt.%. However, beyond this concentration, agglomeration led to reduced mechanical performance and flexibility due to restricted polymer chain movement. In our study, as shown in Fig. 5 and Table 6, the addition of CB to TPU membranes resulted in a notable increase in both maximum force (from 0.22 ± 0.017 to 2.09 ± 0.06) and elongation at break (from 41 ± 0.0002 to 137 ± 0.0003). Unlike MWCNTs, CB did not significantly alter the viscosity of the solution, allowing for a uniform membrane formation and effective electrospinning. Conversely, when MWCNTs were added to the system, the increased viscosity obstructed the electrospinning process, preventing the fabrication of uniform membranes and, consequently, the evaluation of their mechanical properties. This highlights the advantage of CB as a conductive filler in electrospun TPU membranes, as it maintains processability while enhancing mechanical performance. This comparison reinforces the importance

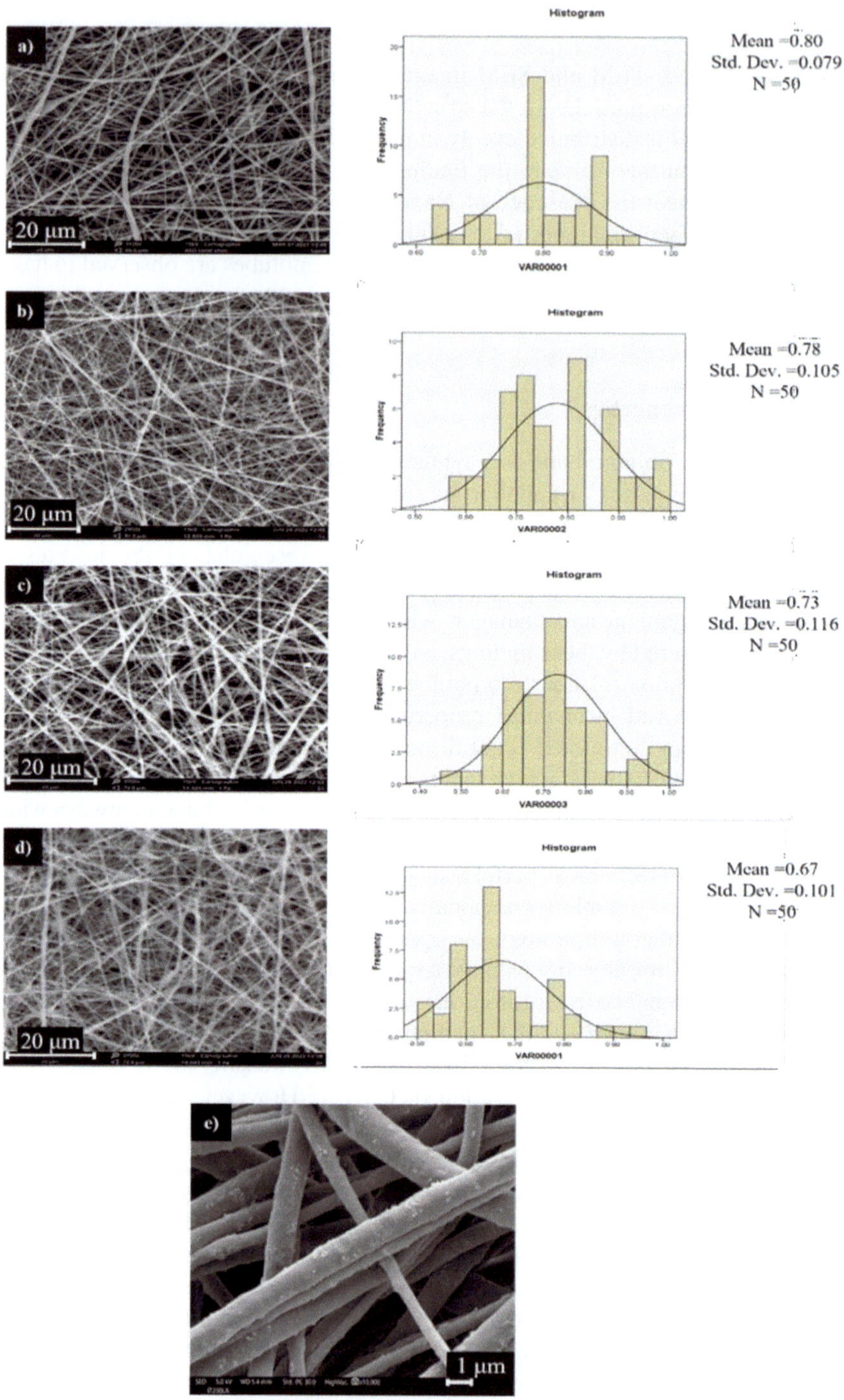

Fig. 3. SEM images of a) TPU/2 wt% CB, b) TPU/3 wt.%. CB, c) TPU/5 wt.% CB, d) TPU/7 wt.% CB by Phenom microscope, and e) TPU/3 wt.% CB using Jeol microscope.

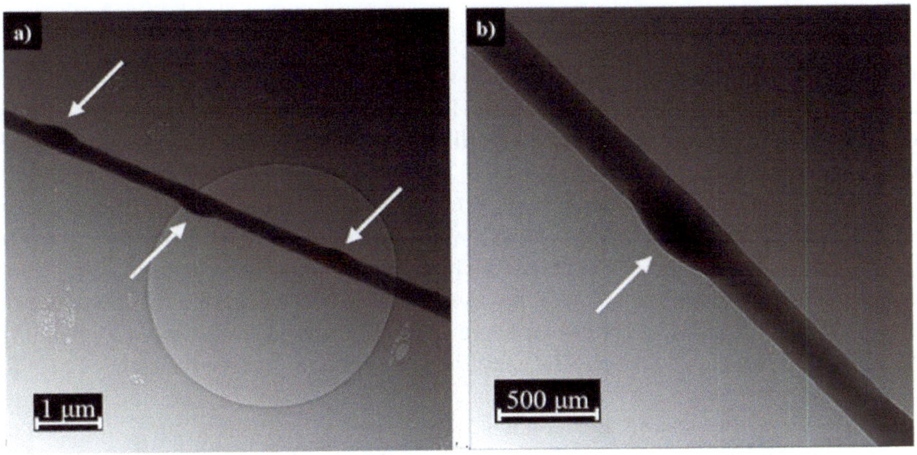

Fig. 4. TEM images of TPU nanofibers containing 3 wt.%. of CB

Table 6. The mechanical properties of the samples

Samples	Average of max force at break (N)	Average of max elongation at break (%)
TPU	0.22 ± 0.017	41 ± 0.0002
TPU/3 CB	2.09 ± 0.06	137 ± 0.0003

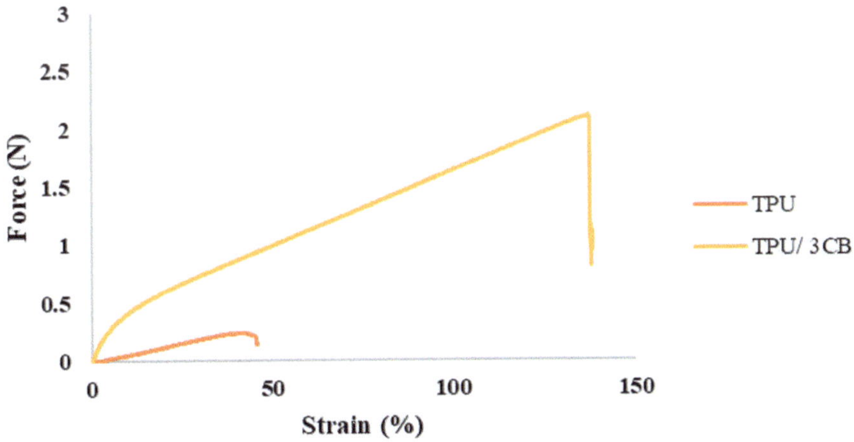

Fig. 5. The force-strain curve of TPU and TPU/3 wt.% CB

of selecting an appropriate conductive filler to balance mechanical reinforcement with solution processability in electrospinning applications [10].

5.4 Electrical Properties

In this study, first, a sheet resistance device was used to measure the resistance of TPU nanofibers comprising MWCNTs and CB. However, no values were obtained using this method. To compare the results with nanofibers, films from polymeric solutions were prepared containing

MWCNTs and heated in an oven at 100 °C for 2 h. Using an ohmmeter, resistance values of 264.2 kΩ, 33 kΩ, and 2.48 kΩ were measured for films including 2 wt.%, 5 wt.%, and 10 wt.% MWCNTs, respectively. Based on these observations, it was concluded that the sheet resistance method was not suitable for measuring the resistance of nanofibers. Additionally, it should be noted that Almuhamed et al. reported obtaining the volume resistance of PAN/CNTs nanofibers, rather than surface resistance [12]. To measure both surface and volume resistance, homemade and industrial ring electrodes were applied. Before testing, the samples were heated in an oven at 40 °C for 24 h to remove any solvent residue. Nonetheless, the resistance values obtained using the ring electrodes were not stable and increased over time. Thus, it is believed that this may have been due to the damage to the particles caused by the pressure applied by the device.

Furthermore, an impedance card was used to measure the impedance at a frequency of 50 KHz, but no resistance or capacitance was observed for the samples. To achieve surface conductivity, the concentration of MWCNTs was increased while decreasing the concentration of polymer to balance the viscosity. Specifically, 5 wt.% and 10 wt.% MWCNTs were added to the polymeric solution by decreasing the concentration of TPU from 20 to 10 and 5 wt.%. However, the electrospinning mechanism did not work well, and nonuniform nanofiber morphology was detected due to high or low viscosity and blockage of the needle.

Previous studies have used various methods, such as immersion, spray, and in-situ polymerization, to make the surface of nanofibers conductive [21, 22]. Since this study aims to use the TPU membranes for air filtration, the immersion method was deemed unsuitable since it blocked the membrane pores. Therefore, the particles were integrated into the structure of TPU membranes to maintain the porosity of the samples, but no surface conductivity was observed

6 Conclusions

In this study, nanofibers and electrospinning were chosen as structures and methods to produce textile sensors. To achieve a flexible system and facilitate interconnection into an electronic device, flexible polyurethane polymer was used. The type of solvents, polymeric concentration, and electrospinning parameters were optimized. After adding MWCNTs and CB to make the samples conductive, the viscosity and conductivity of solutions were studied. SEM images showed that the nanofibers were smooth, uniform, bead-free, and homogeneous. TEM images confirmed the presence of carbon particles in the structure of membranes. According to mechanical tests, the maximum force and maximum elongation at the break of TPU membranes increased from 0.22 ± 0.017 to 2.09 ± 0.06 and from 41 ± 0.0002 to 137 ± 0.0003 by adding CB to the system.

Although the solution comprising carbon particles showed conductivity, and TEM images confirmed the presence of particles within and on the surface of the nanofibers, no conductivity was seen due to the lack of a conductive pathway on the surface of the samples. As a result, the immersion method and film preparation are used for sensor applications, but the porosity of nanofibers could be blocked, which disturbs the passage of air in the air filtration system Xiong et al. announced that the non- conductivity of CB/polyurethane webs could be the consequence of three reasons. i) the pores of fibers, which avoid conductive pathway formation, ii) the long distance between particles, which is caused by the high elongation rate of electrospinning, and iii) the inhomogeneous distribution of CB between filaments and lack of bond between nanofibers due to fast evaporation of solvents during electrospinning [23].

Although in this study, different conductive particles with various percentages were integrated into the TPU nanofiber structure and their presence was confirmed by morphological and mechanical characterization, no surface conductivity was detected. The immersion method and surface treatment were recommended but could not be used in this study, as the objective was to develop strain sensors for detecting air filter clogging. Since the samples would be tested in a ventilation tunnel, the nanofiber porosity needed to be preserved. Therefore, the printing method was used to deposit carbon particles on the sample surfaces in different geometries while maintaining porosity, and the results were presented in our work [1].

References

1. Mohamadi, P., Mohsenzadeh, E., Cochrane, C., Koncar, V.: Highly sensitive electrospun sensory membrane for low-speed air flow detection. IEEE Sensors J. **23**(24), 27755–27763 (2023)
2. P. Mohamadi, E. Mohsenzadeh, C. Cochrane, and V. Koncar, "Investigation of conductive printed thermoplastic polyurethane nanofibers to detect the clogging of air filters," In: IOP Conference Series: Materials Science and Engineering, 2023, vol. 1266, no. 1: IOP Publishing, p. 012005
3. Mohamadi, P., Zaman, S.u., Mohsenzadeh, E., Cochrane, C., Koncar, V.: Preparation and reliability performance evaluation of electro-spun strain sensing membrane. Smart Mater. Struct. **33**(9), 095046 (2024)
4. Khaldi, A., et al.: Development of multi-layer electrochromic polymer display for remotely tunable green-cyan contact lens. AIP Adv. **14**(6), 065003–065012 (2024)
5. Mohamadi, P., Zaman, S.U., Mohsenzadeh, E., Cochrane, C., Koncar, V.: Advanced electrospun membrane for comprehensive strain detection. In: 2024 International Conference on the Challenges, Opportunities, Innovations and Applications in Electronic Textiles (E-Textiles), pp. 345–349. IEEE, Piscataway, NJ (2024)
6. Kim, H., et al.: In situ formation of Ag nanoparticles for fiber strain sensors: Toward textile-based wearable applications. ACS Appl. Mater. Interfaces. **13**(33), 39868–39879 (2021)
7. Ouyang, Z., et al.: Fabrication, characterization and sensor application of electrospun polyurethane nanofibers filled with carbon nanotubes and silver nanoparticles. J. Mater. Chem. B. **1**(18), 2415–2424 (2013)
8. Tang, J., Wu, Y., Ma, S., Yan, T., Pan, Z.: Flexible strain sensor based on CNT/TPU composite nanofiber yarn for smart sports bandage. Compos. Part B. **232**, 109605 (2022)
9. Wang, Y., et al.: Fabrication of high-performance wearable strain sensors by using CNTs-coated electrospun polyurethane nanofibers. J. Mater. Sci. **55**, 12592–12606 (2020)
10. Alizadeh Darbandi, S.M., Nouri, M., Mokhtari, J.: Electrospun nanostructures based on polyurethane/MWCNTs for strain sensing applications. Fibers Polym. **13**, 1126–1131 (2012)

11. Gao, J., et al.: Electrically conductive and fluorine free superhydrophobic strain sensors based on SiO2/graphene-decorated electrospun nanofibers for human motion monitoring. Chem. Eng. J. **373**, 298–306 (2019)
12. Almuhamed, S., Khenoussi, N., Schacher, L., Adolphe, D., Balard, H.: Measuring of electrical properties of MWNT-reinforced PAN nanocomposites. J. Nanomater. **2012**(1), 750698 (2012)
13. Zadeh, Z.E., Solouk, A., Shafieian, M., Nazarpak, M.H.: Electrospun polyurethane/carbon nanotube composites with different amounts of carbon nanotubes and almost the same fiber diameter for biomedical applications. Mater. Sci. Eng. C. **118**, 111403 (2021)
14. Mohamadi, P., Mohsenzadeh, E., Cochrane, C., Koncar, V.: Morphological and mechanical properties of electrospun polyurethane nanofibers—air-filtering application. In: Kumar, A., Ramakrishna, S. (eds.) Electrospun Nanofibers: Principles, Technology and Novel Applications, pp. 417–450. Springer, Cham (2022)
15. Grossiord, N., Regev, O., Loos, J., Meuldijk, J., Koning, C.E.: Time- dependent study of the exfoliation process of carbon nanotubes in aqueous dispersions by using UV−Visible spectroscopy. Anal. Chem. **77**(16), 5135–5139 (2005)
16. Tondnevis, F., Keshvari, H., Mohandesi, J.A.: Physico-mechanical and in vitro characterization of electrically conductive electrospun nanofibers of poly urethane/single walled carbon nano tube by great endothelial cells adhesion for vascular tissue engineering. J. Polym. Res. **26**, 1–16 (2019)
17. Haider, M.K., et al.: Polyacrylonitrile/Carbon Black nanoparticle/Nano-Hydroxyapatite (PAN/nCB/HA) composite nanofibrous matrix as a potential biomaterial scaffold for bone regenerative applications. Mater. Today Commun. **27**, 102259 (2021)
18. Datsyuk, V., et al.: Thermal properties enhancement of epoxy resins by incorporating polybenzimidazole nanofibers filled with graphene and carbon nanotubes as reinforcing material. Polym. Test. **82**, 106317 (2020)
19. Li, Y., Yin, X., Yu, J., Ding, B.: Electrospun nanofibers for high- performance air filtration. Compos. Commun. **15**, 6–19 (2019)
20. Wang, X., Liu, X., Schubert, D.W.: Highly sensitive ultrathin flexible thermoplastic polyurethane/carbon black fibrous film strain sensor with adjustable scaffold networks. Nano-Micro Lett. **13**, 1–19 (2021)
21. Ren, M., et al.: Highly stretchable and durable strain sensor based on carbon nanotubes decorated thermoplastic polyurethane fibrous network with aligned wave-like structure. Chem. Eng. J. **360**, 762–777 (2019)
22. Wang, L., et al.: Highly stretchable, anti-corrosive and wearable strain sensors based on the PDMS/CNTs decorated elastomer nanofiber composite. Chem. Eng. J. **362**, 89–98 (2019)
23. Xiong, Z.-Y., Kong, X.-Y., Guo, Z.-X., Yu, J.: Poly (ethylene terephthalate)/carbon black composite fibers prepared by electrospinning. Chin. J. Polym. Sci. **33**, 1234–1244 (2015)

Absorbing Meta-Textile: An Innovative Barrier Against the Carcinogenic Effects of Electromagnetic Waves

Hanen Zribi(✉) and Amine Hadj Taieb

Tunisia Textile Engineering Research Laboratory, Higher Institute of Arts and Crafts, University of Sfax, LGTex, Kssar Hellal, Tunisia
zribihanen41@gmail.com

Abstract. With the increasing presence of electromagnetic (EM) waves in our environment—primarily driven by the extensive adoption of wireless technologies and smart devices—concerns have intensified over the possible health implications, particularly in relation to long-term exposure and cancer development. While scientific findings on this matter remain debated, the precautionary principle promotes the pursuit of effective protective measures. This paper investigates the use of absorbing meta-textile based on woven meta material as a promising approach to reducing EM wave exposure. These innovative textile-based structures exhibit distinct electromagnetic characteristics that allow them to target and attenuate specific frequency ranges through selective absorption or reflection. The article presents the fundamental design and functionality of meta-textile, discusses their incorporation into intelligent fabrics, and explores their potential use in shielding applications aimed at health preservation. Special attention is given to their possible role in cancer prevention and their interdisciplinary significance across materials science, electromagnetics, and biomedical engineering.

Keywords: Meta-textile · electromagnetic waves · cancer · barrier effect

1 Introduction

In recent times, the widespread integration of wireless communication tools and smart technologies has led to a notable rise in the ambient levels of electromagnetic (EM) radiation. Everyday devices such as smartphones, Wi-Fi routers, connected home systems, and industrial sensors contribute to a constantly active digital environment saturated with EM waves. While these advancements have greatly enhanced our ability to connect and share information, they have simultaneously sparked growing concern about the potential health implications of long-term EM exposure. Of particular interest is the debated association between continuous EM radiation and the emergence of certain cancers, a topic that continues to draw attention from the scientific community [1].

Despite the lack of conclusive evidence, cautionary approaches support the pursuit of solutions that could help minimize unnecessary or prolonged exposure to electromagnetic fields. Within this framework, the innovation of advanced shielding materials is

M. Hamdaoui et al. (Eds.): ITDM'3 2025, PTT, pp. 99–108, 2026.
https://doi.org/10.1007/978-981-95-7704-0_9

gaining momentum. A particularly promising approach lies in the development of meta-textiles—intelligent fabrics incorporating woven metamaterials designed to interact with electromagnetic waves in precise and predictable ways. These specialized textiles can be engineered to absorb or reflect targeted frequency ranges, potentially serving as a barrier to harmful radiation [2]. This study focuses on exploring the structural characteristics, electromagnetic behavior, and practical applications of absorbing meta-textiles for EM shielding purposes. It examines their potential integration into wearable technologies and protective infrastructures, with a specific emphasis on their relevance in preventive health measures—especially in reducing risks potentially linked to cancer. By bridging disciplines such as material science, electromagnetism, and biomedical engineering, meta-textiles emerge as a pioneering solution in the development of smart, health-conscious fabrics for the digital age.

2 Electromagnetic Waves and Their Impact on Health

Certain frequencies of electromagnetic (EM) waves carry potential danger, as their harmfulness is directly linked to their frequency. Indeed, the higher the frequency, the more energetic the wave, which facilitates its deep absorption by body tissues. Among high-frequency waves, we find gamma rays and X-rays, commonly used in radiology but posing a notable danger to biological tissues. Ultraviolet rays alter the DNA of skin cells, promoting the development of skin cancers (Rozières, 2016). On the other hand, visible light and infrared rays are much less harmful, suggesting that reducing their frequency can help minimize risks. The radiofrequency range is commonly used in many applications, such as mobile phones and wireless internet routers, and this energy is sufficient to excite water molecules, causing heating of exposed body areas, much like how a microwave oven works [3].

2.1 Sources and Propagation of Electromagnetic Waves

Electromagnetic (EM) waves are energy transmissions that result from the interaction of electric and magnetic fields, and they can travel through both vacuum and matter. These waves are categorized based on their frequency and wavelength, which define their position on the electromagnetic spectrum. This spectrum extends from low frequency radio waves to high-energy gamma rays, encompassing microwaves, infrared, visible light, ultraviolet, and X-rays [4]. Within the scope of everyday technology and public health, radiofrequency (RF) and microwave radiation are particularly noteworthy, as they are frequently used in modern wireless communication systems. Modern lifestyles are increasingly surrounded by devices that emit EM waves. Among the most prevalent sources are:

- Smartphones: These are major emitters of radiofrequency waves, especially when establishing communication with cellular base stations. The emission levels can vary depending on signal strength and usage patterns.
- Wi-Fi routers: Operating on frequency bands such as 2.4 GHz and 5 GHz, these devices emit EM waves continuously to provide wireless internet access.

- Mobile network antennas: Deployed to enable mobile connectivity across vast areas, these installations emit EM waves non-stop to maintain network service.
- Connected objects and smart technologies: Devices like smartwatches, voice assistants, and home automation systems add to the ambient EM radiation in indoor and outdoor environments.

The widespread integration of wireless technologies into homes, workplaces, and public spaces has led to a marked increase in EM wave exposure. Developments such as the rollout of 5G networks, the expansion of IoT (Internet of Things), and the shift toward remote and connected lifestyles have all contributed to this growing exposure [4]. Even though current exposure levels are generally within regulatory limits, the accumulation of low-level radiation over time has prompted ongoing scientific inquiry and public concern, particularly regarding its possible link to long-term health effects such as cancer.

2.2 Biological Effects of Electromagnetic Waves

The standards governing the use of everyday devices are set with consideration for the thermal effects of the radiofrequencies emitted and received. In 1981, a NASA study highlighted the harmful effects of human exposure to powerful electromagnetic waves, causing symptoms such as headaches, fatigue, nervous system disruption, muscle pain, and even cases of cancer. The challenge lies in finding a concrete link between mobile phone use and brain cancer, a topic at the heart of many current studies. Indeed, some studies suggest that using a mobile phone for half an hour a day can quadruple the risk of developing a tumor [5].

Our daily environment is filled with electromagnetic waves that enable technological innovation but also present harmful effects on our health. Professions at risk include those in the following sectors:

- Telecommunications sector: Radio or television broadcast antennas, mobile phone antennas, radar stations; corresponding to intermediate frequency ranges (or radio frequencies), high frequencies (VHF), and microwaves (UHF).
- Industrial sectors: Industrial microwave ovens, arc welding, dielectric heaters, high-frequency presses, etc.; corresponding to high frequencies and microwaves. Electrolytic cells use a direct current at zero frequency.
- Power distribution sector: High and very high voltage power lines and substations, corresponding to low frequency ranges (ELF).
- Medical sector: Including MRI (continuous electric field at 0 Hz) based on static magnetic fields or electrocoagulation (thermal effect obtained by high frequency current >300 kHz).
- Higher education and office sectors: This exposure is the same as that of the general public but comes from more sources and with more frequent use of the following devices: computers, Wi-Fi, RFID chips, sirens, remote controls, mobile phones, etc.

The World Health Organization (WHO) asserts that waves emitted by television antennas and mobile phones can cause hyperthermia and electric shock, disrupting the body's proper functioning. These waves are classified as carcinogenic agents for humans.

Symptoms of electromagnetic sensitivity in individuals exposed to these waves include fatigue, anxiety, headaches, dizziness, concentration issues, digestive problems, and dermatological issues [6].

The contestation of these effects is difficult because living beings are impacted by electromagnetic waves. However, not all humans are equally affected, and children are the primary focus and object of concern. Indeed, children's bodies are more exposed to electromagnetic waves due to the resonance effect, as they are smaller than adults. The electromagnetic field easily reaches the internal organs of children, as they have a lower body mass, making the structure and function of their organs more vulnerable, since they have not yet reached full maturity [7].

Furthermore, children grow into dependent and responsible adolescents. They seek knowledge, testing, confrontation, and transgression... Their belief in being free with their mobile phones and talking endlessly has become difficult, as they are still controlled by their parents who must ensure:

- Clear explanation of Bluetooth devices and how they work.
- Limitation of time spent using Wi-Fi-enabled digital devices.
- Equipping adolescents with wired computers and mice.
- Engaging with other parents to raise awareness among teaching staff (obligation for students to deposit their mobile phones, turned off, in a common box or personal locker during lessons, etc.).
- Encouraging good practices to ensure a healthy life.

To protect ourselves and stay safe from electromagnetic waves, we must innovate a product or material capable of blocking these waves, which is where the concept of woven barrier-effect metamaterial comes in. It is also necessary to raise awareness among the connected population, whether children or adults. In the next section, we will discuss the woven barrier-effect metamaterial, mentioning its two types and the various means and tools for raising awareness, emphasizing the role of different design approaches [6].

In 2019, Bernard Veyret, with a scaling factor of 100, led a study involving several sets of 90 rats and mice, subjected to intensive exposure to various levels of electromagnetic radiation. No cases of tumors were found among the control group, which was not exposed. In the various exposed groups, the prevalence of brain tumors varied between 1.1% and 3.3%. Notably, the control group, which was not exposed, showed no tumor development, which is statistically anomalous.

Researchers noted that the observed rate of 2.2% to 3.3% was slightly higher than the historical rate of the control group. They suggested that this could be attributed to the fact that the control group had a shorter lifespan, as tumors tend to manifest later in life. It is also interesting to note that females appeared less affected. In September 2020, a dozen people died in France, three on the Bezons site and nine on the Clayes-sous-Bois site, due to complications linked to cancer caused by prolonged exposure to electromagnetic fields. This issue was revived by an Inserm study, which established a link between frequent mobile phone use and the risk of brain tumors. While intensive users have already raised this correlation, the Inserm unit in Bordeaux now provides nearly mathematical precision on the risks involved [7].

When we talk about 896 h of calls over a lifetime, this leads to an increased risk of brain tumors. Over a period of 30 years, this amounts to only two hours per month. In this context, the idea of unlimited telecom packages could be questioned. On average, Belgians use their mobile phones for 2 h and 15 min per month for personal purposes and 3 h per month for professional needs. Adolescents, on the other hand, can use their phones for up to 30 min per day, with peaks of two hours on the phone. According to Inserm, the risk of tumors appears after 24 years of use [16].

3 Meta-Textile: An Innovative Solution

3.1 Definition and Characteristics of Meta-Textile

Understanding Metamaterials and Their Interaction with Electromagnetic Waves

Metamaterials are engineered materials specifically designed to manipulate electromagnetic waves in ways that traditional materials cannot. Their distinctive behavior arises not from their chemical composition, but from their intricate internal structure—typically composed of repeating elements smaller than the wavelength of the electromagnetic waves they influence. Thanks to these structural features, metamaterials can control wave behavior in unique ways: they can redirect electromagnetic radiation, filter out specific frequencies, or absorb energy. This makes them highly valuable in various domains, including telecommunications, optics, and health-related radiation shielding [8].

- Meta-textile: Functionality and Areas of Use meta-textile, is a specialized form of meta-material designed to act as filters for electromagnetic waves. They are made up of periodic arrangements—such as grids, rings, or other geometrical shapes—that can either block, transmit, or absorb certain frequency bands depending on their configuration. These surfaces operate through resonant effects, selectively interacting with waves based on frequency. This enables precise control over which signals pass through and which are attenuated. Meta-textile is widely used in antenna systems, stealth technology, and signal management, as well as in environments where controlling EM radiation is crucial (Fig. 1).

- Incorporating meta-materials into Textile Materials for Enhanced EM Protection

By embedding meta-textile structures into fabrics, researchers have developed woven metamaterials—flexible, wearable solutions that combine electromagnetic shielding with textile functionality. The integration can be achieved using methods such as conductive thread weaving, inkjet printing, or embroidery with metallic components. These smart fabrics serve a dual purpose: they provide effective absorption or reflection of unwanted electromagnetic radiation and remain adaptable for various forms, including wearable garments, interior furnishings, or protective covers for sensitive equipment. Their lightweight and adaptable nature makes them a promising option for applications in both personal and industrial health protection. As barriers against EM exposure, these materials can help reduce risks linked to prolonged radiation, particularly in environments where such exposure is unavoidable [9] (Fig. 2).

Fig. 1. Meta-textile with periodic cross patches

Fig. 2. Mini collection of wearable meta-textile garments

3.2 Electromagnetic Wave Absorption Mechanisms

Woven meta-textile offer a cutting-edge approach to controlling electromagnetic (EM) waves. These advanced structures are specifically designed to interact with certain frequencies, allowing them to absorb electromagnetic energy with precision and efficiency.

- How meta-textile Structures Dissipate Electromagnetic Energy

Meta-textile is made of periodic metallic patterns placed on a dielectric substrate. When exposed to EM waves, these patterns resonate at designated frequencies, enabling them to either absorb the energy and convert it into heat or redirect it through scattering. In the context of absorptive meta-textiles, the goal is to minimize both wave reflection and transmission to ensure maximum energy dissipation. Meta-textile designs can be tailored to function primarily in reflective, transmissive, or absorptive modes depending on material composition, geometry, and structural layering. For protective textile applications, the absorption mode is optimized to reduce the penetration of harmful radiation [10] (Fig. 3).

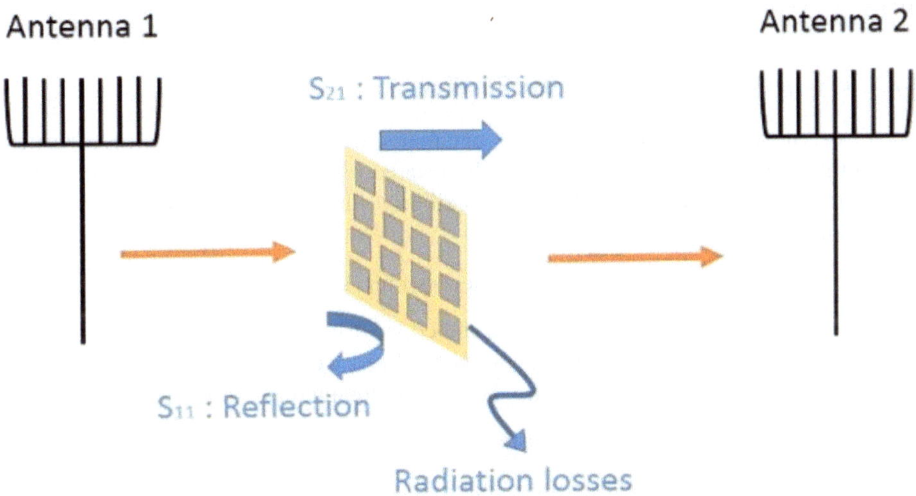

Fig. 3. Meta-textile and antennas configuration

- Comparison with Traditional EM Shielding Materials

Compared to conventional electromagnetic shielding solutions—such as solid metal enclosures or conductive films—meta-textiles offer frequency-specific control, allowing them to block only unwanted radiation without interfering with essential communication signals. While traditional shielding materials often block a broad spectrum of frequencies indiscriminately, FSS structures allow for precision targeting of harmful frequencies. Moreover, meta-textiles are lightweight, pliable, and easily integrated into various everyday items like clothing, wall coverings, or curtains— making them suitable for both wearable and environmental protection, unlike the rigid nature of metallic barriers [11].

- Performance of meta-textile in EM Exposure Reduction

Experimental data shows that meta-textiles can achieve high absorption efficiency, often exceeding 90% within specific frequency ranges. This high level of performance positions them as a viable option for reducing electromagnetic exposure, especially in densely connected environments like homes, offices, and public spaces. Their adaptability and selective shielding capacity make them especially valuable in settings that require both protection and usability. As such, woven meta-materials are emerging as a promising solution in the effort to mitigate potential health risks linked to prolonged EM wave exposure.

4 Use of Meta-Textile in Cancer Prevention

The incorporation of meta-textile into textile applications represents a significant step forward in preventive health strategies, particularly regarding the reduction of long-term exposure to potentially carcinogenic electromagnetic (EM) waves. Owing to their ability to manipulate specific frequencies, these advanced materials offer targeted shielding solutions [12].

4.1 Development of EM-Protective Textiles

One of the most practical uses for these metamaterials lies in the fabrication of protective clothing tailored to environments with elevated EM wave exposure. These textiles are engineered to maintain comfort and flexibility while actively mitigating radiation absorption.

- Everyday integration in garments and accessories

Protective layers of FSS fabrics can be embedded in everyday apparel such as jackets, hats, gloves, and even bag linings. Using specialized weaving methods, these textiles offer EM attenuation without affecting the look or breathability of the material.

- Use cases for high-risk or vulnerable individuals

Protective wear incorporating these materials is particularly useful for:

- Telecommunications professionals who work near antennas and transmitters.
- Expectant mothers, to shield developing fetuses from electromagnetic exposure.
- Electromagnetically hypersensitive individuals, who report symptoms linked to EM fields and benefit from additional shielding.
- Research and standards compliance

Numerous experimental studies have validated the efficiency of these textiles in blocking radiation. Independent certification bodies have measured up to 90% attenuation across specific bands. Performance evaluations often reference standards such as EN 62209–1 (related to Specific Absorption Rate or SAR) [13].

4.2 Reducing EM Exposure in Living Spaces

Beyond wearables, meta-textiles are increasingly being deployed in residential and workplace environments to mitigate constant low-level EM radiation emitted by connected devices.

- Applications in interior fabrics and surfaces

Smart shielding fabrics are being developed for use in curtains, drapes, wall panels, and wallpapers. These elements are designed to filter harmful frequencies without altering room aesthetics or interfering with visible light transmission.

- Contribution to health risk reduction

By limiting everyday exposure to EM radiation, these materials could play a role in preventing health effects such as chronic fatigue, headaches, and possibly certain cancers linked to prolonged EM field exposure [14].

- Ongoing industry and research efforts

Numerous innovation-driven companies and research institutes are actively developing and testing these technologies. Initiatives like SmartProtect and smart-building programs across Europe aim to implement these materials in infrastructure such as schools, hospitals, and office buildings for long-term public health benefits [15].

5 Conclusion

The meta-textile with cross patches structure incorporating cross-shaped patches shows strong compatibility with a wide range of frequency bands, highlighting its effective electromagnetic wave absorption and functional versatility. However, the highlighted values in reveal that the meta-textile with cross patches model performs poorly in the 5 to 8 GHz range. As a result, it cannot be classified as a high-performance absorbing woven metamaterial for that frequency spectrum. In contrast, the metamaterial equipped with cross resonators demonstrates significant absorption in the 0 to 5 GHz and 8 to 14 GHz bands, making it suitable for attenuating microwaves, which lie between infrared and broadcast waves on the electromagnetic spectrum.

Moreover, microwaves—commonly used in wireless technologies—are suspected of having adverse effects on human health. Prolonged exposure to this type of radiation may lead to thermal effects, cellular disturbances, and potentially increase the risk of certain diseases, including brain cancers. The demonstrated efficiency of this meta textile in critical frequency ranges suggests promising applications in personal protection systems, aligning with a preventive approach to public health.

References

1. Organisation Mondiale de la Santé: Radiation: Champs électromagnétiques. Organisation Mondiale de la Santé, Genève (2016)
2. Feng, Y., Liang, M., You, R.: Fabrication and modulation of flexible electromagnetic metamaterials. Microsyst. Nanoeng. **11**(1), 1–20 (2024)
3. Kundi, M.: Extremely low frequency electromagnetic fields and cancer: The need for research. Environ. Res. **178**, 108688 (2019)
4. Yang, Y., Wang, J., Liu, X.: Electromagnetic shielding using flexible embroidery metamaterial absorbers: Design, analysis and experiments. Mater. Des. **223**, 111164 (2022)
5. Feng, Y., Liang, M., You, R.: Fabrication and modulation of flexible electromagnetic metamaterials. Microsyst. Nanoeng. **10**, 1–12 (2024)
6. Yan, Z.L., Chen, S.H., Zhao, K.F.: Effects of electromagnetic radiation on health and immune function of operators. Chin. J. Ind. Hyg. Occup. Dis. **31**(8), 602–605 (2013)
7. National Research Council (US): Possible Health Effects of Ground Wave Emergency Network (GWEN). Washington, DC, National Academies Press. https://www.ncbi.nlm.nih.gov/books/NBK208983/

8. Feng, Y., Liang, M., You, R.: Fabrication and modulation of flexible electromagnetic metamaterials. Microsyst. Nanoeng. **11**(1), Article 14 (2024)
9. Yang, Y., Wang, J., Liu, X.: Electromagnetic shielding using flexible embroidery metamaterial absorbers: Design, analysis and experiments. Mater. Des. **223**, 111164 (2022)
10. Feng, Y., Liang, M., You, R.: Fabrication and modulation of flexible electromagnetic metamaterials. Microsyst. Nanoeng. **11**(1), Article 14 (2024)
11. Kundi, M.: Extremely low frequency electromagnetic fields and cancer: The need for research. Environ. Res. **178**, 108688 (2019)
12. Abdeddaim R., Métamatériaux pour l'IRM à ultra-haut champ, Joliot, 17 octobre 2018, https://joliot.cea.fr/drf/joliot/pages/actualites/actualites/actualites_scientifiques/2018/irm-antennes-metamateriaux.aspx
13. Alonso-Gonzalez, L., Ver-Hoeye, S., Fernandez-Garcia, M., Vazquez-Antuna, C.: From threads to smarttextile: Parametric characterization and electromagnetic analysis of woven structures. IEEE Access. **7**, 1486–1501 (2019)
14. Burokur, S.N.: Structures périodiques et méta-matériaux: Antennes et dispositifs électromagnétiques. Université Paris-Sud, France (2013)
15. Caloz, C., Itoh, T.: Electromagnetic Metamaterials: Transmission Line Theory and Microwave Applications. Wiley IEEE Press, Hoboken, NJ (2006)
16. Verset, J.-C., Ondes électromagnétiques: la mort après 900 heures d'appel? 19 février 2015, https://www.rtbf.be/article/ondeselectromagnetiques-la-mort-apres-900-heures-d-appel-8911010

Re-Designing the Living: Rethinking Sustainable Materials Through Biofabrication—A Case Study

Mariem Fakhfakh[1,2](✉)

[1] Higher Institute of Arts and Crafts of Sfax, University of Sfax, 3002 Sfax, Tunisia
mariefahfakh1@gmail.com
[2] University of Monastir, LGTex, LR11ES42, 5070 Ksar-Hellal, Tunisia

Abstract. This paper explores the emerging intersection of design and synthetic biology within the context of sustainable biomaterials and biotextile innovation. As the textile industry and materials manufacturing face increasing scrutiny for their environmental impact, designers are turning to living materials and biofabrication as strategies to regenerate ecological systems rather than depleting them. Drawing on a biomimetic approach, this study examines how the creative reconfiguration of biological processes enables the development of hybrid materials that blur the boundaries between the natural and the artificial. This article is based on a qualitative analysis of projects led by designer-biologists. It investigates how design can creatively harness biological intelligence to shape new sustainable biotextiles and biomaterials. These practices challenge conventional manufacturing paradigms and propose alternative design models that are grounded in the integration of living organisms and the systemic logic of nature.

Keywords: Sustainable design · Biofabrication · Biotextiles · Synthetic biology · Design-Biohacking · Living materials

1 Introduction

The textile industry and materials manufacturing are among the most polluting and degenerative sectors, whether relying on plant-based materials or petroleum-derived products. In response to ecological crises - such as climate change and biodiversity loss-manufacturing practices are increasingly shifting towards approaches that promote the repair and restoration of natural systems. Within this dynamic, designers are increasingly exploring living materials, including biomaterials and biotextiles, as part of a renewed vision for sustainable design, while also drawing upon knowledge from biological sciences. This approach reflects a new relationship with the living world, in which creators question the boundaries between nature and artificiality, while experimenting with processes that artificially recreate living systems. This positioning aligns with a biomimetic logic, wherein the duality between living and artificial elements guides designers toward biofabrication and the use of emerging, disruptive biotechnologies. These practices leverage tools from synthetic biology, enabling not only the decoding of the intelligence of

M. Hamdaoui et al. (Eds.): ITDM'3 2025, PTT, pp. 109–119, 2026.
https://doi.org/10.1007/978-981-95-7704-0_10

living organisms but also the ability to "re design" life itself to develop hybrid materials that merge organic compounds with industrial elements. Such hybridization illustrates a design approach of living or with living, where biological intervention allows for the adaptation of natural properties to specific physico-chemical uses within the realm of biotextiles. In this context, several key questions arise: How can design creatively integrate the potential of biology into biotextile manufacturing processes? How can the convergence of biomimicry and synthetic biology foster new paths toward sustainable biomaterials innovation? Prospectively and speculatively, could scientists be considered the designers of the future? This article presents a qualitative study of the work of designer-biologists who explore this synergy and examine the integration of synthetic biology into design to develop sustainable biomaterials and biotextiles. Their projects offer concrete alternatives, inspired by the logic of living systems and grounded in the potential of synthetic biology.

2 Synthetic Biology: A New Paradigm of Bio-Design

Design has always been rooted in the observation, analysis, and understanding of the surrounding world. Since its origins, it has shown a fascination for the complexity of nature and the intelligence of its systems. Today, we are witnessing an unprecedented convergence between nature and human creation, made possible by the rapid advances in biotechnology and synthetic biology. The hybridization of scientific knowledge and design practices has led to the emergence of a new perspective within bio-design. This approach offers a relevant response to contemporary challenges by harnessing the properties of living systems to address technical, environmental, or societal issues. Bio-design is thus evolving in close collaboration with life sciences, placing synthetic biology at the heart of the design process. This raises a central question: how does synthetic biology contribute to enriching design practices in the development of alternative and sustainable bio-textiles and biomaterials?

2.1 From Studio to Laboratory: The Evolution of Design

Design is characterized by its porous nature and its ability to engage in dialogue with other disciplines. It is an intellectual, creative, multidisciplinary, and deeply humanistic process, aiming to address complex social and environmental issues. In this context of hybridization and interdisciplinarity, design integrates advancements in biology to explore both the material and immaterial aspects of life. Synthetic biology opens new horizons. It allows designers to harness the tools of synthetic biology to create living and sustainable biomaterials cultivated systematically through microorganisms such as bacteria, yeast, algae, or stem cells. This approach involves replacing inert materials with living organisms and designing functional relationships between matter and organism to meet specific human needs. According to Thierry Marcou, a philosopher specializing in science and technology, this has led to the concept of synthetic nature, which aims to transpose "living" into a new material to be explored to find alternative solutions based on its vital intelligence and potential [1]. The biological and biotechnology revolution has given rise to a new generation of designer-researchers who shift their approach towards scientific thinking, enabling them to grasp the complexity of living systems. Textile designers and engineers are exploring the contributions of synthetic biology to biofabrication They explore various ways of interacting with life, questioning "how they manipulate the living -to give it form, decompose and recompose it, make it evolve- and how the living can itself be set to work, shaped to fulfill specific purposes, to produce forms, energy, colors, or materials" [2]. Biology has introduced a new paradigm based on the use of living organisms as a resource capable of contributing to the development of a new industry. This has led designers to take on a key role within scientific laboratories, exploring alternative production techniques and innovative materials through their creative processes. This progressive evolution has resulted in "a shift from object design to scientific creation, from the design agency to the laboratory" [3]. This transition from object-centered design to scientific practice is accompanied by a move toward a more interdisciplinary approach that prioritizes experimentation and rigorous scientific reasoning. The laboratory thus becomes a privileged space for designers engaging in scientific projects, encouraging the cross-pollination of knowledge between art, science, and technology.

2.2 The Designer as Biohacker

From this perspective, biomimicry promotes the hybridization of knowledge, not only to extract principles from nature but also to integrate the living into the design process. This phenomenon strengthens the convergence between design and biological systems by introducing a new paradigm of "bio-design." Advances in biotechnology now allow designers to collaborate with living organisms, developing innovative biological systems and functions tailored to human needs. This emerging movement merges elements of design, chemistry, and synthetic biology, establishing new relationships with living matters and contributing to the rise of a new paradigm: the design of the living. As illustrated in the adjacent diagram, bio-design is now being approached from new angles grounded in scientific rationality (Fig. 1).

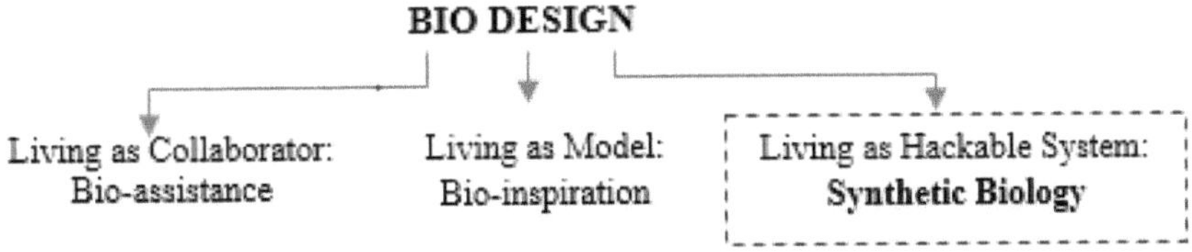

Fig. 1. The different roles of living organisms in the bio-design approach (Created by author) [4]

This methodology allows us to "draw inspiration from it as a model (bio-inspiration and mimicry), to use it as a material (bio-utilization), to benefit from it as a process (bio-assistance), or even to engineer it from scratch (synthetic biology)" [5]. The designer integrates living matter into both the design and manufacturing processes to generate new production alternatives located at the interface between the natural and the artificial. This approach is being applied across various domains such as habitat, furniture, food design, textiles, fashion, and energy [6]. Within this framework, the designer envisions a future "where everything is programmable—even life itself" while also acknowledging the possibility of a world where nature remains authentically natural [7]. Designers thus work in close collaboration with biologists and engineers to envision the objects, interfaces, and environments of tomorrow through the reprogramming of life. This approach is rooted in the study of interactions between design and synthetic biology, reflecting a synthetic and reconfigured form of nature. Design helps to objectify scientific research while interacting with living systems in a responsible and creative manner, thereby influencing their naturality and functionalities. In this context, Carole Collet emphasizes that living design can be used to create new materials, including synthetic "living" textiles. Designers, often referred to as "biohackers", can harness genetically reprogrammed nature synthetic nature to endow it with new functions, forms, and characteristics.

2.3 The Emergence of "Scientific Design"

Design today is evolving into a powerful tool for exploration, capable of establishing a fertile dialogue with biological disciplines and profoundly transforming both the research directions and manufacturing practices associated with them. In its creative process, design now operates within a rigorous conceptual and interpretive framework, conducive to the creation of new living organisms or functional biological systems [8]. This synergy between design and science has given rise to what is known as "scientific design", an approach that appropriates knowledge from the experimental sciences to work with innovative materials and processes [9]. Scientific design is driven by curiosity, investigation, research, experimentation, and creativity, fully engaging the designer in a dynamic of research-creation. In this context, designer and researcher Neri Oxman developed Krebs' cycle of creativity, a transdisciplinary model that articulates engineering, design, social sciences, and art (see Fig. 2).

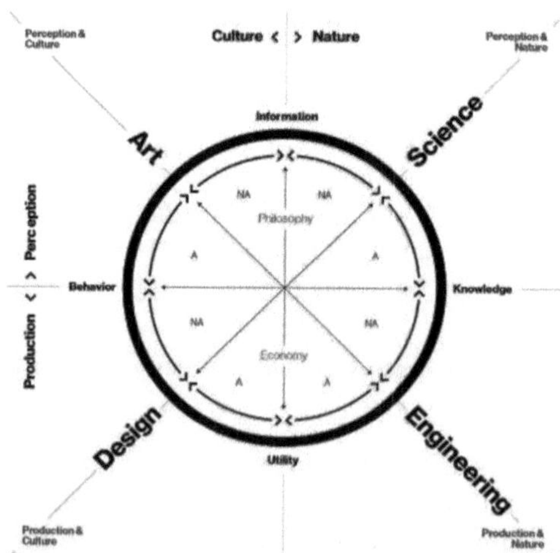

Fig. 2. Neri Oxman's krebs cycle of creativity [9]

This cycle highlights a circular and collaborative dynamic in which:

- **Science** plays an explanatory and predictive role, converting information into knowledge.
- **Engineering** applies this knowledge to address concrete problems, turning knowledge into a utility.
- **Design** materializes these solutions to maximize functionality and enrich human experience, converting utility into behavior.
- **Art** then questions these behaviors and heightens our perception of the world, transforming behavior into a new form of information [10].

This paradigm of scientific design redefines traditional disciplinary boundaries by emphasizing a collaborative approach based on co-creation and co-fabrication. It establishes a reciprocal relationship between disciplines, generating a continuous cycle of cooperation, innovation, and alternative sustainable practices.

3 Re-Designing the Living for a Regenerative Approach

By working directly with living systems, design no longer views nature and life as mere natural resources but as active partners in the creation of materials. Drawing on the diverse reflections of bio designer-researchers, living design plays a fundamental role in systemic and holistic regeneration by developing natural, synthetic, and/or hybrid natures. This new conception of nature fosters the development of alternative resources within a multi-species approach. By adopting this perspective, design contributes to maximizing our collective potential by designing not only for humans but for all species. This opens the way for a harmonious and mutually beneficial coexistence between human and non-human beings. Through the reconciliation of emerging biotechnologies and nature, living design introduces synthetic and hybrid concepts that have the potential to incorporate living matter, envisioning materials as active entities within their ecosystems.

3.1 "Biolace": Genetically Programmed Nature for a Sustainable Biotextile

Most current design projects still view natural resources as mere tools to create products or textiles. In contrast, regenerative design considers the product as the outcome of a living process of regeneration, one that aims to restore ecosystems and enhance biodiversity. It is within this regenerative framework that Carole Collet's speculative project Biolace emerges. In Biolace, Collet imagines a future set in the year 2050, shaped by global overpopulation and the urgent need for radical sustainability. This speculative scenario explores how future advances in agriculture and synthetic biology might allow for new models of sustainable production suited to a densely populated planet. She conceptualizes four fictional, genetically modified plants:

Strawberry Noir, Basil n 5, Tomato Factor 60, and GoldNano Spinach. Inspired by current advances in synthetic biology, these plants are programmed to produce both enriched food and textile fibers, redefining plants as multifunctional "living factories" [11]. Figure 3 illustrates the Strawberry Noir concept, where a genetically reprogrammed strawberry plant produces black lace from its roots while also yielding vitamin C-rich, antioxidant-laden black strawberries.

Fig. 3. "Strawberry Noir": A plant from the "Biolace" project, designed by Carole Collet, presented at the "Alive, New Design Frontiers" Exhibition, Paris, 2013 [11].

According to Collet, the goal of this project is to "highlight the potential of emerging living technologies and challenge the advantages and risks of such extreme genetic engineering" [12]. As a biohacker-designer, she leverages the tools of synthetic biology - including DNA sequencing- to transform plants into hybrid production systems capable of generating both food and textiles. Through genetic reprogramming, living organisms become both conceptual media and productive materials, giving rise to novel forms and alternative biotextiles. The lace produced by the plant then be used as a material to create garments and accessories. In this context, Collet explores synthetic biology and biotechnology as catalysts for design innovation, integrating biological principles into futuristic biofabrication practices [13]. Her goal is not simply to mimic the rhythms of life, but to incorporate them into the creative process. She extends the vital dimensions of life itself into the heart of design. Through this hybrid approach, rooted in genetically modified living organisms, Collet investigates the possibilities of biofabrication to 'grow' textiles while contributing to nature's regeneration. In this way, design becomes a meaningful response to today's environmental challenges by offering alternative production techniques that support a viable future for the planet.

3.2 "BioCouture": Towards a Cohabitation Approach

This new vision of design encourages us to rethink our concepts and production technologies, to develop more ethical and sustainable approaches that preserve biodiversity and contribute to the regeneration of natural ecosystems. This approach urges us to imagine manufacturing methods that are not only ethical, sustainable, and regenerative but also capable of limiting their environmental impact and actively contributing to ecosystem restoration. In response to resource depletion, designers are exploring biofabricated materials as sustainable alternatives. Multi-species design plays a key role in this transition, encouraging mutualistic interactions with the living world. This perspective expands design beyond an anthropocentric view of design to incorporate non-human organisms as co-actors in the creative process [14]. Suzanne Lee's "BioCouture" project fully illustrates this transition (see Fig. 4). In collaboration with biologist researchers and chemists, Lee has been growing microbial materials for fashion since 2003 and explores the potential of synthetic biology and microbial fermentation to offer a radical alternative to conventional textiles [14].

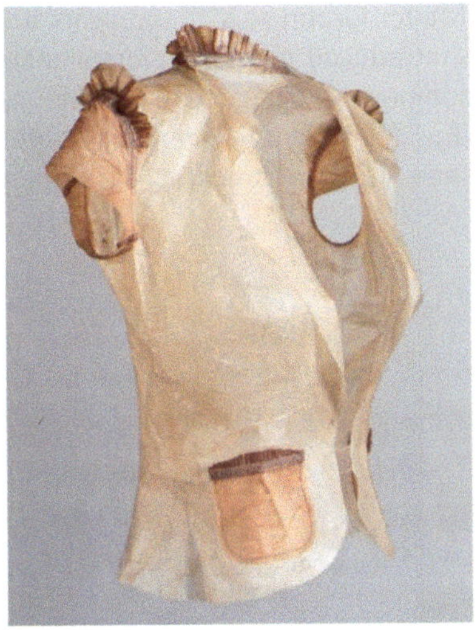

Fig. 4. Example of the "BioCouture" project in microbial cellulose, created by Suzanne Lee, 2010 [15]

Her goal is to cultivate clothing using bacterial cellulose, produced through the fermentation of a sweetened green tea solution [15]. The process involves the interaction of bacteria (Gluconacetobacter xylinus) and yeast, which feed on sugar and produce pure cellulose nanofibers. These fibers form layers that float on the surface of the liquid. In large containers, at temperatures below 30 °C, Suzanne Lee mixes 30 l of green tea, 2 k of sugar, yeast, bacteria, and acetic acid — a component of vinegar. After two weeks, a cellulose membrane forms. Once harvested, this membrane is shaped, dried, cut, and sewn. The result is a flexible textile whose texture is like leather. The longer it dries in the sun, the thinner and stiffer the fabric becomes [16]. This process does not involve intensive agriculture, nor the use of animal or petrochemical resources, significantly reducing the ecological footprint of textile production. Furthermore, the dyes used are plant-based: vegetable peels, organic indigo, or other natural pigments, reinforcing the sustainable ethics of the project. This collaboration aims to develop living entities and alternative natural resources, thus paving the way for the creation of biomaterials that present properties specifically suited not only to our needs as humans but also to those of nature itself. In this context, biological design stands out for its goal of transcending the conventional limits of design by creatively translating the potentials of life, not just as a passive actor in the natural ecosystem, but as a positive transformative agent for the well-being of biodiversity and our planet. Design engages in a close collaboration with living organisms to give birth to new manifestations of nature, fostering a shift towards a more sustainable way of life and cohabitation with nature.

3.3 "Aguahoja": Biocomposites from Organic Waste, Towards Material Ecology

In the search for alternatives to conventional materials with a high environmental impact, both in terms of resource extraction and production, the Aguahoja project, led by Neri Oxman and the Mediated Matter Group at MIT, presents itself as a pioneering exploration. In 2014, the group introduced an experimental material library composed of bioplastics made from biocomposites, designed at the intersection of synthetic biology and digital fabrication (see Fig. 5). This project aims to rethink the relationship between material, living organisms, and the environment, positioning nature as an active partner in the design process.

Fig. 5. "Aguahoja Artifacts", created by Neri Oxman and the MIT Mediated Matter Group, SFMOMA, California, 2015 [17].

Through several creations, Oxman and her team demonstrate that bio-fabrication, combined with digital technologies, allows for the imagination of materials and structures in symbiosis with natural systems. The approach is based on a material library composed exclusively of organic waste, such as fallen leaves, apple peels, and shrimp shells. The originality of this approach lies not only in the use of renewable biological resources but also in the design of intelligent materials capable of interacting with their environment. This project is embedded in her philosophy of material ecology, which offers a speculative and proactive interpretation of the relationships between materials, natural environments, and emerging technologies [18]. In this context, laboratory experiments have led to the creation of biocomposites with physicochemical properties close to those of conventional plastics. These biocomposites are made from natural polymers, such as: Pectin (extracted from fruit peels), Cellulose (found in plant cell walls), Chitosan (derived from chitin in shrimp shells). For example, the manufacturing process derived from shrimp shells begins with the transformation of chitin into chitosan using acetic acid. In powder form, chitosan is mixed with water at concentrations ranging from 1%

to 12%, forming a hydrogel [19]. This hydrogel is then used in an additive manufacturing process driven by digital technologies. A KUKA robot deposits these hydrogels in varying concentrations, generating a range of textures from translucent, fluid materials at low concentrations (3%, with viscosity close to honey) to opaque, rigid structures like natural rubber at high concentrations. This modulation of concentrations allows for the control of a wide range of optical and mechanical properties: opacity, translucency, rigidity, flexibility, and even dynamic chromatic properties. Thanks to digital fabrication, it is possible to generate complex geometries and fine textures, optimized for developing bio-inspired, flexible, and biodegradable polymers. Water plays a crucial role in this co-fabrication process: it facilitates the assembly of panels during printing and makes the materials responsive to environmental stimuli such as light, humidity, or temperature [20]. This responsiveness allows the structures to change their shape, mechanical behavior, or color depending on the ambient conditions, giving rise to intelligent and adaptive artifacts. These biopolymers also exhibit thermochromic properties, changing color according to temperature, and can return to their original hue at 30 °C, demonstrating a true dynamic connection with their environment [21]. Made from organic materials, produced by robots, and self-assembled by water, this hybrid manufacturing system tends toward a fusion of the cultivated and the manufactured. This union paves the way for a new approach to digital and synthetic fabrication, capable of controlling not only material properties but also processes of programmed degradation. As demonstrated by the MIT group, exposure to sunlight triggers selective degradation: pectin molecules dissolve rapidly, allowing cellulose and chitosan molecules to deform and degrade in a programmed manner. These materials then return to the ecosystem, nourishing soil microorganisms and reintegrating into natural biological cycles [22]. This material fully embodies a circular and regenerative logic, restoring ecological balances during its life cycle. Through their controlled growth and natural degradation, these artifacts represent a material ecology where organism-becomes-shelter and shelter-becomes-organism [22]. By harnessing natural processes, this approach transforms the degenerative impact of manufacturing into a regenerative action, integrating design, material science, biology, and ecology. It illustrates an advanced form of speculative biomimicry, referred to as neo-materialism, which is not simply about mimicking the form (or function) of an organism in a material object, but understanding form and material as co-emergent in a growth process [23]. This approach opens possibilities for a future that is synthetically natural, where materials are co-fabricated by and for ecosystems.

A future where solutions to climate crises arise from a deep understanding of the relationships between matter, environment, and biology.

4 Conclusion

A new living materiality is emerging at the intersection of the natural and the synthetic, questioning humanity's place within its environment. Based on a transdisciplinary approach, contemporary design is increasingly turning to biofabrication, relying on various techniques derived from synthetic biology to create sustainable biomaterials and enable the development of sustainable biotextiles and biomaterials. In this context, designer-biologists are challenging the boundaries between living and artificial by experimenting with chemical and biological processes on living organisms. The projects presented illustrate a true co-creation with life itself, grounded in a biomimetic and transdisciplinary approach. These initiatives aim to invent naturally synthetic life forms or to imagine new modes of cohabitation with them.

This qualitative study has highlighted scientific, environmental, and ethical issues, shedding light on design's potential to adopt tools from the biological sciences. Raising awareness among designers, engineers, and citizens about the urgent need for alternatives to conventional materials is essential. The goal is to shift from degenerative manufacturing processes to regenerative approaches, in harmony with natural systems. Each project examined presents its own unique characteristics, but they all share the same ambition: to harness the potential of synthetic biology to create biotextiles grounded in a speculative and salutogenic vision, which anticipates environmental degradation and seeks to restore biodiversity. By working directly with living organisms as alternative resources and leveraging biotechnologies and digital tools, these approaches enable the development of intelligent, innovative, and durable materials with adaptive physicochemical properties. This field paves the way for a future where the cultivated and the manufactured merge, where systems designed by and for nature offer credible and necessary solutions to contemporary ecological challenges. Based on deep scientific knowledge, this approach aims to rethink our relationship with living, ensuring a hybridization and reconciliation between the natural and the artificial. The living then becomes a central actor in the creative process, giving birth to a new form of nature – hybrid and synthetic. From the agency to the laboratory, designers have taken on a central role in biological research, contributing to a gradual shift from nature-inspired design to nature-inspired by design.

References

1. Thierry, M.: Vers un design de la vie synthétique. InternetActu.net (2014)
2. Labex Transfers: La vie à l'œuvre: Explorer les potentialités du bioart et du biodesign [online]. http://www.transfers.ens.fr
3. Koch, E.: Design biologique: l'innovante Neri Oxman. Maze [online]. https://maze.fr/art/05/2016/design-biologique-linnovante-neri-oxman (2016). Accessed: 10 Dec 2023
4. Fakhfakh, M.: Le design du vivant au cœur d'une approche régénérative durable: matériaux co-fabriqués avec la nature pour un écosystème sain et résilient. PhD thesis, Institut Supérieur des Arts et Métiers de Sfax (2024), 257 pp.
5. Holocene Design Gallery: Guillian Graves, designer en bio-design. Holocene Design Gallery [online] http://holocene-design-gallery.com/guillian-graves- bio-design
6. Huffpost: En vie: une expo sur le vivant à la fondation EDF [online]. http://www.huffingtonpost.fr/2013/04/28/en-vie-une-expo-sur-le-vivant-fondationedf_n_3163010.html
7. Semavoine, C.: Design du vivant: salut ou ruine de l'humanité? Prospective Design [online]. http://www.prospectivedesign-leblog.com (2013)
8. Léchot Hirt, L.: RECHERCHE-CRÉATION EN DESIGN Modèles pour une pratique expérimentale. Vues Densemble.
9. Oxman, N.: Official website. https://oxman.com/mission
10. En Vie/Alive: Biofacturing Textiles: Biolace [online]. http://thisisalive.com/fr/biolace/
11. Collet, C.: En Vie, Alive: aux frontières du design. Fondation EDF, Paris (2013)

12. WGSN Insider: Regenerative Design: Repairing the World with Carole Collet, https://www.wgsn.com/en/blogs/regenerative-design-repairing-world-carole-collet-0
13. Fakhfakh, M., Taieb, A.H.: The living design: The soul of regenerative design and the key to an ecological transition. In: Abdessalem, S.B., Hamdaoui, M., Baffoun, A., Elamri, A. (eds.) Proceedings of the Second International Conference of Innovative Textiles and Developed Materials ITDM'2, pp. 125–136. Springer, Singapore (2024)
14. Lee, S.: Biofabricate. Our Why. Biofabricate. https://www.biofabricate.co (2020)
15. Scarpitti, C.: Fashion-oriented bio textiles: the new speculative aesthetics of biocouture. PAD. Pages on Arts and Design V22 (2022)
16. Lee, S.: BioCouture: Growing Clothing. TED Talk. https://www.ted.com/talks/suzanne_lee_grow_your_own_clothes (2011)
17. Benayer, A.: La Fabrique du Vivant: Mutations / Créations 3. Paris, IRCAM, Centre Pompidou (2019), p.19
18. Stream: Interview: Neri Oxman, Material Ecology. https://www.pca-stream.com/fr/articles/material-ecology-19 (2014)
19. Oxman, N.: Biopolymer Aguahoja III Pavilion Shows How "We Can Begin to Redesign Our Built Structures as If They Were Grown". Dezeen. https://www.dezeen.com/2021/11/19/neri-oxman-aguahoja-iii-pavilion-robotically-fabricated/ (2021)
20. Oxman, N.: Platform: Water-based digital fabrication. https://oxman.com/projects/water-based-digital-fabrication
21. Neville, M.: Abstract: The Art of Design – Neri Oxman: Bio Architecture. Netflix documentary, USA (2019), 45 minutes
22. Oxman, N. Project: Aguahoja. https://oxman.com/projects/aguahoja
23. Kamili, L.: Biomimicry and bio-inspiration: New techniques, new ethics?. Techniques & Culture, Varia (2022), p.9

Piezoelectric Smart Textiles for Energy Harvesting

Ichrak Trabelsi[1(✉)], Adel Ghith[1], and Faten Fayala[2,3]

[1] Université de Monastir, ENIM, Unité de Recherche Matériaux et Procédés Textiles (MPTex), UR17ES33, 5000 Monastir, Tunisie
ichrak.trabelsi@enim.u-monastir.tn

[2] University of Monastir, Laboratory of Thermal and Energetic Systems Studies (LESTE), 5000, Monastir, Tunisia

[3] Monastir Textile Science & Technology Research Center, 5000 Monastir, Tunisia

Abstract. The growing demand for sustainable and autonomous energy solutions has driven significant research into energy harvesting technologies. Among these, piezoelectric materials have emerged as a promising solution for converting mechanical energy into electrical power due to their high electromechanical coupling efficiency and adaptability. With the rapid advancement of wearable electronics and smart textiles, the integration of piezoelectric materials into textile structures offers a self-sustaining energy source by harnessing mechanical deformations from human-body movements and environmental vibrations. This review explores recent developments in piezoelectric textiles, highlighting the fundamental principles of piezoelectricity, material selection, fabrication techniques, and performance evaluation. Additionally, key applications are discussed. The insights provided in this review underscore the potential of piezoelectric textiles in enabling next-generation self-powered wearable systems.

Keywords: Piezoelectricity · Smart textiles · Nanogenerators · Self-powered systems

1 Introduction

The extensive reliance on non-renewable energy sources has led to critical environmental concerns, including pollution and resource depletion. Traditional batteries, although indispensable for powering electronic devices, present limitations in terms of lifespan, environmental impact, and waste management. For instance, lithium-ion battery production is highly energy-intensive and contributes to environmental degradation [1]. In response to these challenges, researchers have increasingly explored ambient energy harvesting, particularly thermal and vibrational energy. Among the various energy conversion mechanisms, piezoelectricity has emerged as a highly efficient approach due to its ability to convert mechanical vibrations into electrical energy [2]. Piezoelectric transduction exhibits a high electromechanical coupling factor and piezoelectric coefficient compared to other transduction mechanisms such as electrostatic, electromagnetic, and

The original version of the chapter has been revised. A correction to this chapter can be found at https://doi.org/10.1007/978-981-95-7704-0_25

M. Hamdaoui et al. (Eds.): ITDM'3 2025, PTT, pp. 120–133, 2026.
https://doi.org/10.1007/978-981-95-7704-0_11

triboelectric [3]. Moreover, unlike other energy harvesting methods, piezoelectric generators are not significantly affected by environmental factors, making them particularly suitable for various applications [4].

With the rapid advancement of wearable electronics and smart textiles, integrating piezoelectric materials into textile structures has gained increasing interest. Smart textiles embedded with piezoelectric materials can harness mechanical energy from human body movements or environmental vibrations, thereby providing a continuous and self-sustaining power source [5]. This technology has promising applications in self-powered healthcare monitoring, portable electronics, and interactive textiles. Additionally, the miniaturization of electronic components has further driven the need for flexible and lightweight energy solutions, reinforcing the relevance of piezoelectric textiles [6]. This review aims to explore recent advancements in the integration of piezoelectric materials into smart textiles.

The paper begins with the fundamental principles of piezoelectricity, followed by an exploration of various piezoelectric materials and their suitability for textile applications. Key characterization parameters are then discussed, along with the design and functioning of piezoelectric nanogenerators. The structural configurations, energy conversion mechanisms, and applications of piezoelectric nanogenerators are explored

Finally, we present the challenges and the potential future research directions in this field.

2 Piezoelectricity: Fundamental Principles

Piezoelectricity was first discovered in 1880 by Pierre and Jacques Curie, who observed that some crystalline materials generate an electric charge when subjected to mechanical stress [7].

This phenomenon, known as the direct piezoelectric effect, occurs due to the displacement of electric dipoles within a non-centrosymmetric crystal structure leading to an asymmetric charge distribution. As a result, electrical charges accumulate at opposite ends of the material, forming an electric dipole. Conversely, the inverse piezoelectric effect occurs when these materials deform in response to an applied electric field [8].

The constitutive equations of direct and converse piezoelectric effect are as follows [3]:

$$D = dT + \varepsilon E \text{ (Direct effect)} \tag{1}$$

$$X = sT + dE \text{ (Converse effect)} \tag{2}$$

Where:

- D is electrical displacement;
- d is the piezoelectric coefficient;
- T is the mechanical stress;
- is thepermittivity of the material;
- E is the electric field;
- X is the mechanical strain;
- s is the mechanical compliance.

3 Piezoelectric Materials

Piezoelectricity was initially explored in natural crystals like quartz and Rochelle salt. In the mid-20th century, the development of synthetic piezoelectric ceramics, particularly barium titanate ($BaTiO_3$) and lead zirconate titanate (PZT) marked a significant technological breakthrough, enabling the commercialization of high-performance piezoelectric devices [9].

However, due to environmental concerns, research has increasingly focused on lead-free piezoelectric materials, such as potassium sodium niobate (KNN) and bismuth ferrite ($BiFeO_3$) [10]. These materials offer promising alternatives while maintaining good piezoelectric performance.

In addition to sustainability considerations, the growing demand for flexible and wearable electronics has driven research towards soft, polymer-based, and textile-integrated piezoelectric materials. These materials offer mechanical flexibility, lightweight properties, and adaptability to curved surfaces, opening new possibilities for applications in health monitoring, smart clothing, and self-powered sensors.

Piezoelectric materials can be classified into four main categories: piezoelectric crystals, piezoelectric ceramics, piezoelectric polymers and piezoelectric composites. Each category exhibits distinct structural, mechanical, and electrical characteristics, making them suitable for specific applications.

3.1 Piezoelectric Crystals

Piezoelectric crystals are naturally occurring or synthetically grown materials that exhibit intrinsic piezoelectric properties due to their non-centrosymmetric crystal structures.

The most well-known natural piezoelectric crystal is quartz (SiO_2), which has been widely used in frequency control and precision sensing such as in oscillators, resonators, and precision timing devices. However, synthetic crystals such as lithium niobate ($LiNbO_3$) and lithium tantalate ($LiTaO_3$) offer enhanced piezoelectric coefficients and broader application potential such as acoustic wave devices and optoelectronic applications [11].

Compared to other piezoelectric materials, crystals generally exhibit high thermal and chemical stability, making them suitable for harsh environments. However, their relatively high stiffness, brittle nature and complex manufacturing process limit their application in flexible and wearable electronics [12].

3.2 Piezoelectric Ceramics

Piezoelectric ceramics feature strong piezoelectric properties, good resistance to acids, bases and humidity. However, their intrinsic rigid nature restricts their applications in flexible devices [4, 12]. From 2000 to 2016, PZT dominated piezoelectric energy harvesting applications due to its high piezoelectric coefficient and excellent electromechanical coupling [13].

However, lead (Pb) is a toxic element that threatens both the environment and human health. Therefore, governments have restricted its use in many products. Consequently, researchers have devoted significant efforts to develop high-performance lead-free piezoelectric materials, leading to the emergence of several new piezoceramics. Among the most notable are potassium sodium niobate (KNN) and sodium bismuth titanate (NBT). These materials exhibit promising properties but still face challenges in matching the high performance of PZT [3].

3.3 Piezoelectric Polymers

Piezoelectric polymers are a class of materials characterized by their intrinsic flexibility, lightweight nature, and ease of processing, which make them particularly attractive for integration into wearable and deformable systems. The most well-known piezoelectric polymer is polyvinylidene fluoride (PVDF) and its copolymers, such as polyvinylidene fluoride-trifluoroethylene (PVDF-TrFE), which exhibit enhanced piezoelectric response due to their strong molecular dipole alignment.

In addition to PVDF-based materials, other polymers have shown potential piezoelectric behavior. These include polylactic acid (PLA), a biodegradable and biocompatible polymer often used in biomedical applications and polyamide-11 (PA11), which can also exhibit piezoelectric properties when processed appropriately. Although these polymers generally present lower piezoelectric coefficients than ceramics, however, their flexibility and adaptability are ideal for textile integration [3, 4].

3.4 Piezoelectric Composites

Piezoelectric composites combine piezoelectric ceramics or crystals with polymer matrices to balance flexibility and high energy conversion efficiency. These materials are particularly promising for applications in smart textiles, where both mechanical flexibility and energy harvesting are needed. For example, PVDF-based composites, which combine PVDF with piezoelectric ceramics like PZT or $BaTiO_3$, offer enhanced piezoelectric output and good flexibility. These composites have been widely used in wearable devices for energy harvesting and pressure sensing [3].

Epoxy-based composites, embedding piezoelectric powders in epoxy resin, offer mechanical strength, making them suitable for automotive and industrial sensors. Despite their advantages, challenges remain in optimizing the interface between the materials and improving scalability, cost-effectiveness, and durability for real-world textile applications.

4 Characterization Parameters of Piezoelectric Materials

To optimize the performance of piezoelectric materials for a specefic application, it is essential to evaluate their fundamental properties through specific characterization parameters. These parameters provide insights into the material's ability to generate and maintain electrical signals under mechanical stress and vice versa.

The key characterization parameters includes [14]:

- Piezoelectric charge coefficient (d): also referred to be the piezoelectric modulus, it describes the change in volume on application of an electric field. There can be many piezoelectric coefficients, d_{xy}, where x refers the polarisation direction, and y refers to the applied mechanical stress direction.
- Piezoelectric voltage constant (g): it represents the electric field generated by a piezoelectric material per unit of applied mechanical stress or, alternatively, is the mechanical strain produced per unit of electric displacement.

  Higher g-values indicate better energy conversion efficiency.
- Permittivity (ε): also known as the dielectric constant, it quantifies the dielectric displacement obtained per unit electric field applied. εT and εS denote the permittivity at constant stress and strain, respectively.

  A higher permittivity generally enhances charge storage capabilities but may also lead to increased dielectric losses.
- Young's modulus (Y): It is a ratio of applied stress to the resulting strain. It measures the elasticity, or the stiffness, of a material. Materials with lower Y-values present good flexibility which is an essential characteristic for stretchable and flexible piezoelectric devices.
- Elastic compliance (s): it is the reciprocal of Young's modulus. It measures the strain produced per unit of applied mechanical stress. sD and sE are two types of elastic compliances, which denote, respectively, compliance under constant electric displacement and constant electric field.
- Piezoelectric coupling factor (k): this parameter defines the efficiency of energy conversion between electrical and mechanical forms. It is the square root of the ratio of electrical energy output to the mechanical energy input.

  k always remains below unity due to the inability of complete energy.
- Dielectric Dissipation Factor (DDF): also known as the dielectric loss tangent (tan δ), it quantifies energy losses within the material due to dielectric polarization effects. It is determined by the ratio of effective conductance to effective susceptance in a parallel circuit.

  Lower dielectric loss values are more suitable for high-efficiency piezoelectric materials.

5 Piezoelectric Nanogenerators

Piezoelectric nanogenerators (PENGs) are energy harvesting devices that convert nanoscale mechanical energy into electrical energy through the piezoelectric effect [15].

5.1 Structural Configurations of Piezoelectric Nanogenerators

PENGs can adopt various structural configurations to optimize energy conversion efficiency. The fundamental construction of a PENG includes two fundamental components: a piezoelectric material and metallic electrodes.

The piezoelectric material generates electrical charges and, to facilitate the collection and transportation of the produced electric charges, metallic electrodes are placed on both the upper and lower surfaces of the piezoelectric layer [15].

- Metal-Insulator-Metal (MIM) sandwich structure

PENGs can adopt a metal-insulator-metal (MIM) sandwich structure, where a piezoelectric material is placed between two conductive electrodes. This configuration ensures efficient charge collection and enhances the conversion of mechanical energy into electrical energy.

When subjected to mechanical deformation, the piezoelectric layer generates an electric potential. The top and bottom metal electrodes then act as charge collectors. This layered configuration not only optimizes charge separation but also enhances mechanical flexibility, making it particularly suitable for wearable and stretchable electronic devices [15].

Recent advancements in PENG technology had led to the incorporation of nanomaterials within the piezoelectric layer to improve its performance. For instance, ZnO nanowires, $BaTiO_3$ nanoparticles, and carbon-based materials have been integrated into the piezoelectric matrix to enhance its dielectric and piezoelectric properties, thereby increasing the output voltage and current of the device [15].

Recent research has demonstrated significant advancements in MIM sandwich-structured materials, which can be effectively used in smart textiles. Several studies have explored innovative designs and material compositions to optimize the performance of such structures. For instance, a review by Wu & et al. [16] highlights the development of advanced sandwich structures with customized core designs, enhancing mechanical properties and lightweight characteristics. Similarly, research on polyurethane foam-based sandwich composites by Khan & et al. [17] provides insights into improved flexibility and durability, which are critical for wearable applications. Furthermore, studies on acoustic properties by Ciaburro & et al. [18] emphasize the potential of sandwich textiles in noise reduction, while Militký & et al. [19] demonstrated the effectiveness of these structures in heat management for wearable technologies.

The integration of such structures in smart textile applications has been further explored in recent research on post- pandemic advancements in functional textiles by Ivanoska-Dacikj & et al. [20] demonstrating their potential in protective and responsive fabric technologies. These collective contributions pave the way for the development of novel textile-integrated piezoelectric nanogenerators, leveraging the MIM configuration for enhanced energy conversion and wearability.

- Cantilever-based structure

Cantilevers are fundamental structural elements in engineering and physics, characterized by being fixed at one side, while the other side remains free. When a force is applied to the free end, the cantilever bends, generating tensile and compressive stresses along its length. These stresses result in mechanical deformation [15].

A seismic mass is usually attached to the free end of the cantilever to adjust its resonant frequency to the available environment frequency, which is often below 100 Hz [3]. The resonant frequency () can be calculated using the following equation where k is the stiffness of the cantilever and m is the effective mass [15]:

$$F_r = \frac{1}{2\pi}\sqrt{\frac{k}{m}}$$

One of the main advantages of the cantilever configuration is its ability to achieve high strain levels with relatively low input forces, making it suitable for wearable electronics, self-powered sensors, and biomedical devices. Additionally, the mechanical resonance of the cantilever beam can be optimized to enhance energy output by tuning its dimensions and material properties [13, 21].

Several studies have explored the integration of cantilever-based piezoelectric nanogenerators into textiles for energy harvesting and biomechanical sensing applications. Wan & et al. [22] developed a fully textile-based piezoelectric nanogenerator utsing a 3D knitted fabric electrode, demonstrating enhanced flexibility and breathability for wearable applications. Similarly, Song & et al. [23] designed a hybrid large-area textile nanogenerator that combines piezoelectric and triboelectric units in a cascading structure, enabling efficient mechanical energy harvesting for self-powered electronics. In another study, Javed & et al. [24] integrated PVDF nanofibers into a knitted fabric to develop a piezoelectric sensor which allows monitoring human joint angles, highlighting the potential of smart textiles for motion tracking.

- Stacked-layer structure

Stacked-layer configurations consist of multiple thin piezoelectric layers separated by conductive electrodes and stacked along the direction of the electric field. When mechanical stress is applied, each piezoelectric layer undergoes deformation, leading to the cumulative generation of electrical charge across the layers. This multilayer approach significantly enhances the output voltage and current compared to single-layer devices [13].

Compared to cantilever structure of similar weight and size, stacked-layer structures exhibit considerably higher electrical power and power density in both resonance and off-resonance modes due to the increased effective piezoelectric volume and their charge collection [13].

One of the key advantages of stacked-layer PENGs is their ability to operate under low mechanical strain while still achieving high energy output. This feature makes them particularly suitable for applications in self-powered sensors, biomedical implants, and wearable electronics. Additionally, the stacked configuration is well-suited for high-load applications due to its robust mechanical design. However, the complex fabrication process, especially in confined spaces, remains a significant challenge in implementing this structure effectively [13].

Kim & et al. [25] developed a simplified method for manufacturing piezoelectric textiles based on the stacked-layer structure integrating a 70-µm-thick ferroelectric polymer film between two conductive fabric layers. The result was a flexible and durable textile able to harvest piezoelectric energy.

Additionally, recent studies have demonstrated the effectiveness of stacked-layer configurations in the development of smart textiles. For instance, He & et al. [26] integrated ZnO nanowires into textile fibers to create PENGs able to convert mechanical energy into electrical energy. A study by Wan & et al. [22] on flexible PENGs based on PVDF nanofibers has shown promising results for wearable applications, as these nanogenerators can efficiently harvest biomechanical energy from body movements.

These advancements highlight the potential of stacked-layer structures in developing energy-harvesting in smart textiles.

- Cymbal structure

The cymbal structure consists of two concave metallic endcaps with a piezoelectric disk placed between them. The metal endcaps act as a mechanical transformer focusing and concentrating mechanical stress on the piezoelectric material, amplifying the piezoelectric coefficient. Thus, a portion of the incident axial stress is converted into radial and tangential stresses which amplify the piezoelectric coefficient of the generator [13].

The main advantages of the cymbal structure is its ability to capture low-amplitude vibrations and efficiently convert them into electrical energy The design allows energy harvesting from a broad range of vibration frequencies, making it highly versatile compared to traditional cantilever configurations. Additionally, the metallic shells provide mechanical stability and durability, making the structure suitable for long-term wearable electronics [27].

However, the fabrication of the Cymbal structure can be more complex than simpler designs like cantilever beams, as it requires precise alignment of the metallic shells and piezoelectric materials [13].

Many research has explored the integration of piezoelectric structures, such as the cymbal configuration, into smart textiles for energy harvesting and sensing applications. For instance, Zhao & et al. [28] conducted a finite element analysis of cymbal piezoelectric transducers, demonstrating their potential for energy harvesting. Although this study focused on infrastructure, the insights gained are applicable to the development of smart textiles energy- harvesting. Additionally, Kim & Yun [29] developed a stretchable power-generating sensor array using piezoelectric functional threads with hemispherical dome structures, highlighting their feasibility for wearable applications.

Furthermore, recent studies have investigated the use of cymbal-type piezoelectric transducers in smart textiles for energy harvesting and sensing. For example, Shim & et al. [30] demonstrated that cymbal transducers exhibit broadband characteristics when arranged in arrays. This configuration significantly enhances the efficiency of mechanical-to-electrical energy conversion, which is crucial for smart textile applications.

5.2 Energy Conversion Mechanism

The working principle of PENGs is based on the conversion of mechanical energy into electrical energy through the deformation of a piezoelectric material. Fig. 1 illustrates the electrical generation process and the variation of its electrical characteristics during one pressing and releasing movement [4].

This process follows a cyclic sequence of four distinct states:

- Initial state (i): At rest, the charge centers of cations and anions coincide, resulting in no polarization within the material. Consequently, no electrical potential or current is generated.
- Application of mechanical stress (ii): When a pressing force is applied, the piezoelectric material deforms, generating an internal dipole moment. This leads to the formation of a piezoelectric potential, a piezopotential, across the electrodes. If the electrodes are connected to an external circuit, electrons start flowing to partially compensate for this potential difference, resulting in an electric current.

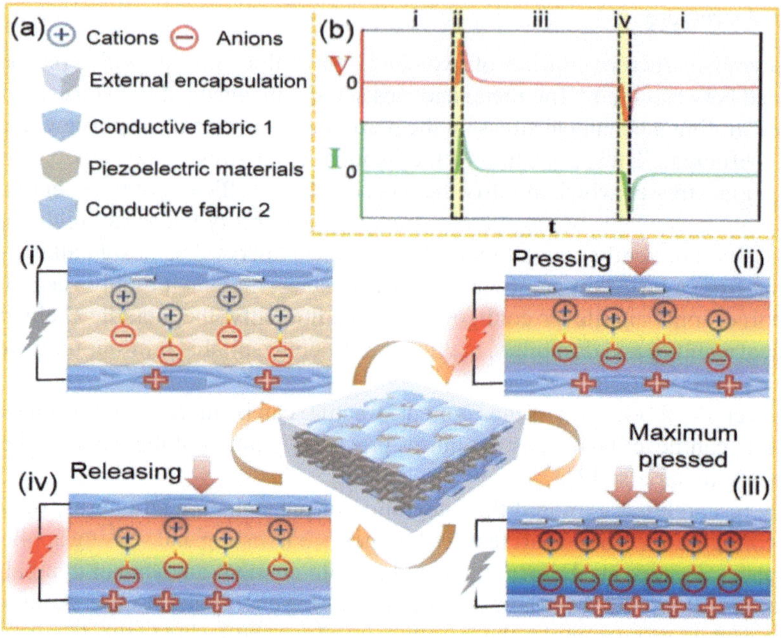

Fig. 1. Working mechanisms of textile-based PENGs. a) Schematic illustration of piezoelectric charge generation process. b) Typical open-circuit voltage (Voc) and short-circuit current (Isc) versus time curves, in which four parts corresponded to the four movement processes in (a). Reproduced from Ref. [4] with permission from the publisher.

- Maximum compression state (iii): As the material reaches its maximum deformation, the polarization is at its peak, meaning the charge distribution is stabilized. At this moment, the output voltage is at its highest value, but no current flows because there is no further variation in the electric field.
- Release of mechanical stress (iv): When the external force is removed, the material returns to its original shape, causing the charge distribution to shift again. This results in an opposite potential difference, generating a reverse current through the external circuit. If the mechanical deformation is cyclically applied, a continuous alternating pulse current is produced.

5.3 Applications of Piezoelectric Nanogenerators

Piezoelectric nanogenerators (PENGs) have gained significant attention for their potential to harvest mechanical energy and convert it into electrical power for various applications.

Wearable Electronics

PENGs are increasingly integrated into wearable systems to enable self-powered devices for real-time health monitoring and motion tracking. Due to their ability to convert biomechanical movements into electrical energy, they have been embedded in textiles and garments to develop smart wearable devices. For instance, hybrid piezoelectric fibers composed of barium titanate (BT) nanoparticles and PVDF have demonstrated significantly enhanced energy generation capabilities. These knitted textiles can generate up to 4 V and a power density of 87 μW.cm^{-3}. Such technology has been successfully integrated into knee sleeves for real-time health monitoring, showing promising potential for large-scale production of smart textiles [6].

Walking Activity Energy Harvesting

Human walking generates multiple forms of mechanical energy, including heel strikes, leg swings, and compressive forces. Researchers have developed PENG-based shoe insoles and watch straps permitting to harvest energy. A piezoelectric cantilever beam generator integrated into a shoe successfully increased its output voltage from 1.06 V to 3.55 V as walking speed increased from 2 km/h to 8 km/h, with power output rising from 0.03 mW to 0.35 mW [31]. Similarly, a PVDF-based curved PENG was designed to harvest low-frequency biomechanical energy, achieving a power density of 3.9 mW/cm^2 that is sufficient to power 476 LED bulbs. When integrated into a shoe insole, it generated an average output voltage of 14 V, and when attached to a watch strap, it produced 22 V [32].

Implantable and Biomedical Sensors

PENGs have been widely investigated for implantable medical devices due to their ability to harvest biomechanical energy. The flexibility and sensitivity of these devices allow for real-time monitoring of physiological activities such as heart rate, respiration and body motion. Flexible PENGs have been integrated into insoles to monitor footstep pressure and gait, enabling real-time biomechanical analysis for rehabilitation applications [33]. They have also been used as physiological sensors to detect weak physiological vibrations, such as eye blinks and pulse variations [34].

In cardiac energy harvesting, a novel helical-structured piezoelectric harvester has been developed to be integrated with pacemaker leads, eliminating direct contact of the harvesting device with the heart while efficiently scavenging mechanical energy. The developed harvesting device eliminates the need for additional surgeries since it is integrated with the pacemaker lead and is compatible with the pacemaker lead implantation approach [35].

Moreover, PENGs have been used to power lead-free pacemakers. A beam structure was designed to harness heartbeat vibrations, generating over 10 μW of power, which is sufficient to power an autonomous lead-free pacemaker. This approach eliminates the need for battery replacements, reducing surgical interventions [36].

Environmental Energy Harvesting

PENGs can be used for large-scale environmental energy harvesting, particularly from wind. In fact, flexible piezoelectric devices have been used to collect wind energy. Orrego & et al. [37] investigated the wind energy harvesting performance of inverted piezoelectric flags under controlled and ambient wind conditions. Inverted flag configuration offered multiple benefits such as providing sustained power generation over a wide range of wind speed. A peak power output of 1–5 mW/cm^3 has been achieved at wind speeds of 5–9 m/s, while at lower wind speeds of 2.5–4.5 m/s, power generation remains at 0.1–0.4 mW/cm^3.

Smart Home Applications
PENGs have been integrated into smart home technologies. A self-powered floor tile based on PZNxC thick film was developed to generate an output voltage of 42 V and a current of 52 μA from the footsteps of an average person (50–80 kg). This energy was sufficient to operate a wireless transmitter sensor node and a receiver switch module, demonstrating the feasibility of using PENGs for energy-efficient smart home systems [39]

Limitations of PENGs in Smart Textiles
Despite the considerable potential of PENGs in smart textiles, several challenges remain. For instance, the energy harvested from human motion or environmental vibrations is often insufficient to power high-demand devices, limiting the practical applications in wearable electronics and healthcare monitoring systems. A study demonstrated that PENGs integrated into textiles can harvest energy from body movements, but their power density remains below the practical requirements for real-time applications [39].

Additionally, the durability of piezoelectric materials under constant mechanical stress can affect the long-term performance of PENG-integrated textiles. Bagla & et al. [40] has demonstrated that while some piezoelectric materials such as PVDF-TrFE exhibit good flexibility, their performance may decrease after prolonged deformation.

Furthermore, the scalability of manufacturing processes and the integration of energy-harvesting systems into smart textiles present economic and technical challenges, particularly when considering the cost-efficiency and flexibility required for mass production. These limitations underscore the need for further research and development to enhance the performance, stability, and integration of PENGs in real-world applications.

6 Conclusion

Piezoelectricity has emerged as a key technology for sustainable and self-powered energy solutions, enabling the direct conversion of mechanical energy into electrical power. Its integration into textile structures represents a major advancement, giving rise to smart textiles able to harvest energy from human-body movements and environmental vibrations. These textile-based systems offer flexibility, wearability, and adaptability, making them ideal for next-generation applications in healthcare, sports, and interactive clothing.

Although some limitations remain, including low energy output and material durability, the field is rapidly evolving. With ongoing research and development, piezoelectric textiles stand as a highly promising domain, expected to play a crucial role in the future of self-powered wearable systems and smart fabrics.

References

1. Arnaud Delubac « What is the environmental impact of a battery? ». https://greenly.earth/en-us/blog/industries/carbon-footprint-battery (2025). Accessed 28 Feb 2025

2. Shaukat, H., Ali, A., Bibi, S., Altabey, W.A., Noori, M., Kouritem, S.A.: A Review of the Recent Advances in Piezoelectric Materials, Energy Harvester Structures, and Their Applications in Analytical Chemistry. Appl. Sci. **13**(3), 1300 (janv. 2023). https://doi.org/10.3390/app13031300
3. Sezer, N., Koç, M.: A comprehensive review on the state-of-the-art of piezoelectric energy harvesting. Nano Energy. **80**, 105567 (févr. 2021). https://doi.org/10.1016/j.nanoen.2020.105567
4. Dong, K., Peng, X., Wang, Z.L.: Fiber/Fabric-based piezoelectric and triboelectric nanogenerators for flexible/stretchable and wearable electronics and artificial intelligence. Adv. Mater. **32**, 1902549 (2020). https://doi.org/10.1002/adma.201902549
5. Maestri, G., Ferreira, L.B., Bachmann, P., Paim, A.A., Merlini, C., Steffens, F.: Recent advances in piezoelectric textile materials: A brief literature review. J. Eng. Fibers Fabr. **18**, 151242 (janv. 2023). https://doi.org/10.1177/15589250231151242
6. Mokhtari, F., et al.: Wearable Electronic Textiles from Nanostructured Piezoelectric Fibers. Adv. Mater. Technol. **5**(4), 1900900 (avr. 2020). https://doi.org/10.1002/admt.201900900
7. Curie, J., Curie, P.: Développement par compression de l'électricité polaire dans les cristaux hémièdres à faces inclinées. Bull. Minéralogie. **3**(4), 90–93 (1880). https://doi.org/10.3406/bulmi.1880.1564
8. Jaffe, H.: Piezoelectric Ceramics. J. Am. Ceram. Soc. **41**(11), 494–498 (1958). https://doi.org/10.1111/j.1151-2916.1958.tb12903.x
9. Haertling, G.H.: Ferroelectric Ceramics: History and Technology. J. Am. Ceram. Soc. **82**(4), 797–818 (1999). https://doi.org/10.1111/j.1151-2916.1999.tb01840.x
10. Shrout, T.R., Zhang, S.J.: Lead-free piezoelectric ceramics: Alternatives for PZT? J. Electroceram. **19**(1), 113–126 (sept. 2007). https://doi.org/10.1007/s10832-007-9047-0
11. Shur, V.Y.: Lithium niobate and lithium tantalate-based piezoelectric materials. In: Advanced Piezoelectric Materials, Woodhead Publishing, p. 204–238 (2010). https://doi.org/10.1533/9781845699758.1.204
12. Zhang, C., Fan, W., Wang, S., Wang, Q., Zhang, Y., Dong, K.: Recent progress of wearable piezoelectric nanogenerators. ACS Appl. Electron. Mater. **3**(6), 2449–2467 (juin 2021). https://doi.org/10.1021/acsaelm.1c00165
13. Li, L., Xu, J., Liu, J., Gao, F.: Recent progress on piezoelectric energy harvesting: structures and materials. Adv. Compos. Hybrid Mater. **1**(3), 478–505 (sept. 2018). https://doi.org/10.1007/s42114-018-0046-1
14. « Piezoelectric Charge Constants | APC International », americanpiezo. Accessed on: 10 Mar 2025 [Online]. Available on : https://www.americanpiezo.com/knowledge-center/piezo-theory/piezoelectric-constants/
15. Jean, F., Khan, M.U., Alazzam, A., Mohammad, B.: Advancement in piezoelectric nanogenerators for acoustic energy harvesting. Microsyst. Nanoeng. **10**(1), 197 (déc. 2024). https://doi.org/10.1038/s41378-024-00811-4
16. Sahu, S.K., Sreekanth, P.S.R., Reddy, S.V.K.: A Brief Review on Advanced Sandwich Structures with Customized Design Core and Composite Face Sheet. Polymers. **14**(20), 4267 (oct. 2022). https://doi.org/10.3390/polym14204267
17. Khan, T., Acar, V., Aydin, M.R., Hülagü, B., Akbulut, H., Seydibeyoğlu, M.Ö.: A review on recent advances in sandwich structures based on polyurethane foam cores. Polym. Compos. **41**(6), 2355–2400 (2020). https://doi.org/10.1002/pc.25543
18. Ciaburro, G., Romero, V.P., Iannace, G., Bravo Moncayo, L.: Improving Acoustic Properties of Sandwich Structures Using Recycled Membrane and HoneyComb Composite (RMHCC). Buildings. **14**(9), 2878 (sept. 2024). https://doi.org/10.3390/buildings14092878
19. Militký, J., Křemenáková, D., Venkataraman, M., Večerník, J., Martínková, L., Marek, J.: Sandwich Structures Reflecting Thermal Radiation Produced by the Human Body. Polymers. **13**(19), 3309 (sept. 2021). https://doi.org/10.3390/polym13193309

20. Ivanoska-Dacikj, A., et al.: Advanced and smart textiles during and after the COVID-19 pandemic: Issues, challenges, and innovations. Healthcare. **11**(8), 1115 (avr. 2023). https://doi.org/10.3390/healthcare11081115
21. Xu, Q., Gao, A., Li, Y., Jin, Y.: Design and optimization of piezoelectric cantilever beam vibration energy harvester. Micromachines. **13**(5), 675 (avr. 2022). https://doi.org/10.3390/mi13050675
22. Wan, X., et al.: All-textile piezoelectric nanogenerator based on 3D knitted fabric electrode for wearable applications. ACS Sens. **9**(6), 2989–2998 (juin 2024). https://doi.org/10.1021/acssensors.4c00158
23. Song, J., et al.: Highly Flexible, Large-Area, and Facile Textile-Based Hybrid Nanogenerator with Cascaded Piezoelectric and Triboelectric Units for Mechanical Energy Harvesting. Adv. Mater. Technol. **3**(6), 1800016 (2018). https://doi.org/10.1002/admt.201800016
24. Javed, Z., et al.: Piezoelectric nanogenerator for bio-mechanical strain measurement. Beilstein J. Nanotechnol. **13**, 192–200 (2022). https://doi.org/10.3762/bjnano.13.14
25. Kim, J., et al.: Cost-effective and strongly integrated fabric-based wearable piezoelectric energy harvester. Nano Energy. **75**, 104992 (sept. 2020). https://doi.org/10.1016/j.nanoen.2020.104992
26. He, Z., Mohsenzadeh, E., Zhang, S., Rault, F., Salaün, F.: Development of high-sensitive piezoelectric nanogenerators of all-organic PVDF multilayer nanofibrous membrane with innovative 3D structure via electrohydrodynamic processes. J. Polym. Res. **30**(10), 385 (oct. 2023). https://doi.org/10.1007/s10965-023-03776-6
27. J. Palosaari, « Energy harvesting from walking using piezoelectric cymbal and diaphragm type structures ». (2017).
28. Zhao, H., Yu, J., Ling, J.: Finite element analysis of Cymbal piezoelectric transducers for harvesting energy from asphalt pavement. J. Ceram. Soc. Jpn. **118**(1382), 909–915 (2010). https://doi.org/10.2109/jcersj2.118.909
29. Kim, K., Yun, K.-S.: Stretchable Power-Generating Sensor Array in Textile Structure Using Piezoelectric Functional Threads with Hemispherical Dome Structures. Int. J. Precis. Eng. Manuf.-Green Technol. **6**(4), 699–710 (août 2019). https://doi.org/10.1007/s40684-019-00127-z
30. Shim, H., Kim, K., Seo, H., Roh, Y.: Equivalent Circuit to Analyze the Transmitting Characteristics of a Cymbal Array. Sensors. **22**(22), 8743 (nov. 2022). https://doi.org/10.3390/s22228743
31. Fan, K., Liu, Z., Liu, H., Wang, L., Zhu, Y., Yu, B.: Scavenging energy from human walking through a shoe- mounted piezoelectric harvester. Appl. Phys. Lett. **110**(14), 143902 (avr. 2017). https://doi.org/10.1063/1.4979832
32. Jung, W.-S., et al.: Powerful curved piezoelectric generator for wearable applications. Nano Energy. **13**, 174–181 (avr. 2015). https://doi.org/10.1016/j.nanoen.2015.01.051
33. Matsouka, D., Vassiliadis, S., Bayramol, D.V.: Piezoelectric textile fibres for wearable energy harvesting systems. Mater. Res. Express. **5**(6), 065508 (juin 2018). https://doi.org/10.1088/2053-1591/aac928
34. Ani Melfa Roji, M., Jiji, G., Ajith Bosco Raj, T.: A retrospect on the role of piezoelectric nanogenerators in the development of the green world. RSC Adv. **7**(53), 33642–33670 (2017). https://doi.org/10.1039/C7RA05256A
35. Dong, L., et al.: In vivo cardiac power generation enabled by an integrated helical piezoelectric pacemaker lead. Nano Energy. **66**, 104085 (déc. 2019). https://doi.org/10.1016/j.nanoen.2019.104085
36. Ansari, M.H., Karami, M.A.: Piezoelectric energy harvesting from heartbeat vibrations for leadless pacemakers. J. Phys. Conf. Ser. **660**, 012121 (déc. 2015). https://doi.org/10.1088/1742-6596/660/1/012121

37. Orrego, S., et al.: Harvesting ambient wind energy with an inverted piezoelectric flag. Appl. Energy. **194**, 212–222 (mai 2017). https://doi.org/10.1016/j.apenergy.2017.03.016
38. Kim, K.-B., et al.: Optimized composite piezoelectric energy harvesting floor tile for smart home energy management. Energy Convers. Manag. **171**, 31–37 (sept. 2018). https://doi.org/10.1016/j.enconman.2018.05.031
39. Bagla, A., et al.: ZnO@C/PVDF electrospun membrane as piezoelectric nanogenerator for wearable applications. J. Phys. Chem. C. **129**, 5808 (2025)
40. Whiter, R.A., Narayan, V., Kar-Narayan, S.: A Scalable nanogenerator based on self-poled piezoelectric polymer nanowires with high energy conversion efficiency. Adv. Energy Mater. **4**(18), 1400519 (déc. 2014). https://doi.org/10.1002/aenm.201400519

Multisensory in the Metaverse: Designing Textile Data and Informative Feedback

Manel Abdelkefi(✉) and Amine Hadj Taieb

Higher Institute of Arts and Crafts, Sfax University, Sfax, Tunisia
Manel.abdelkefi@isams.usf.tn

Abstract. The rise of the metaverse is unlocking exciting new ways for us to connect with technology, where our very bodies and what we wear are poised to become active players. This article dives into the potential of smart textiles to create a truly embodied, almost intuitive way to interact within these immersive digital worlds. Imagine clothing that not only senses what's happening around us and within us but also provides feedback that feels natural and integrated. While smart textiles hold incredible promise for capturing our senses and giving us information back, making them feel like a seamless and effective part of the metaverse experience still presents some interesting design puzzles. This study shines a light on how crucial thoughtful information design and, especially, well-crafted feedback are to solving these puzzles. We're laying out a way of thinking about designing smart textile interfaces for the metaverse that puts these design considerations front and center. Our goal is to figure out how to create interactions that feel fluid, meaningful, and truly two-way. We'll be pinpointing key design ideas that keep you, the user, and the virtual environment in mind. We'll also be exploring creative ways to show information and provide sensory feedback that work with the unique qualities of smart textiles. By looking closely at what's holding us back right now and suggesting a design approach that focuses on clear information and intuitive feedback, this research aims to inspire new paths for creating wearable tech that genuinely feels like a natural extension of ourselves within the immersive and interactive future of the metaverse

Keywords: Metaverse · Smart Textiles · Interaction Design · Informative Feedback

1 Introduction

The metaverse, this ever-evolving, immersive digital space, is moving beyond just a virtual version of our social, professional, and recreational interactions. It's becoming a whole new realm for engaging our senses and our bodies [13]. In this emerging landscape, the human body—rooted as it is in the physical world—and its most intimate extension, our clothing, are shifting from mainly symbolic roles to becoming active interactive interfaces. Now, while everyday clothing seems like a natural fit to take center stage in the metaverse, its inability to really communicate back and forth with the digital environment—you know, through analog-digital interfaces (ADIs)—poses a pretty significant hurdle to its full integration [5].

M. Hamdaoui et al. (Eds.): ITDM'3 2025, PTT, pp. 134–143, 2026.
https://doi.org/10.1007/978-981-95-7704-0_12

That's where the Human Digital Twin (HDT) and smart textiles come into play. They're like the technological catalysts, packing in sensors, actuators, and transmission modules that can grab real-time physiological and environmental data [10, 20], paving the way for totally new kinds of human-machine interaction [15, 16]. But going beyond just their functional role as interfaces, these souped-up textiles are really catching the eye of designers because of their poetic potential. How can these tech-infused clothes and fabrics tell stories, convey emotions, send out alerts, offer reassurance when things are uncertain, or even capture and bring back a personal bodily experience within this immersive digital world? From this angle, textiles aren't just seen as simple data carriers anymore; they're becoming a sensitive medium, with their own unique way of expressing things, right at the intersection of our lived bodies, the code that makes them interactive, and the personal or shared narratives they can weave.

However, the challenge isn't just about hitting some technical performance benchmark. The real trick lies in designing a smooth, intuitive interactive experience—the UX—where what our bodies sense (the input) and the digital feedback we get (the output) are all orchestrated by a clear and meaningful feedback system. While current research often zooms in on individual components of smart textiles [5], it's becoming crucial to take a step back and look at the whole system, guided by information design and, especially, feedback design. This two-pronged approach—dealing with both the information and how it feels—plays a vital role in turning raw data streams into experiences that we can actually perceive, engage with, and use. In this two-way street, it's not just about 'wearing' data; it's about embodying it, feeling it, and understanding it. This brings up a key question: What role do information design and feedback design play in creating effective and usable human-machine interfaces (HMIs) based on smart textiles, so we can have that smooth and meaningful two-way interaction within the metaverse?

2 Conceptual Foundations

This section lays the conceptual groundwork for our project, right at the sweet spot where interaction design, smart textiles, and immersive environments meet. Our goal is to think about textiles as a sensitive interface and the metaverse as a stage for sensory interaction, all to sketch out a framework for designing informative multisensory feedback within these cyber-physical spaces.

2.1 Smart Textiles and Augmented Interaction

Smart textile refers to hybrid materials that can interact with their surroundings by picking up, processing, or sending out signals. What makes them special is how they're built—they've got sensors, conductive fibers, flexible electronic components, or what?

we call "reactive" materials woven right in. This lets these textile surfaces be both perceptive and expressive [6, 19].

Contemporary thinking sees smart textiles as advanced bodily interfaces, potentially able to extend or even transform our skin by embedding digital functionalities. These textiles can gather a wide range of data, including things like our heart rate, posture, environmental info, and even behavioral cues like pressure and muscle tension. This makes them rich channels of information flowing between the user and digital environments. This sensing capability opens the door to what we call "bottom-up" interaction: our bodies become an active source of data, which is then translated into immersive environments to trigger events or change how the user experiences things [12]. Conversely, some textile devices can deliver sensory feedback—think warmth, vibrations, or light—directly to the body, creating a "top-down" interaction. This two- way loop of interaction fits into the model of embodied interaction [8], where connected objects play a part in our situated cognition. This is illustrated in Fig. 1 below.

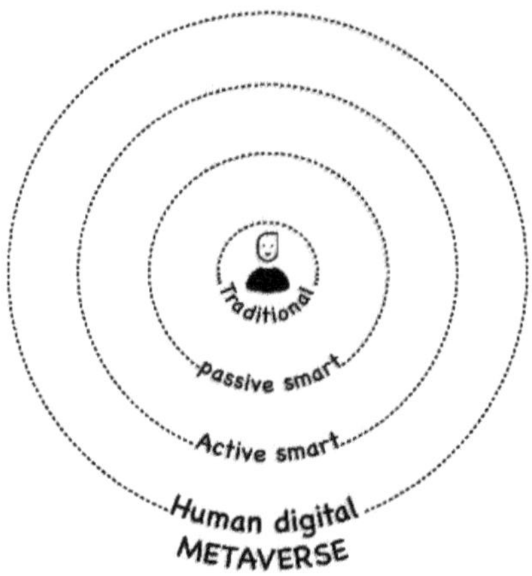

Fig. 1. Concentric Architecture of Smart Textile Integration in the Metaverse, Centered on the Human Digital Twin

Building on this exploration of how HMI textiles can fit into the metaverse, design research is increasingly fascinated by the poetic side of these augmented textiles. The big question becomes: how can these tech-enhanced clothes and fabrics go beyond just being useful tools to actually tell stories, convey emotions, send out alerts, offer comfort when we're uncertain, or even capture and bring back a unique bodily experience? From this perspective, textiles aren't just seen as simple carriers of raw data anymore; they're turning into a sensitive medium, with their own way of expressing things, right at the intersection of our lived bodies, the computer code that makes them interactive, and the personal or shared stories they can weave.

It's precisely within this dynamic tension—between the technical functionality and this poetic expressiveness—that the heart of our exploration lies. Designing smart textiles for the metaverse isn't just about collecting and sending data; it absolutely needs to engage the user on a deep sensory and emotional level. The answer to this challenge potentially lies in a carefully considered approach to feedback, one that can transform raw information into an experience that's not just meaningful and engaging, but also rich in significance and feeling.

2.2 The Metaverse: An Immersive Cyber-Physical Space

The metaverse is commonly defined as a collective, persistent, and immersive virtual environment, accessible through avatars and integrating layers of social, spatial, and economic interaction [6, 7]. It constitutes an extension of the paradigms of virtual reality (VR) and augmented reality (AR), while opening up a field of experimentation for the design of hybrid experiences. In this cyber-physical space, which merges physical and digital realities, the user interacts with virtual elements through physical devices and perceives digital information via their bodily senses. In this context, the user is no longer a simple spectator but a full-fledged actor, called upon to interact and act within immersive environments. Traditional interaction devices (keyboard, mouse, screen), while fundamental, present intrinsic limitations for deep immersion: their non- corporeal nature and indirect sensory feedback can hinder the naturalness of spatial and gestural interactions, and thus diminish the feeling of presence in the virtual space. Bodily interfaces, and notably smart textiles, offer a promising alternative to overcome these limitations, by enabling intuitive interaction and direct, integrated sensory feedback to the body [1, 2].

Figure 2 illustrates the diverse capabilities of e-textiles, encompassing the sensing of physiological signals (ECG, EMG, EEG), movement (constraint, pressure), environmental conditions (temperature, gas), and visual display (OLED, PLED, electrochromic), highlighting their potential as truly multimodal interfaces for the metaverse. By activating tangible sensory feedback, these textiles can embody a new grammar of immersive interaction. We're talking about informative feedback that bypasses the visual and verbal, instead using touch, temperature, vibration, or tension [14]. This novel grammar allows for intuitive and expressive interactions: for example, the sensation of holding a virtual object could be simulated by subtle pressure and gentle vibrations felt through smart gloves, while the texture of a virtual surface could be communicated by variations in roughness or temperature perceived by a haptic garment. This type of feedback, whether tactile or proprioceptive, can enrich the understanding of virtual events (feeling the resistance of a virtual door when pushing it) or further enhance the feeling of presence in the digital environment (having the sensation of being touched by an avatar) [17].

The design of these interactions draws inspiration from an experience-centered design approach, where feedback is conceived as a sensory dialogue between the textile material, the user's body, and the digital data [21]. Design thus becomes a vector of "interactional intelligence," capable of modulating the quality of experience based on the context, the user's intentions, and the temporality of use. Finally, the integration of smart textiles into the metaverse invites us to rethink the classic paradigms of HMI design: How do we design for non-screen interfaces that engage the entire body? How do we translate complex numerical data into pertinent, legible, and memorable tactile sensations? And how do we orchestrate the back-and-forth of information between the user's intentions, their bodily actions, and the virtual system's reactions?

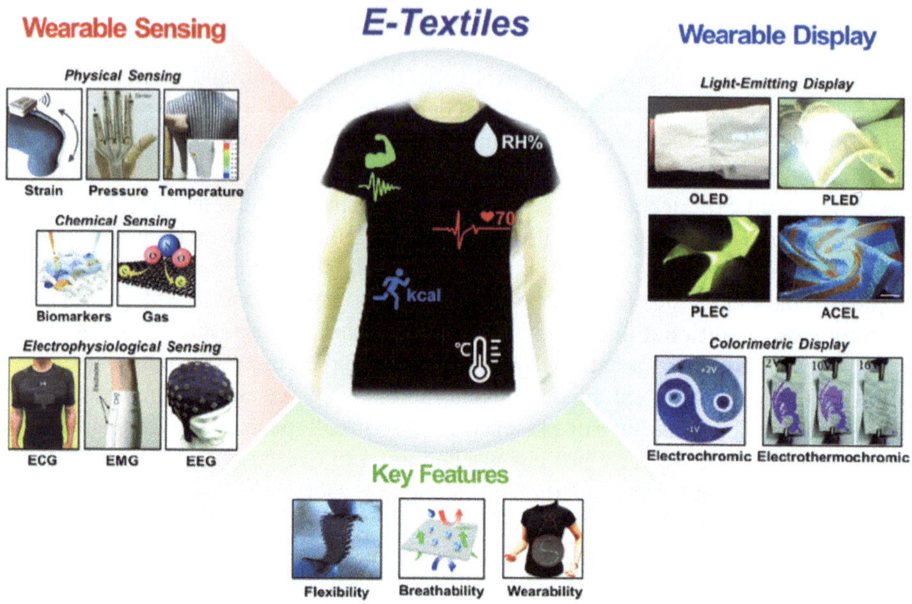

Fig. 2. e-textiles with Wearable Sensing and Wearable Display capabilities, highlighting key features of flexibility, conformability, and wearability. Adapted from Cho et al., 2022.

3 Designing Multisensory Informative Feedback

To further explore the mechanisms by which smart textiles can serve as immersive and intuitive interfaces within the metaverse, it is essential to consider the body itself as the starting point for this interaction. This section thus addresses the notion of the body as interface, drawing upon fundamental theories of human-machine interaction to understand how we can effectively design multisensory informative feedback through smart textiles.

3.1 Body as interface

Within the framework of our exploration of smart textiles as an interface for immersion and augmented perception in metaverse environments, it appears essential to invoke certain theories from human-machine interaction (HMI), notably the contributions of Donald Norman and work on haptic and proprioceptive feedback in wearable interfaces. These approaches allow for a better understanding of the role of feedback in the interaction loop and clarify the cognitive and sensory value of responses generated by bodily interfaces. Donald Norman (2010) proposes a seven-stage model that decomposes interaction with a technical system into two major phases: the formulation of action (intention to perform an action in the metaverse, planning of the necessary bodily movement, execution of this movement via the textile interface) and the evaluation of the result (perception of sensory feedback from the textile—a pressure, a vibration, a temperature change –, interpretation of this feedback as a consequence of the action in the virtual world, comparison of this perceived result with the initial intention). This is illustrated in Table 1 below.

Table 1. Norman's Seven-Stage Model of Action

Phase	Étapes	Description
Formulation	1. Formulation of the goal	Define what the user wants to accomplish
	2. Formulation of the intention	Translate the goal into anticipated actions
	3. Specification of the Action	Determine how to realize the intention
	4. Execution of the Action	Perform the action with the system
Evaluation	5. Perception of the system State	Observe the system's response
	6. Interpretation of the State	Undesrtand the meaning of the feedback
	7. Evaluation of the outcome	Compare with the initial goal

For example, if the user wishes to interact with a virtual button, the intention (step 2) translates into a hand movement detected by textile sensors (step 4). Slight haptic feedback at the fingertip (step 6) confirms the button's activation, allowing the user to interpret (step 7) the success of the action and compare it to their intention. Haptic and proprioceptive feedback, particularly via smart textiles that envelop the body, offer a rich and direct channel for communicating the state of the virtual world and the consequences of the user's actions, significantly enriching immersion and perception [14].

3.2 Multisensory Feedback

The integration of multisensory feedback is essential for enriching sensory immersion in virtual environments, as highlighted by a recent study on the impact of smell in virtual architectural environments [22], and this is particularly relevant for bodily interfaces such as smart textiles. The synchronized combination of multiple sensory modalities—visual, haptic, auditory (optional), and thermal—permits a more natural and coherent integration of information, leveraging the synergy between the senses for an enhanced bodily perception and heightened user reactivity [9]. This approach aligns with an understanding of user experience (UX) that goes beyond mere functionality to encompass a profound relationship between the body, the interface, and the environment.

In a textile perspective, this implies rethinking feedback as an embodied multisensory experience, where sensory perceptions and emotions are intrinsically linked. As Nancy (2000) emphasizes, perceptual experience is fundamentally corporeal, involving a range of physical actions and contacts. As illustrated in Table 2, a smart textile can thus translate virtual events into a rich palette of sensations: a visual alert (for example, an imminent danger in the metaverse) can manifest simultaneously through localized vibration (haptic), a change in color or a flashing pattern (visual), and a slight increase in temperature (thermal), reinforcing the urgency of the signal and soliciting a more pronounced emotional response [4]. The temporal synchronization of these stimuli is crucial for the coherence of the experience and for establishing an intuitive "feeling of presence" [18].

One of the key principles for successful immersion in virtual environments is that the synchronized integration of visual and haptic feedback leads to a more natural and

Table 2. Examples of Feedback Modalities in Smart Textiles.

Feedback Type	Sensory channel	Exemple in a Smart Textile
Visual	Sight	Color change / dynamic pattern
Haptic	Touch	Vibration, Fabric tension, swelling
Thermal	Temperature	Localized warmth or coolness
Auditory (optional)	Hearing	Soft sound or click upon an action

intuitive experience. Extending this idea to smart textiles is particularly promising for metaverses, where bodily sensations and emotions could become central elements of interaction.

3.3 Typology of Bodily Textile Feedback

We propose considering textile feedback along two axes: the sensory modality and the cognitive function. The intersection of these axes allows for the creation of a typological framework for interactive textile design. This approach draws inspiration from both user-centered design principles and theories of multisensory experience. This is illustrated in Table 3 below.

Table 3. Examples of Feedback Modalities in Smart Textiles.

Modality	**Confirmation**	**Alert/Error**	**Guidance**	**Emotional Simulation**
Visual	Subtle patter change	Rapid blinking abrupt color change	Progressive directional color	Soft gradient effect shooting or stimulating patterns
Haptic	Gentle vibration, texture modification	Intense vibration, sudden pressure	Directional modulation of intensity	Rhythmic pulsing sensation of the presence
Thermal	Stable maintained temperature	Sudden and intense heat or coldness	Progressive and localized heat or coldness	Sensation of localized comfort or discomfort
Auditory (optional)	Soft clic sound integrated with body movment	Rapid strident beep sound, abrupt tonal, variation	Directional tone variation enveloping sound scape	Soft melodies or emotionally evocative sounds

This way of categorizing feedback is all about helping designers get creative with interactive textiles. It gives them a framework to think about how different senses can be used to communicate, evoke emotions, and make the interaction work better. Ultimately, it's about unlocking the expressive power of how textiles feel on our bodies within the metaverse. Because, as sensory design research suggests, when you design feedback as a whole sensory experience, it can turn what we wear into a truly embodied way to interact.

Throughout this section, we've really emphasized how central feedback is to designing interactive textiles, especially when we're thinking about augmented embodiment and that feeling of "being there"—embodied UX. Leaning on Norman's action cycle model and the principles of multisensory feedback [3, 11], it becomes super important to design interaction not just as a mental exchange of information, but as a rich, grounded sensory experience that really connects with the user's body. In this context, smart textiles go beyond just being a simple support for embodied interaction; they become a genuine tool for it. This is illustrated in Fig. 3 below.

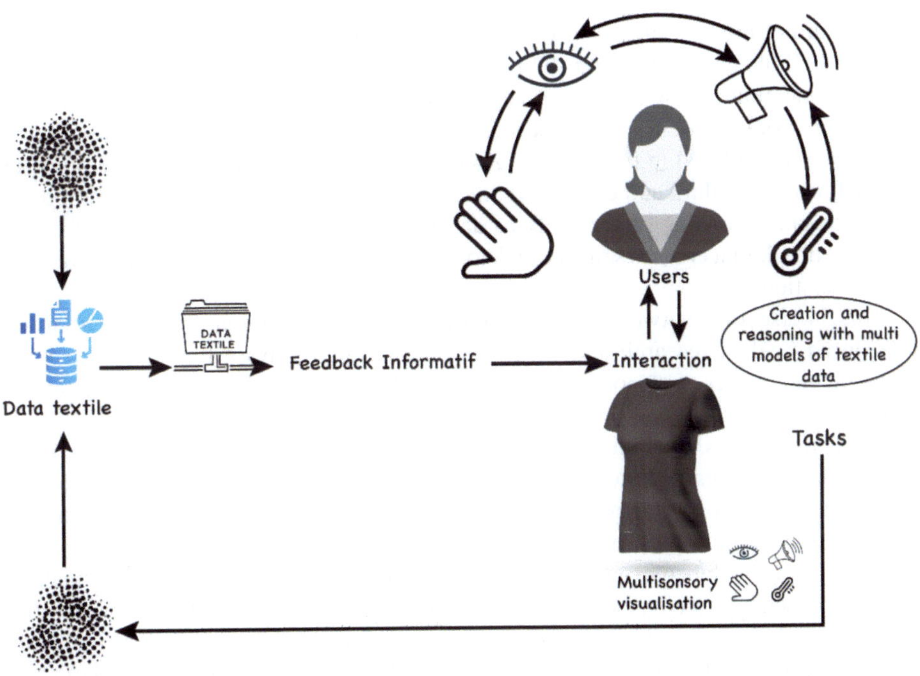

Fig. 3. Schematic of a multisensory visualization system for textile data interaction

This feedback—whether it's visual (like changing colors or dynamic patterns), haptic (through different vibrations, tensions, or pressure), thermal (sensations of warmth or coolness in specific spots), or even auditory (with subtle integrated sound cues)—goes way beyond just passing along raw data. It turns into a powerful way to convey emotions, offer intuitive guidance, and provide contextually relevant meaning. This approach opens up a whole new way of designing interactions that are fundamentally intuitive and expressive, capable of orchestrating and synchronizing different sensory modalities to really boost the user's immersion in hybrid environments, whether they're real, virtual, or a mix of both.

So, thinking about textile feedback means seeing clothing not just as a static object, but as a dynamic and living interface, inherently capable of having a real-time conversation with our bodies, responding to our actions and the environment around us. This shift in perspective lays the groundwork for innovative interaction design, where sophisticated technological experience is intertwined with a deep, perceptive sensitivity. Our ultimate goal is to create a new kind of augmented bodily experience, packed with expressive potential and deeply integrated into how we interact with the digital and physical world around us, especially within the immersive and sensory-rich context of metaverses.

4 Conclusion

In conclusion, the design of smart textile HMIs for the metaverse cannot be approached solely from a technical or functional standpoint. It requires design thinking capable of articulating informational, sensory, and contextual dimensions into a coherent and embodied whole. By mobilizing the principles of information design centered on the human, contextualization, significant visualization, and particularly rich and multimodal sensory feedback—carefully orchestrated to create a truly bidirectional and embodied interaction, as explored in the preceding sections—the smart garment has the potential to transform into a fluid and expressive intelligent interface. This transformation relies on a deep understanding of the body as an interface, leveraging a typology of bodily textile feedback that crosses sensory and cognitive functions for intuitive and significant communication.

This intersection between textile, interaction, and immersion opens up a largely unexplored field of research and creation, where design plays a fundamental role as mediator, interpreter, and poetic technologist. Through this approach, the metaverse ceases to be a purely digital space to become a tangible sensory environment, inhabited by augmented bodies, intelligent textiles, and narrative experiences woven into the fabric of data. However, this ambitious vision raises significant challenges. The technological constraints related to the integration of miniaturized and high- performance sensors and actuators into textiles, questions of social and aesthetic acceptability of these intelligent garments, and the inherent complexities in the design of coherent and personalized multimodal sensory feedback remain crucial research axes for the future. Navigating this sensory design with a human-centered approach and a creative exploration of the expressive potentialities of textiles will be essential to fully realize the potential of smart textile HMIs in the metaverse.

References

1. Billinghurst, M., Kato, H., Kiyokawa, K., Belcher, D., Poupyrev, I.: Experiments with face-To-face collaborative AR interfaces. Virtual Real. **6**(3), 107–121 (2002). https://doi.org/10.1007/s100550200012
2. Cho, S., Chang, T., Yu, T., Lee, C.H.: Smart Electronic Textiles for wearable sensing and display. Biosensors. **12**(4), 222 (2022). https://doi.org/10.3390/bios12040222
3. Cornelio, P., Velasco, C., Obrist, M.: Multisensory integration as per technological advances: A review. Front. Neurosci. **15**, 652611 (2021). https://doi.org/10.3389/fnins.2021.652611

4. Desmet, P. : Measuring emotion: development and application of an instrument to measure emotional responses to products. In: Blythe, M.A., Overbeeke, K., Monk, A.F., Wright, P.C. (eds) Funology. Human-Computer Interaction Series, vol 3. Springer, Dordrecht (2003). https://doi.org/10.1007/1-4020-2967-5_12
5. Diak, V., Diak, A.: Smart Textile as Advanced Human Digital Interface in Metaverse. SSRN. (2025). https://doi.org/10.2139/ssrn.5083525
6. Dias, T.: Electronic Textiles. Woodhead Publishing. Elsevier (2015). ISBN: 9780081002230
7. Dionisio, J.D.N., Iii, W.G.B., Gilbert, R.: 3D Virtual worlds and the metaverse: Current status and future possibilities. ACM Comput. Surv. **45**(3), 1–38 (2013). https://doi.org/10.1145/2480741.2480751
8. Dourish, P.: Where the Action is: The Foundations of Embodied Interaction. MIT Press paperback edn. MIT Press (2004). https://doi.org/10.7551/mitpress/7221.001.0001
9. Ernst, M.O., Bülthoff, H.H.: Merging the senses into a robust percept. Trends Cogn. Sci. **8**(4), 162–169 (2004). https://doi.org/10.1016/j.tics.2004.02.002
10. Heikenfeld, J., Jajack, A., Rogers, J., et al.: Wearable sensors: Modalities, challenges, and prospects. Lab Chip. **18**(2), 217–248 (2018). https://doi.org/10.1039/C7LC00914C
11. Lécuyer, A.: Simulating haptic feedback using vision: A survey of research and applications of pseudo-haptic feedback. Presence-Teleoper. Virtual Environ. **18**(1), 39–53 (2009). https://doi.org/10.1162/pres.18.1.39
12. McCann, J., Bryson, D.: Smart Clothes and Wearable Technology. Elsevier (2009). ISBN: 978-1-84569-357-2
13. Mu, X., Zhang, H., Shi, J., Hou, J., Ma, J., Yang, Y.: Fashion intelligence in the Metaverse: Promise and future prospects. Artif. Intell. Rev. **57**(3), 67 (2024). https://doi.org/10.1007/s10462-024-10703-8
14. Pacchierotti, C., Sinclair, S., Solazzi, M., Frisoli, A., Hayward, V., Prattichizzo, D.: Wearable Haptic Systems for the Fingertip and the Hand: Taxonomy, Review, and Perspectives. IEEE Trans. Haptics. **10**(4), 580–600 (2017). https://doi.org/10.1109/TOH.2017.2689006
15. Pouta, E., Mikkonen, J.V.: Woven eTextiles in HCI—A literature review. In: Proceedings of the ACM Designing Interactive Systems Conference, pp. 1099–1118 (2022). https://doi.org/10.1145/3532106.3533566
16. Ruckdashel, R.R., Khadse, N., Park, J.H.: Smart E-Textiles: Overview of Components and Outlook. Sensors. **22**(16), 6055 (2022). https://doi.org/10.3390/s22166055
17. Slater, M., Wilbur, S.: A framework for immersive virtual environments (FIVE): Speculations on the role of presence in virtual environments. Presence-Teleoper. Virtual Environ. **6**(6), 603–616 (1997). https://doi.org/10.1162/pres.1997.6.6.603
18. Stein, B.E., Meredith, M.A.: The merging of the senses (Cognitive Neuroscience). Bradford Books (1993). ISBN : 978-0262693011
19. Stoppa, M., Chiolerio, A.: Wearable Electronics and Smart Textiles: A Critical Review. Sensors. **14**(7), 11957–11992 (2014). https://doi.org/10.3390/s140711957
20. Wang, H., et al.: A smart mechanical-energy harvesting and self-heating textile device for photo-thermal energy utilization. EcoMat. **5**(5), e12337 (2023). https://doi.org/10.1002/eom2.12337
21. Wright, P., Wallace, J., McCarthy, J.: Aesthetics and experience-centered design. ACM Trans. Comput.-Hum. Interact. **15**(4), 1–21 (2008). https://doi.org/10.1145/1460355.1460360
22. Yildirim, M., Globa, A., Gocer, O., Brambilla, A.: Multisensory Virtual Reality Environments in Architecture: The Impact of Smell on Presence, Engagement, and User Experience. Available at SSRN: https://ssrn.com/abstract=5098840 or http://dx.doi.org/10.2139/ssrn.5098840

Breathability Performance of Disposable Baby Diapers

Wala Kallala[1(✉)], Sofien Benltoufa[1], Faten Fayala[1,2], and Adel Ghith[3]

[1] University of Monastir, Laboratory of Thermal and Energetic Systems Studies (LESTE), 5000 Monastir, Tunisia
walakallala21@gmail.com

[2] Monastir Textile Science & Technology Research Center, 5000 Monastir, Tunisia

[3] Université de Monastir, ENIM, Unité de Recherche Matériaux et Procédés Textiles (MPTex), UR17ES33, 5019 Monastir, Tunisie

Abstract. Disposable baby diapers undergo continuous innovation to improve absorption, comfort, and breathability. Despite short use, they incorporate advanced technologies and materials tailored to infants' physiological needs. Their performance relies on absorbing and retaining liquids while minimizing skin irritation, achieved through a complex interaction of layers. Diaper design prioritizes absorption, impermeability, and breathability.

This study analyzes five baby diaper brands available in Tunisia to assess their breathability and comfort properties based on physical and morphological parameters. Measurements include total mass, thickness, density, and microscopic observations of topsheet, high loft, and backsheet. Air and water vapor permeability tests are also conducted to assess breathability.

The results indicate that the density and morphological structure of the layers directly influence their breathability. High density and compact structure, as observed for topsheet D (0.42 g/cm^3), limit the circulation of air and water vapor. This topsheet has the lowest air permeability (3903 mm/s), a reduced water vapor transmission rate (RWVP%) of 78.92%, and a high evaporative thermal resistance (RET) of 4.61 m^2-Pa/W, reflecting a low capacity to evacuate heat and moisture.

Keywords: Breathability · Baby Diapers · Comfort

1 Introduction

Disposable baby diapers have become an essential part of modern baby hygiene. Their development has radically transformed parenting practices, facilitating baby hygiene and reducing the amount of work involved in diapering. Since their commercial appearance in the 1940s, they have evolved through major technological advances, improving their absorbency, comfort and environmental impact. However, their production and disposal still pose ecological and health challenges, motivating the search for more sustainable solutions that respect the health of babies.

Before the invention of disposable diapers, babies were mainly dressed in cloth diapers, requiring frequent washing. The first modern disposable diapers were introduced in

M. Hamdaoui et al. (Eds.): ITDM'3 2025, PTT, pp. 144–153, 2026.
https://doi.org/10.1007/978-981-95-7704-0_13

the 1950s by the Johnson & Johnson brand, followed by Pampers in 1961, which revolutionized the market with a more absorbent and ergonomic design [1]. Since then, diapers have evolved to include superabsorbent polymers (SAP), leak barriers and improved fastening systems. Their conception is based on a multi-diaper structure integrating various materials, each with specific roles in humidity management, prevention of irritation and optimization of adjustment.

In a typical diaper as shown in Fig. 1, the first layer is the topsheet, which is in contact with the skin. It is an essential part of baby diaper conception, providing a soft layer that maintains the integrity of the structure while facilitating the passage of moisture to the absorbent core.

To ensure comfortable contact with the skin, it is made from polypropylene (PP), a material widely used today in two main forms: carded short fibers (staple fibers) and spunbonded continuous filaments. Each of these technologies has its own distinct characteristics. Carded fabrics, composed of crimped fibers, are characterized by their greater volume and slightly rough texture. Spunbonded fabrics, on the other hand, offer a silky softness and are generally stronger, especially in the cross-machine direction, enabling the use of less dense and therefore more economical materials. Thanks to this cost advantage, spunbonded fabrics have gradually gained ground over thermobonded carded fabrics. In Europe, most topsheets are manufactured from homopolymer polypropylene, a naturally hydrophobic material. To guarantee effective absorption, surface treatments are applied to allow moisture to pass through the sheet while limiting its return to the surface. The aim is to optimize liquid strike- through and minimize re-wet. These treatments evolve towards permanent hydrophilicity, guaranteeing that the absorbent effect does not dissipate over time and that the material remains high-performance, even after several successive flows [2]. This study [3] highlights that Air-Through technology in the conception of disposable diaper topsheets offers the best permeability and breathability, effectively reducing moisture in contact with the skin and minimizing the risk of dermatitis. Spunbonded sheets, on the other hand, although more resistant and economical, have a denser structure, limiting air circulation and potentially favoring moisture accumulation. The SMS group (Spunbond-Meltblown-Spunbond) represents an intermediate alternative, combining good absorption capacity with enhanced protection against leaks. The choice of material therefore has a direct influence on the comfort and hygiene of diapers, and future improvements could incorporate innovative materials to further optimize this performance.

Directly below the topsheet, we often find the acquisition layer also called high loft, this layer is essential for improving absorption by speeding up liquid acquisition, optimizing retention to limit the rewet effect (return of moisture to the surface) and enabling more uniform distribution of liquid throughout the layer, making more efficient use of its absorption capacity [4].

These acquisition layers are generally made up of carded staple fibers, assembled using a variety of techniques, resin bonding (adhesive) being the most used. Depending on the layer concept, other processes such as thermal air bonding, needling or thermobonded spunbonded fibers may also be used. Innovation in the field of diapers is constant, with the regular introduction of new models [2].

Then there is the absorbent core, which is the most important component of the diaper. It's made from a combination of materials specially designed to absorb liquids quickly and hold them in place, preventing leakage and irritation. Its main components include cellulose fibers derived from wood pulp and a superabsorbent polymer (SAP)

Superabsorbent polymers (SAP) are hydrophilic structures capable of absorbing and retaining very large quantities of water or aqueous solutions, for example, they can absorb more than 30 times their own weight in liquid [5].

The positioning of SAP in the composite structure has a significant impact on its efficiency. In addition, particle size, location and proportions play a key role in optimizing absorption. When the swelling of the superabsorbent is delayed in the wetting zone of certain diaper models, this allows better distribution of urine in the diaper. By distributing the liquid more evenly, core saturation in the wetting zone is reduced, improving absorption capacity [6].

Fluff pulp is a cellulose pulp made up of cellulose fibers forming a capillary network that favors the absorption and retention of liquids. Thanks to its high porosity and low density, he capacity of pulp is around 10 cm^3 of water per gram when "free swelling" but drops to less than 2 cm^3 under a pressure of 5 Kpa [7].

Finally, the backsheet is usually a nonwoven/polyethylene (PE) film composite. This layer helps to ensure the product's performance by preventing the transfer of liquids to the outside, thus ensuring the containment of the whole structure. As the first material in contact with the consumer, its softness is an important criterion. It must also be sufficiently resistant to ensure its effectiveness [8].

In addition to the main components mentioned previously, the baby diaper includes other elements, such as the fastening system, consisting of a comfort band at the front, side and back fasteners and elastic between the legs, as well as leakage barriers.

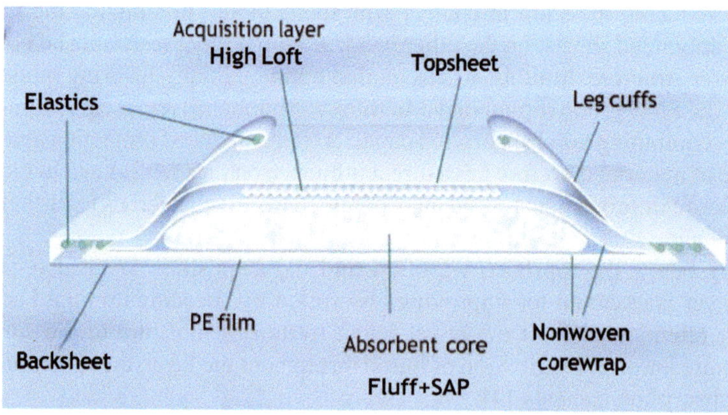

Fig. 1. Disposable Baby diapers composition.

In this study, we carried out a comparison between five brands of baby diapers available on the Tunisian market. We specifically separated out certain non-woven components, such as topsheet, high loft and backsheet, to better understand their impact on overall diaper performance. Tests were carried out on these components, including measurements of air and water vapor permeability, to identify those offering the best characteristics in terms of comfort, absorption efficiency and moisture management.

2 Materials and Methods

2.1 Materials

To carry out this study, we selected and tested five brands of disposable baby diapers from different origins and available on the Tunisian market. These diapers were anonymous and identified as A, B, C and D. Diapers A, B and C are from Tunisia, while diaper D originates from Egypt and diaper C from Algeria. Each diaper has been carefully separated into its parts, and its specific properties evaluated through a series of tests. All the diapers selected for this study are size 5. This choice was made to ensure a homogeneous comparison between the different brands and to evaluate their performance under similar conditions.

2.2 Methods

We carried out dimensional measurements of five brands of baby diapers, including total mass and thickness. We also measured non-woven components such as topsheet, high loft and backsheet, recording their mass and thickness, and then calculated the density of each. These calculations enabled us to better understand the physical properties of each layer and analyze their impact on the global performance of the diapers.

Microscope observation was carried out to ensure accurate characterization of the various components of baby diapers. For this purpose, an optical microscope with several levels of magnification was used. Non-woven structures were observed at 0.63X magnification, enabling analysis of material morphology and fiber distribution within the samples.

Air and water vapor permeability is a key parameter in assessing the breathability and comfort of baby diapers. To analyze this property, an air permeability tester was used in accordance with international standard ISO 9237 (1995). The test consists in measuring the airflow speed through a 20 cm^2 circular sample, placed perpendicular to the airflow, at a constant pressure of 200 Pa. Results are expressed in mm/s. Water vapor permeability is determined using the Permetest apparatus. This method consists of diffusing a precise quantity of water using a syringe. The device then transforms this water into vapor, which passes through the sample to be tested. Air and water vapor permeability were applied to all diapers and their main non-woven components, including topsheet, high loft and backsheet.

3 Results and Discussions

3.1 Dimensional Measurements

All the diapers tested share similar compositions. Firstly, each has a topsheet, high loft, absorbent core, which contains a blend of cellulose pulp and superabsorbent polymer (SAP), and a backsheet. Although the composition of these diapers is essentially identical between the different models tested, differences can still be observed in the placement of the components within each diaper, as well as in the quantity of each material used.

The results for mass (g), thickness (mm) and density (g/cm3) of baby diapers are shown in Table 1.

The results of the diapers masses shown in Table 1 indicate a significant variation between the mass values. Diaper A has the highest mass (48.54 g). On the other hand, Diaper E has a significantly lower mass, at around half that of diaper A. More precisely, the mass of diapers E is the lowest of all the diapers, at around 25.09 g.

Table 1. Mass, thickness and density values for baby diapers.

	A	B	C	D	E
Muti-layer					
Mass (g)	48.54 ± 0.7	43.28 ± 1.91	41.09 ± 0.61	38.02 ± 0.59	25.09 ± 0.22
Topsheet					
Mass (g)	1.14 ± 0.01	0.81 ± 0.04	0.94 ± 0.04	1.37 ± 0.01	0.97 ± 0.05
Thickness (mm)	0.19	0.19	0.2	0.33	0.17
Density (g/cm^3)	0.084	0.073	0.065	0.42	0.094
High Loft					
Mass (g)	0.69 ± 0.06	0.66 ± 0.01	0.55 ± 0.01	0.68 ± 0.03	1.1 ± 0.05
Thickness (mm)	0.6	0.57	0.55	0.6	0.78
Density (g/cm^3)	0.055	0.061	0.064	0.87	0.049
Backsheet					
Mass (g)	1.44 ± 0.15	1.22 ± 0.23	1.25 ± 0.1	1.8 ± 0.1	1.29 ± 0.1
Thickness (mm)	0.16	0.18	0.16	0.25	0.14
Density (g/cm^3)	0.088	0.083	0.094	0.072	0.086

The difference in mass observed between baby diapers can be attributed to variations in the quantities of materials used in their manufacturing, which can influence essential parameters such as comfort and breathability.

Topsheets (TP) masses result range from 0.81 g to 1.37 g, indicating a significant variation. Diaper B has the lowest topsheet (TP) mass, while diaper D has the highest.

Thickness of the topsheets varies between 0.17 mm and 0.33 mm. The topsheet E has the lowest thickness, while topsheet D has the highest. Density of the topsheets varies between 0.065 and 0.42 g/cm^3. Topsheet D has the highest density, while topsheet E has the lowest.

High Loft (HL) masses range from 0.55 g to 1.1 g, which represents a difference. Diaper E has the highest High Loft mass, while diaper C has the lowest. High Loft thickness varies between 0.55 mm and 0.78 mm. High Loft of diaper D is the thickest, while that of diaper C is the thinnest. Density of the high loft varies between 0.049 and 0.087 g/cm^3. High Loft D has the highest density, while high loft E has the lowest.

Backsheet (BK) masses range from 1.22 g to 1.8 g, with relatively little variation. Diaper D has the highest backsheet mass, while diaper B has the lowest. The thickness of the backsheet varies from 0.14 mm to 0.25 mm. Backsheet E is the thinnest, while backsheet D is the thickest. Density of the backsheets varies between 0.072 and 0.094 g/cm^3. backsheet C has the highest density, while the backsheet D has the lowest.

The difference between the values found for physical properties such as total mass, thickness and density of the various non-woven components (TP, HL and BK) used in baby diapers can be a factor influencing the breathability and comfort performance of the diapers tested.

3.2 Microscopy

Figure 2 shows microscopy images of the topsheet, high loft and backsheet of the various used diapers.

Figure 2 shows some differences between the structures of the five topsheets, high loft and backsheet. Regarding topsheet, we observe that Topsheets a and b have relatively similar structures. Their structure is mostly fibrous, consisting of interlaced fibers with oval fusion points formed by thermobonding. Topsheet c shows fusion points in its structure, although they are not as clearly visible as in the other topsheets. The structure is made of randomly interlaced fibers, with a loose distribution and an increased presence of fiber spaces. Topsheet d has a compact structure with reduced porosity, limiting the passage of liquids while ensuring comfort for the skin. It is also characterized by the presence of circular melting points, randomly distributed over the surface.

Concerning the High Loft, we observe that High loft c structure presents fibers interlaced in violet and arranged randomly, forming a porous and irregular structure.

Concerning the Backsheet, backsheet a, b and c feature a relatively homogeneous interlaced fiber structure. These fibers form a light, partially porous network. In addition, these backsheets are characterized by the presence of fusion points, generally oval, resulting from the thermobonding process.

3.3 Air Permeability

Figure 3 shows the values of the air permeability test carried out on the three nonwovens from the different diapers tested.

Figure 3 shows that Topsheet air permeability values range from 3903 to 4702 mm/s. Topsheet D showed the lowest air permeability, suggesting a dense internal structure. This hypothesis was confirmed by microscopic observation, which revealed a compact structure with a low porosity rate. This high density limits the passage of air through the material. In contrast, topsheet D showed the highest air permeability of the samples tested, indicating a more open structure. This observation is consistent with the microscopic results, which showed a less dense distribution of fibers and more porosity, facilitating air circulation.

High Loft permeability values range from 5152 mm/s to 9000 mm/s. High Loft D has the lowest value, which can be attributed to its structure and high density. Indeed, higher density can reduce pore size and limit air circulation. Conversely, High Loft C has the highest permeability value, indicating better air permeability. This suggests that its structure is more porous, and its density is relatively low, facilitating the passage of air through the material.

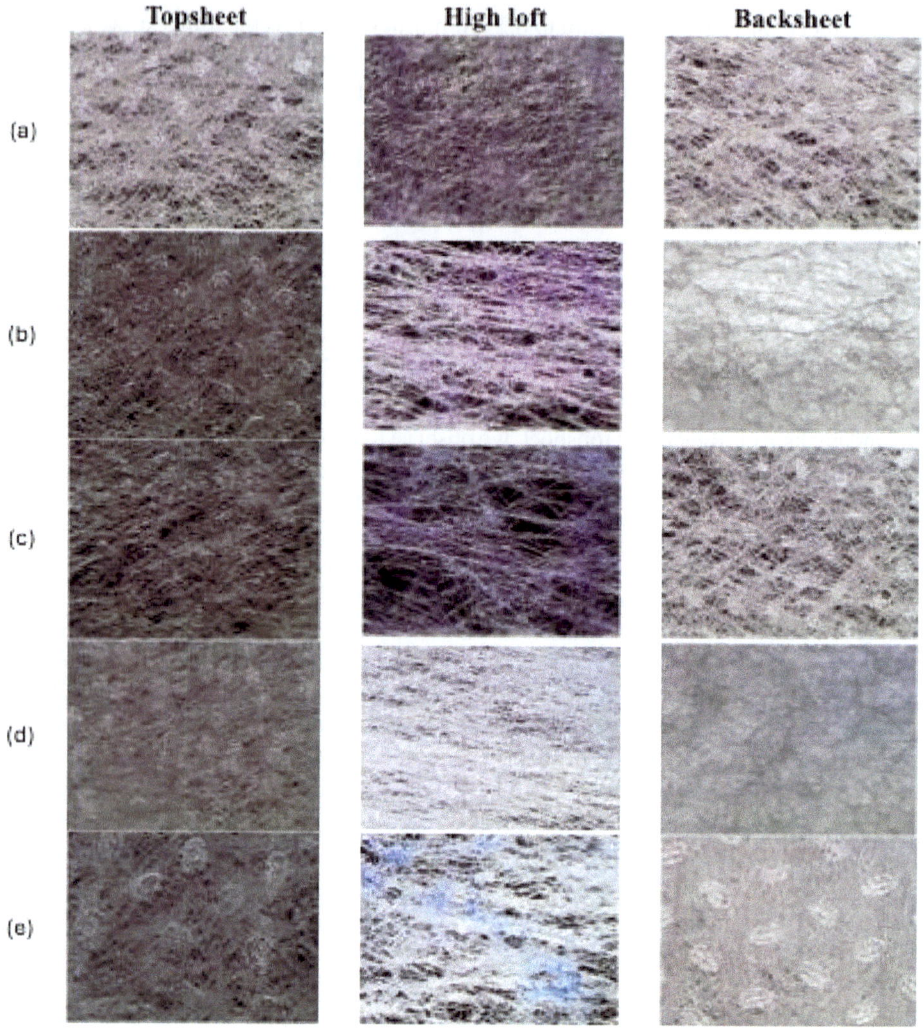

Fig. 2. Microscope images of different diaper layers (0.63 X).

Backsheet permeability values range from 2726 mm/s to 7563 mm/s, which represents a significant difference. Backsheet D and E show the lowest permeability values. This low permeability is mainly due to their compact structure and high density, which limits air circulation. Indeed, higher density reduces pore size, which hinders the passage of air through the material. In contrast, Backsheet A and C have the highest permeability values. This is due to their lower density and more open, porous structure, which promotes better air circulation and thus improves the material's breathability.

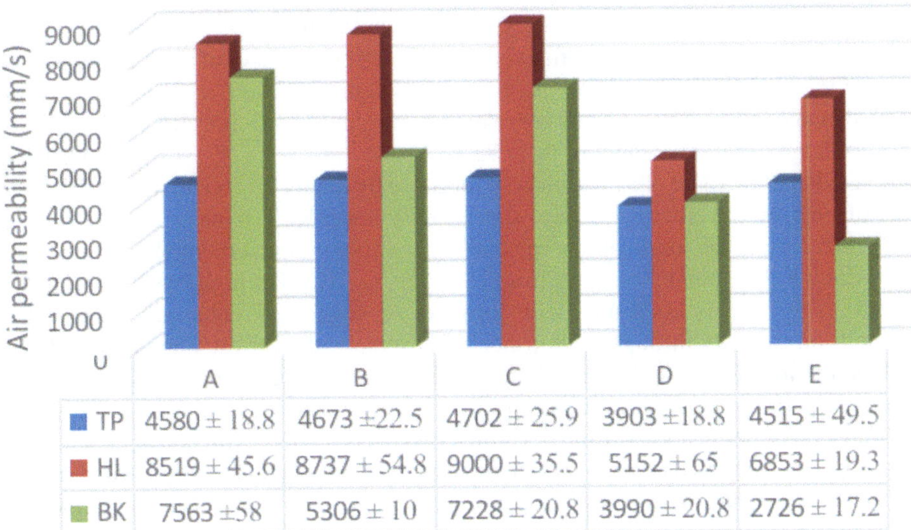

Fig. 3. Air permeability values of topsheet, high loft and backsheet (mm/s)

3.4 Water Vapour Diffusion

The Relative Water Vapor Permeability RWVP (expressed in %) and Thermal Evaporative Resistance RET (expressed in m^2-Pa/W) of five brands of baby diapers are shown in Table 2.

Table 2. Water vapour permeability values for baby diapers.

	A	B	C	D	E
Muti-layer					
RWVP %	11.95 ± 0.46	11.94 ± 1.07	10.3 ± 0.42	3.51 ± 0.86	4 ± 0.02
RET (m^2·Pa/W)	102.92 ± 18.66	79.2 ± 2	73.83 ± 25.72	308.12 ± 75.91	260.46 ± 2.73
Topsheet					
RWVP %	93.31 ± 3.71	92.48 ± 1.52	85.07 ± 9.03	78.92 ± 4.15	85.86 ± 1.27
RET (m^2·Pa/W)	1.06 ± 0.87	0.71 ± 0.41	0.44 ± 0.16	4.61 ± 3.21	1.49 ± 0.29
Backsheet					
RWVP %	93.57 ± 4.63	90.89 ± 0.01	87.72 ± 0.31	79.01 ± 1.82	89.3 ± 2.96
RET (m^2·Pa/W)	1.02 ± 1.07	1.33 ± 0.46	1.88 ± 0.70	3.13 ± 1.37	1.19 ± 0.21

The results presented in Table 2 show a notable difference between the values obtained. Relative water vapour permeability (RWVP) percentages of baby diapers range from 3.51% to 11.95%. The RWVP percentages of diapers D and E are the lowest, while the percentages of the other diapers are relatively similar. As concerns RET values, there is a very large difference, ranging from 73.83 to 308.12(m^2. Pa /W). Diaper D has the highest value (308.12 m^2. Pa /W), closely followed by diaper E (260.46 m^2. Pa/W). The other diapers, on the other hand, have significantly lower RET values.

In reference to the topsheet, the relative water vapor permeability (RWVP) percentages exhibit a close range, spanning from 78.2% to 93.31%. The highest value, recorded at 93.31 ± 3.71%, is associated with topsheet A, suggesting enhanced water vapor transmission capacity through the topsheet. Conversely, the lowest value of 78.92% is attributed to topsheet D, indicating diminished permeability and consequently reduced breathability of the topsheet in this diaper.

Water vapor permeability results exhibit a strong correlation with nonwoven density, as well as with morphological characteristics observed through microscopy. Microscopic analysis reveals that high-density structures consist of more compact fibers, thereby reducing the available space for vapor passage. This structure leads to a substantial reduction in water vapor permeability. This phenomenon is particularly pronounced in topsheet D, which, due to its high density, possesses the lowest values for relative water vapor permeability (RWVP) and the highest values for thermal evaporative resistance (RET), indicating limited breathability. These observations corroborate that the density and microstructure of nonwovens are pivotal parameters in evaluating the comfort of baby diapers.

In contrast, topsheet C, characterized by its medium density and more porous structure, offers superior breathability performance. It exhibits average RWVP% values, as well as a lower RET value, indicating an enhanced ability to facilitate thermal exchange between the baby's skin and the external environment.

The results obtained for the backsheet exhibit a comparable range of values. The relative water vapor permeability (RWVP) percentages for the backsheets ranged from 79.01% to 93.57%. The maximum value, 93.57%, corresponds to backsheet A, while the minimum value, 79.01%, is associated with backsheet D.

The thermal evaporative resistance (RET) values span a range of 1.02 to 3.13 m^2-Pa/W. Backsheet D holds the highest value, while backsheet A exhibits the lowest.

The results demonstrate density values that closely align with those previously measured for the different backsheets.

4 Conclusion

This study assessed the comfort and breathability characteristics of baby diapers available on the Tunisian market. The findings indicate that the physical properties and morphological attributes of materials, particularly nonwovens, significantly influence diaper comfort and breathability.

Through measurements of mass, thickness, density, microscopic examinations, and tests of air and water vapor permeability, we compared the various layers. It was discovered that a more porous and well-structured nonwoven fabric facilitates the seamless circulation of air and vapor, enabling the efficient wicking of moisture and maintaining skin dryness. This approach mitigates the risk of irritation and enhances comfort. Conversely, a denser, closed structure restricts these exchanges, promoting moisture accumulation and resulting in a sensation of discomfort.

These properties are therefore crucial for ensuring the quality of baby diapers. Future research will explore novel materials, particularly biodegradable options, to optimize the comfort, breathability, and environmental impact of diapers.

References

1. Histoire de la couche-Culotte Diaper Alliance Foundation. https://diaperalliancefoundation.com/decouvrir/histoire-de-la-couche-culotte/ (2025). Accessed 20 Apr 2025.
2. Nonwovens in Hygiene applications—Perini Journal. http://www.perinijournal.it/Items/en-US/Articoli/PJL-20/Nonwovens-in-Hygiene (2025). Accessed 20 Apr 2025.
3. Yoshida, J., Shintaku, S., Matsudaira, M.: Structural characteristics for top sheet of disposable diapers. Sen'i Gakkaishi. **61**(12), 317–322 (2005)
4. Cohen, R. Acquisition/distribution layer (United States Brevet US20050267429A1). https://patents.google.com/patent/US20050267429A1/en. (2005).
5. Dey, S., Kenneally, D., Odio, M., Hatzopoulos, I.: Modern diaper performance: Construction, materials, and safety review. Int. J. Dermatol. **55**(S1), 18–20 (2016)
6. Ajmeri, J. R., & Ajmeri, C. J.. 5—Nonwoven personal hygiene materials and products. In R. A. Chapman (Éd.), Applications of Nonwovens in Technical Textiles (p. 85–102). Woodhead Publishing (2010).
7. Sasikumar, G., Senthil, M., Visagavel, K., Zubar, H.A., Dheenathayalan, T.: Development of bio-degradable baby diapers. Int J Res Eng Technol. **3**, 2321–7308 (2014)
8. Uyanik, S., Kaynak, H.K.: Permeability properties of breathable and non-breathable baby diaper backsheet. In: International Advanced Researches & Engineering Congress, Osmaniye, Türkiye (16-18 November 2017)

Circularity and Sustainability in Textiles

Enhanced Valorization of Marine and Textile Waste through Eco-Designed Structures for Sustainable Applications

Arwa Elaissi[1,2](✉), Rania Baghdadi[2], Adel Ghith[1], and Mohsen Missaoui[2]

[1] Université de Monastir, ENIM, Unité de Recherche Matériaux et Procédés Textiles (MPTex), UR17ES33, 5019 Monastir, Tunisie
elaissiarwa@gmail.com

[2] Textile Technical Center (CETTEX), Avenue of Industries, Bir El Kassâa industrial zone, Ben Arous, Tunisia

Abstract. This study aims to develop sustainable and functional textile structures by enhancing the valorization of marine and textile waste—specifically Posidonia Oceanica and cotton fibers. Posidonia, a renewable marine biomass with a high cellulose content, was selected for its environmental abundance and potential as an eco-friendly material. The methodology involves morphological and chemical characterization of raw and chemically treated Posidonia fibers, including cationization using polyimethyldiallylammonium-diallylamine chloride (PDDACD), followed by FTIR and microscopy analyses. Results show that Posidonia fibers have comparable chemical composition to cotton but significantly differ in morphology and diameter. The chemical treatment enhanced fiber compatibility, suggesting improved adhesion potential in composite structures. These findings demonstrate the feasibility of using Posidonia as a biodegradable alternative to synthetic fibers, particularly for packaging and reinforcement in composite applications. The study provides original insight into the functional reuse of marine waste through eco-design strategies, contributing to circular economy practices in textile and material engineering.

Keywords: Fibers · Posidonia · cotton · cellulose · non-woven

1 Introduction

The use of cellulosic fibers has intensified over the past decades and market demand has increased due to the advantages they offer in terms of their bio-renewable traits, ubiquitous availability in various forms, and low costs. Among these cellulosic fibers, we find Posidonia. Oceanic Posidonia is a seagrass species endemic to the Mediterranean Sea. This marine plant loses its leaves in autumn and litter deposits are mainly found along sandy coasts, forming wedge-shaped structures, a few centimeters to several meters thick, called "benches".

Beaches and dune systems in Mediterranean countries depend on oceanic Posidonia to protect ecosystems, prevent coastal erosion, and regulate carbon dioxide uptake in

M. Hamdaoui et al. (Eds.): ITDM'3 2025, PTT, pp. 157–163, 2026.
https://doi.org/10.1007/978-981-95-7704-0_14

Mediterranean countries. On the other hand, the accumulation of these fibers along the Mediterranean coast has also led to the need for annual beach cleaning.

Dural et al. [1] estimated the total primary production of oceanic Posidonia at approximately 5 × 106–5 × 107 Tones/year. The use of this easily available and renewable lignocellulosic biomass would therefore constitute an appropriate solution to the cleaning problem. This marine biomass has recently received considerable attention as a potential new and environmentally friendly source of lignocellulosic material. The Mediterranean Sea is considered as a good area and market of waste sources for fiber production and has been exploited in several sectors [2–4]. This biomass is extracted from its leaves and used as animal feed [5]. Since then, it has also been studied as an adsorbent for the removal of dyes or phenol [2] and as a source of cellulose [6].

The most important current applications, such as its use to produce fibers for papermaking, nanofibers and/or cellulose derivatives for various industrial applications, and lignocellulose for fiber-reinforced composite materials, have been described. Aguir et al. have been used Bleached cellulose pulp obtained from Posidonia Oceanica for the production of carboxymethylcellulose (CMC) [6]. Berga et al. have been used Posidonia extracts as bio-stimulants in tomato and cucumber germination tests [7]. Hamdaoui et al. have been exploited Posidonia fibers as an insulating material in Mediterranean buildings [8].

However, the integration of Posidonia Oceanica fibers into technical nonwovens remains underexplored, especially regarding their compatibility with recycled cotton fibers and their chemical surface reactivity. This study aims to address this gap by evaluating the impact of chemical treatments on the structural and chemical properties of these marine fibers, with the objective of developing eco-designed, bio-based nonwoven materials for sustainable applications such as biodegradable packaging.

2 Material and Methods

Posidonia balls, collected on the beach of Monastir, are carefully transformed for various applications.

Reinforcements based on plant fibers are often difficult to disperse in the matrix due to the presence of large quantities of hydroxyl groups. Due to the formation of hydrogen bonds, these groups have a strong tendency to aggregate, thus weakening the interfacial properties. They are also responsible for the high polarity of the fibers, which makes them poorly compatible with most of the matrices conventionally used; they often exhibit apolar characteristics [9, 10].

There are several surface treatment techniques for natural fibers, such as desizing treatment, causticization [11] and cationization [12]. Cationization is a chemical process that involves introducing positive ions, called cations, into a substance. It makes it possible to improve the hydrophilicity of a material by using quaternary ammonium compounds. In our work, a polyimethyldiallylammonium-diallylamine chloride (PDDACD) copolymer was used. The fibers are immersed at 30 °C in a solution of the cationizer. Afterwards, we add 2% sodium hydroxide (NaOH) for 30 min at 50 °C, followed by rinsing with purified cold water. Next, the fibers are dipped in acetic acid solution and then washed with distilled water.

2.1 Axio Scope Projection Microscope

The morphological analysis of the fibers is measured with an Axio Scope projection microscope. A1 (see Fig. 1). This microscope was engineered, manufactured, and tested according to DIN EN 61010–1 (IEC 61010–1) and IEC 61010–2-101, safety regulations for metering, control, and laboratory equipment. They are universal microscopes specially designed for applications in the biological and medical fields and for materials studies. The fibers can be easily differentiated using a standard optical microscope (total magnification 160X), equipped with a 20x magnification objective.

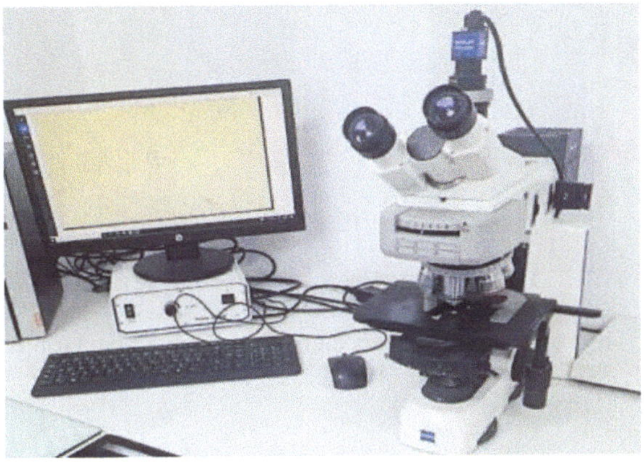

Fig. 1. Optical microscopy

2.2 FTIR Analysis

Fourier transform infrared spectroscopy, known as FTIR analysis, is a commonly used method in chemistry and physics to study the composition of materials. It exploits the fact that atoms absorb infrared light at specific wavelengths, which helps spot chemical bonds present in a substance. The chemical composition of the analyzed material is determined while identifying the different bonds by their vibration wavelength following this analysis technique.

3 Results and Discussions

The fibers were analyzed to determine their properties, which provides a better understanding of their behavior and their potential applications in various fields.

3.1 Microscopic Examination

This method consists of examining the fibers under a microscope, which makes it possible to distinguish their different morphological characteristics and to determine the diameter of different fibers. We tested a sample of raw cotton fibers (see Fig. 2.a) and a sample of raw Posidonia fibers (see Fig. 2.b).

Cotton fibers are distinguished by their characteristic twists, which give them a spiral ribbon appearance. While Posidonia fibers show distinct cracks on their surface. We observed distinct differences between cotton fibers and Posidonia fibers, as well as interesting similarities between Posidonia fibers and linen fibers.

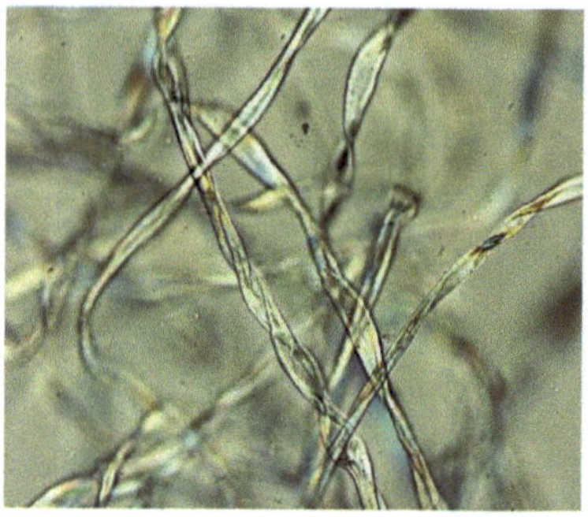

(a)

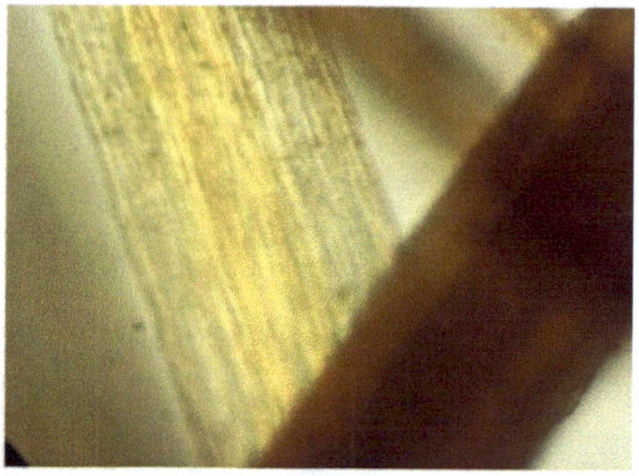

(b)

Fig. 2. Microscopic view of: (a) cotton fibers and (b) Posidonia fibers (objective ×20, magnification ×12.6)

3.2 Fiber Diameter

The diameter of the fibers was measured using a microscope equipped with a 20x magnification objective and a scale of 0.17 μm. The measurements carried out revealed that the average diameter of Posidonia fibers is 155 μm, while that of cotton fibers is 22.40 μm. These results highlight the notable difference in diameter between the two fiber types, with Posidonia fibers being significantly thicker than cotton fibers.

3.3 Chemical Characterization of Fibers

Infrared spectroscopy is a method that allows the functional chemical groups of fibers to be identified. The infrared spectra of cotton and Posidonia fibers are illustrated in Fig 3.

For two fibers, this figure shows a band observed between 3000 and 3500 cm^{-1} indicating the presence of OH groups [13]. Another band centered around 1030 cm^{-1} relates to the vibration of the C-O bond present in the glucose monomers forming cellulose [14]. The band at 1160 cm^{-1} is associated with C–O stretching in cellulose and hemicellulose [15]. The band centered around 557 cm^{-1} is attributed to C–C stretching [16]. For cotton, the band centered around 1108 cm^{-1} corresponds to a C-O and C-C elongation. Another band centered around 1053 cm^{-1}, representing the C-O elongation vibration of cellulose and hemicellulose.

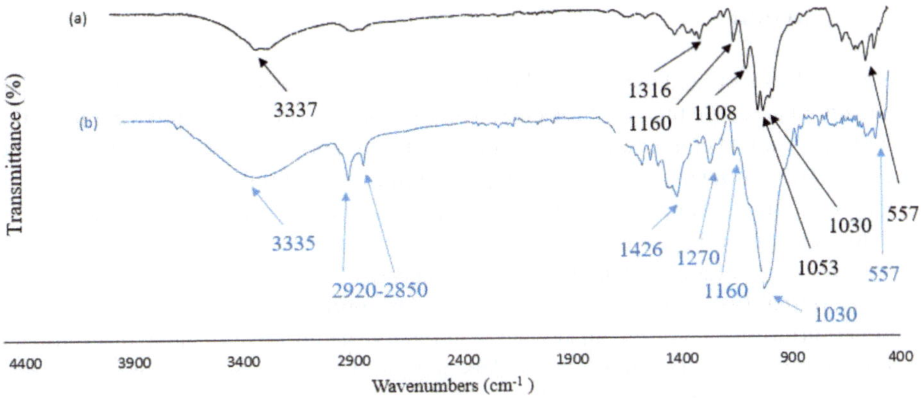

Fig. 3. Spectrum of: (a) cotton fibers and (b) Posidonia fibers

For Posidonia, we note the presence of a band centered around 2920 cm^{-1}, relating to the vibration of the CH bonds of the methyl group. Another band, centered around 1420 cm^{-1}, is attributed to the O–H deformations of the phenolic and aliphatic groups [17]. The band centered around 1270 cm^{-1} shows a C-O elongation in the lignin. Two bands centered around 2920 and 2850 cm^{-1}, due to the asymmetric stretching of the aliphatic C–H groups [18].

Figure 4 shows the spectrum of cationized Posidonia fibers. The spectrum shows a shift from the band centered around 3335 cm^{-1} (OH stretching groups) to the band centered around 3315 cm^{-1}. This confirms the addition of ammonium ions to the PDDACD structure. We can also observe the appearance of a new band centered around 1574 cm^{-1}, attributed to the stretching vibrations of the N-H bonds in the secondary amine (-NH). These results correspond to the reaction of quaternary ammonium salt groups with the cellulose skeleton.

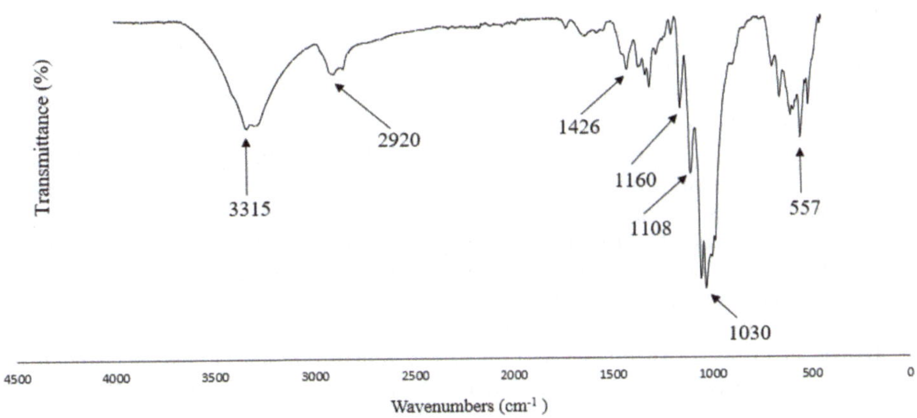

Fig. 4. Spectrum of 6% cationized Posidonia fibers.

4 Conclusion

This study demonstrated the potential of Posidonia oceanica as a sustainable alternative to conventional textile fibers. While chemically similar to cotton, Posidonia exhibits distinct morphological characteristics, including a significantly larger diameter and surface cracking. The cationization treatment improved fiber compatibility by introducing quaternary ammonium groups, as confirmed by FTIR analysis. These modifications enhance the interfacial properties of Posidonia, making it a viable reinforcement in eco-designed composites or as a biodegradable packaging material.

The originality of this work lies in the integrated approach to marine and textile waste valorization through green chemistry and functional structure development. It contributes both theoretically by deepening the understanding of Posidonia's fiber chemistry and practically by proposing novel applications aligned with environmental sustainability and circular economy goals.

Acknowledgments. This project is carried out under the MOBIDOC scheme, funded by the EU through the SWAFY PROJECT and managed by the ANPR ».

The authors thank Textile Technical Center "CETTEX" for their invaluable support and collaboration. Their expertise and resources have significantly contributed to the success of this work.

References

1. Dural, M.U., et al.: Methylene blue adsorption on activated carbon prepared from Posidonia oceanica (L.) dead leaves: Kinetics and equilibrium studies. Chem. Eng. J. **168**(1), 77–85 (2011)
2. Guezguez, I., Dridi-Dhaouadi, S., Mhenni, F.: Sorption of Yellow 59 on Posidonia oceanica, a non-conventional biosorbent: Comparison with activated carbons. Ind. Crop. Prod. **29**(1), 197–204 (2009)
3. Khiari, R., et al.: Chemical composition and pulping of date palm rachis and Posidonia oceanica–A comparison with other wood and non-wood fibre sources. Bioresour. Technol. **101**(2), 775–780 (2010)
4. Khiari, R., et al.: Valorisation of vegetal wastes as a source of cellulose and cellulose derivatives. J. Polym. Environ. **19**, 80–89 (2011)
5. Pergent, G., Pergent-Martini, C., Boudouresque, C.-F.: Utilisation de l'herbier à Posidonia oceanica comme indicateur biologique de la qualité du milieu littoral en Méditerranée: état des connaissances. Mésogée. **54**, 3–27 (1995)
6. Aguir, C., M'Henni, M.F.: Experimental study on carboxymethylation of cellulose extracted from Posidonia oceanica. J. Appl. Polym. Sci. **99**(4), 1808–1816 (2006)
7. Ferrández-Gómez, B., et al.: Valorization of Posidonia oceanica biomass: Role on germination of cucumber and tomato seeds. Waste Manag. **171**, 634–641 (2023)
8. Hamdaoui, O., et al.: Thermophysical characterization of Posidonia Oceanica marine fibers intended to be used as an insulation material in Mediterranean buildings. Constr. Build. Mater. **180**, 68–76 (2018)
9. Khiari, R., Belgacem, M.: Potential for using multiscale Posidonia oceanica waste: Current status and prospects in material science. In: Lignocellulosic Fibre and Biomass-Based Composite Materials, pp. 447–471 (2017)

10. Herrera-Franco, P.J., Valadez-Gonzalez, A.: Mechanical properties of continuous natural fibre-reinforced polymer composites. Compos. Part A Appl. Sci. Manuf. **35**(3), 339–345 (2004)
11. Niu, P., et al.: Study on mechanical properties and thermal stability of polypropylene/hemp fiber composites. J. Reinf. Plast. Compos. **30**(1), 36–44 (2011)
12. Tarbuk, A., Grancaric, A.M., Leskovac, M.: Novel cotton cellulose by cationisation during the mercerisation process—Part 1: Chemical and morphological changes. Cellulose. **21**, 2167–2179 (2014)
13. Zannen, S., et al.: Effect of chemical extraction on physicochemical and mechanical properties of doum palm fibres. Adv. Mater. Phys. Chem. **4**(10), 203 (2014)
14. Pozo, C., et al.: Characterization of temporal biodegradation of radiata pine by Gloeophyllum trabeum through principal component analysis-based two-dimensional correlation FTIR spectroscopy. J. Chil. Chem. Soc. **61**(2), 2878–2883 (2016)
15. Baloyi, R.B., Sithole, B.B., Chunilall, V.: Physicochemical Properties of Cellulose Nanocrystals Extracted from Postconsumer Polyester/Cotton-Blended Fabrics and Their Effects on PVA Composite Films. Polymers. **16**(11), 1495 (2024)
16. Zhou, W., et al.: Cotton-derived cellulose film as a dendrite-inhibiting separator to stabilize the zinc metal anode of aqueous zinc ion batteries. Energy Stor. Mater. **44**, 57–65 (2022)
17. Cocozza, C., et al.: Chemical, physical and spectroscopic characterization of Posidonia oceanica (L.) Del. residues and their possible recycle. Biomass Bioenergy. **35**(2), 799–807 (2011)
18. Cocozza, C., et al.: Comparative management of offshore posidonia residues: composting vs. energy recovery. Waste Manag. **31**(1), 78–84 (2011)

A Gate-to-Gate LCA of the Environmental Footprint of Denim Jeans Production

Maha Abdelileh[1(✉)], Sonia Dridi-Dhaouadi[1,2], and Hatem Dhaouadi[1]

[1] Research Laboratory of Environmental Chemistry and Clean Processes, Faculty of Sciences of Monastir, University of Monastir, 5019 Monastir, Tunisia
maha.abdelileh@gmail.com

[2] Preparatory Institute of Engineering Studies of Monastir (IPEIM), University of Monastir, Monastir, Tunisia

Abstract. This study presents a gate-to-gate Life Cycle Assessment (LCA) to assess the environmental impacts associated with the production of a pair of blue jeans, focusing on the stages from spinning to Jeans washing. The assessment was conducted using The CML 2000 reference method. The results highlight that the dyeing process is the most environmentally burdensome stage overall, contributing significantly to multiple impact categories due to the intensive use of chemicals and energy. Moreover, the washing phase was found to be a major contributor to global warming (40%), abiotic resource depletion (40%), and human toxicity (39%). These impacts are primarily driven by the consumption of non-renewable energy sources, particularly electricity generated from natural gas. The findings underscore the need for more sustainable practices in the textile industry, including the adoption of renewable energy, organic cotton and eco-friendly chemical alternatives.

Keywords: Life Cycle Assessment · gate-to-gate · jeans

1 Introduction

The environmental impact of the textile and fashion industry has become a growing concern due to its significant resource consumption, pollution, and waste generation. This impact stems from various stages of textile production, including raw material extraction, manufacturing, dyeing, and transportation.

Among textile products, blue jeans stand out as one of the most popular and widely produced garments globally. Their production involves multiple stages, including cotton cultivation, fiber processing, dyeing, finishing, and consumer use, all of which contribute to environmental impacts such as water and energy consumption, greenhouse gas emissions, and chemical pollution. According to recent market data, the North America denim jeans market generated a revenue of approximately USD 29.93 billion in 2024 and is projected to grow at a compound annual growth rate (CAGR) of 5.5% from 2025 to 2030 [1].

M. Hamdaoui et al. (Eds.): ITDM'3 2025, PTT, pp. 164–170, 2026.
https://doi.org/10.1007/978-981-95-7704-0_15

Life Cycle Assessment (LCA) is a widely used methodology for evaluating the environmental footprint of a product, process or service across its entire life cycle according to ISO 14040 and ISO 14044 [2, 3]. It provides a comprehensive assessment by considering all stages, from resource extraction to final disposal or recycling.

LCA is recognized as a comprehensive, multi-stage, and multi-criteria approach that assesses various environmental impact categories. It evaluates both local effects, such as toxicity, eutrophication, and acidification, as well as global concerns like climate change and ozone layer depletion [4]. Beyond measuring the environmental performance of a product or service, LCA also serves as a decision-support tool, enabling comparisons between different production systems and the identification of more sustainable alternatives.

This study aims to perform a comprehensive Life Cycle Assessment (LCA) of a pair of jeans, evaluating its environmental impacts and identifying potential improvements to reduce the environmental burdens associated with the manufacturing process.

2 Materials and Methods

Life Cycle Analysis (LCA) consists of four interconnected stages [2]. The first stage, defining the goal and scope, is outlined in Section 2.1. Section 2.2 provides a detailed overview of the inventory data and assumptions related to blue jeans production. The impact assessment methodology employed in this study is detailed in Section 2.3, while the interpretation of the results is presented in Section 3.

2.1 Goal and Scope

The main objective of this study is to assess the potential environmental impacts associated with the production of a pair of jeans and to determine which stages contribute most and least to these impacts. The scope of the assessment is gate-to-gate, with the system boundaries illustrated in Fig. 1. The processes included are spinning, dyeing, weaving, fabric finishing, garmenting, and jeans washing. Cotton cultivation and transportation stages are excluded due to insufficient data. In accordance with the study's goal and scope, the functional unit is defined as one pair of jeans weighing 750 g.

2.2 Life Cycle Inventory (LCI)

The Life Cycle Inventory (LCI) is a stage in LCA, involving the identification, collection, and interpretation of data required to assess the environmental impacts linked to the production of a blue jeans. This study utilized a combination of primary and secondary data sources. Primary data were directly collected from two textile industries one specializing in denim fabric production and the other in jeans manufacturing and washing. Secondary data were obtained from scientific literature and the Ecoinvent database, which provided background data incorporated into the SimaPro 7.1 software. In this section, we will present the various stages in the life cycle of a pair of jeans pants and the data used to model them:

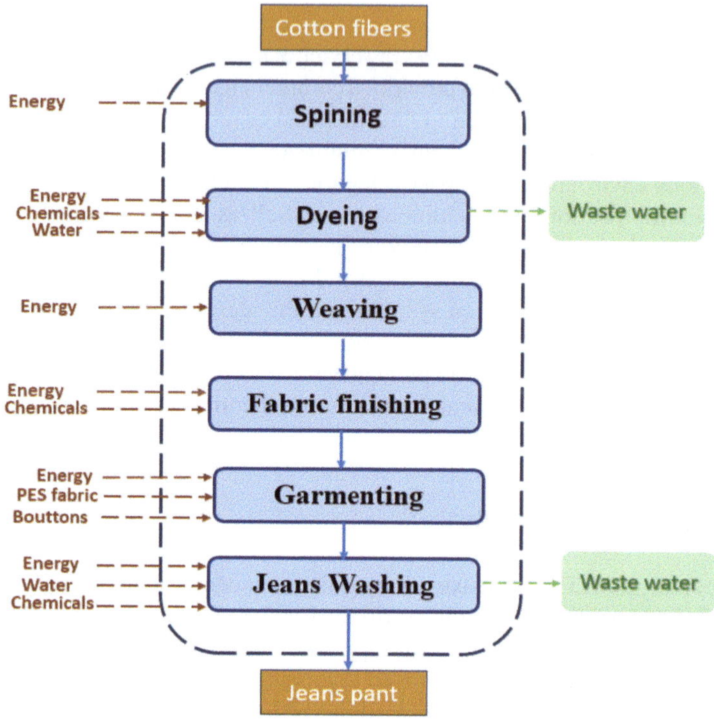

Fig. 1. Flow-chart of the processes included in the life cycle of jeans

Spinning

The first stage of fabric manufacturing is the production of cotton yarn, which is achieved through spinning. The primary inputs in the spinning process include cotton fiber, lubricant, and electricity. Energy consumption is calculated based on the equipment currently used in the industry. 10% of the cotton is lost during the spinning process [5]. This means that to produce 750 g of denim fabric for a pair of jeans, 825 g of raw cotton must be considered for the spinning stage.

Dyeing

In the production of denim fabrics, only the warp yarns undergo the dyeing process. The dyeing of 750 g of cotton yarn was carried out in a dye bath containing sodium dithionite, sodium hydroxide, a pre-wetting agent, a dispersing agent, and indigo. Following the dyeing process, the cotton yarn is rinsed with water to remove excess dye and then dried.

Weaving

The weaving stage consists of the following operations:

- Warping: Preparation of the warp sheet.
- Sizing: The warp yarn is treated with 0.11 Kg sizing agent to enhance its mechanical properties and improve its performance during the weaving process.
- Weaving: The interlacing of warp and weft threads to form the fabric

The warping step is not taken into account due to a lack of data. The electrical energy consumed during the sizing and weaving stages is 4.8 Kwh

Fabric Finishing
The flaming stage has been taken into account and the sub-steps of brushing, weft correction, wet sanforizing, and stenter priming could not be modeled due to a lack of data. However, according to IFTH (Institut français du textile et de l' habillement), the potential environmental impact of these operations is negligible compared to other sub-steps [5].

Garmenting
During this process, various operations such as spreading, cutting, sewing, heat pressing, and ironing are involved. Important considerations here in the environmental impact assessment are the high energy consumption; fabric waste generated during the cutting process (which is often recycled); accessory waste, and rejected garments [6].

Washing
The various treatments considered in this study to give jeans pant a faded appearance are as follows:

- Stonewash (simple): denim pants are machine-washed with pumice stones. The size of the stones and the time spent in the machine determine the fabric's shrinkage, lightening and worn appearance.
- Stonewash with chlorine: to fade denim pants
- Treatment with potassium permanganate
- Vaporization to soften the fabric.

2.3 Life-cycle Impact Assessment

The jeans production was modeled using SimaPro 7 software [7]. Input flow data was obtained from the Ecoinvent life cycle inventory database (Unit Processes) and analyzed using the CML 2000 reference method to evaluate ten environmental impact categories:

- Abiotic Depletion Potential (ADP): Measures the reduction of both renewable and non-renewable abiotic resources. It is expressed in kg Sb eq.
- Acidification Potential (AP): Quantifies the release of acidic substances into the lower atmosphere, which contribute to the acidification of water and soil. It is expressed in kg SO_2 eq.
- Eutrophication Potential (EP): Assesses the impact of nitrogen and phosphorus emissions on aquatic ecosystems, leading to excessive nutrient enrichment. It is expressed in kg PO_4 eq.
- Global Warming Potential (GWP): Evaluates the emissions of greenhouse gases, which are major contributors to climate change. It is expressed in kg CO_2 eq.
- Ozone Depletion Potential (ODP): Measures the impact of chlorinated and brominated compounds that degrade the stratospheric ozone layer, increasing harmful ultraviolet radiation on earth's surface. It is expressed in kg CFC-11 eq.

- Human Toxicity Potential (HTP): Assesses the chronic toxicological effects of carcinogenic and non-carcinogenic substances released into air, water, and soil, impacting human health. It is expressed in kg 1,4-dichlorobenzene (1,4-DCB) eq.
- Freshwater Aquatic Ecotoxicity Potential (FWETP): Represents the harmful effects of toxic substances on freshwater ecosystems. It is expressed in kg 1,4-DCB eq.
- Marine Ecotoxicity Potential (METP): Evaluates the negative impact of toxic pollutants on marine ecosystems. It is expressed in kg 1,4-DCB eq.
- Terrestrial Ecotoxicity Potential (TETP): Measures the impact of toxic substances on terrestrial ecosystems. It is expressed in kg 1,4-DCB eq.
- Photochemical Oxidation Potential (POP): Represents the formation of ground-level ozone due to the reaction of sunlight with volatile organic compounds and other reactive emissions from fossil fuel

3 Results and Discussion

Fig. 2 presents the results of the life-cycle analysis of a denim pant, detailing the impact at each stage across the following categories: Abiotic Depletion, Acidification, Eutrophication, Global Warming, Ozone Depletion, Human Toxicity, Freshwater Aquatic Ecotoxicity, Marine Ecotoxicity, Terrestrial Ecotoxicity, and Photochemical Oxidation.

The results indicate that the dyeing process is the most environmentally impactful stage in jeans production. Dyeing significantly contributes to impact categories such as abiotic depletion (47.97%) and global warming (51.08%). This is largely due to CO_2 emissions resulting from the consumption of fossil fuels (coal, gas, and oil) for electricity generation. Given that Tunisia's electricity production relies predominantly on natural gas, the environmental impact of these stages is further amplified, highlighting the need for more sustainable energy alternatives in the textile manufacturing process. Besides, the dyeing stage accounts for 49.4% of the potential impact on ozone layer depletion. This is primarily due to the intensive electricity consumption, which is largely generated from natural gas. The reliance on fossil fuel-based energy sources underscores the need for adopting more sustainable production methods to mitigate environmental harm.

Additionally, the dyeing stage exhibits the highest acidification potential due to the release of acidifying compounds, such as sulfates formed from the oxidation of sodium dithionite. It can also be noted that the weaving (23%) and the washing (14,77%) stages also play a role, likely due to chemical discharges and the combustion of fossil-based energy sources, such gas.

The weaving step also contributes significantly to both freshwater aquatic ecotoxicity (31.95%) and terrestrial ecotoxicity (47.14%), which are primarily associated with the environmental footprint of cotton.

To summarize, for the studied jeans, the dyeing process is the life cycle phase with the greatest environmental impact across all categories, except for eutrophication. The second most impactful phase is washing, followed by weaving, which ranks third in terms of overall environmental burden.

To mitigate the environmental impact of jeans production, several sustainable solutions can be implemented. Firstly, using organic cotton instead of conventional cotton can significantly reduce the impact on soil and water, as organic farming eliminates the need for synthetic pesticides and fertilizers, thereby lowering freshwater and terrestrial ecotoxicity. Besides the use of recycled cotton can significantly reduce the resource consumption and emissions associated with traditional fiber production.

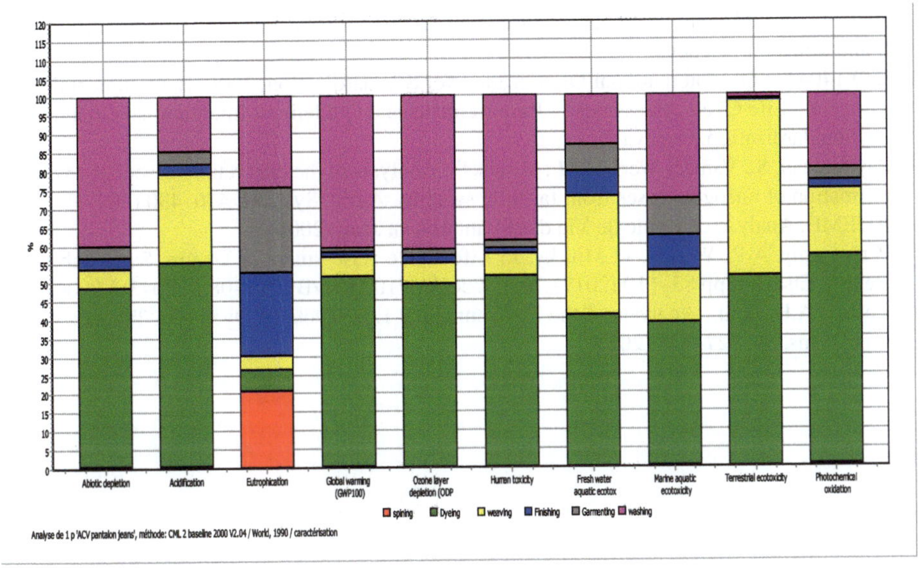

Fig. 2. Relative contribution of manufacturing phases on impact categories

Secondly, shifting to renewable energy sources, such as solar or wind power, for manufacturing processes like dyeing and weaving can help reduce carbon emissions and dependence on fossil fuels, minimizing contributions to global warming and abiotic depletion. Additionally, replacing sodium dithionite, a commonly used reducing agent in the dyeing process, with eco-friendly alternatives such as glucose, or iron-based reducing agents can decrease the release of acidifying compounds like sulfates, thereby reducing acidification potential. Finally, adopting low-water dyeing techniques can further reduce water consumption and wastewater pollution, making jeans production more sustainable.

4 Conclusion

From the study it is clear that the dyeing stage is the most environmentally impactful, contributing significantly across all assessed categories due to its high consumption of chemicals and energy. The negative impacts depend on the two main aspects, energy sources for electricity and heating and the substances used for dyeing and finishing processes. Among the evaluated production stages, jeans garmenting has the lowest environmental impacts of all. Finally, to achieve the best available technologies in the production of jeans, alternative scenarios should be considered, such as the use of organic cotton and the implementation of renewable energy sources

References

1. North America Denim Jeans Market Size & Outlook, 2024-2030. https://www.grandviewresearch.com/horizon/outlook/denim-jeans-market/north-america
2. ISO 14040 ISO – International Organization for Standardization, 2006. Environmental Management – Life Cycle Assessment – Principles and framework. International Organization for Standardization, Geneva
3. ISO 14044 ISO – International Organization for Standardization, 2006. Environmental Management – Life Cycle Assessment – Requirements and Guidelines. International Organization for Standardization, Geneva
4. Payraudeau, S., Van der Werf, H.M., Vertes, F.: Analysis of the uncertainty associated with the estimation of nitrogen losses from farming systems. Agric. Syst. **94**, 416–430 (2007)
5. ADEME, Analyse de Cycle de Vie d'un Pantalon en Jean (2006)
6. Periyasamy, A.P., Wiener, J., Militky, J.: Life-Cycle Assessment of Denim. Sustainability in Denim; Elsevier, pp 83–110 (2017). https://doi.org/10.1016/B978-0-08-102043-2.00004-6
7. Costaa, A.F., et al.: Analysis of the Environmental Life Cycle of Dyeing in Textiles. Chem. Eng. Trans. **86**, 727–732 (2021)

Environmental Impact Assessment of a Pair of Jeans: A Tunisian Case Study

Mouna Haj Nasr[1(✉)], Hassen Hedfi[2], and Ayda Baffoun[1]

[1] University of Monastir, National Engineering School of Monastir, Textile Materials and Processes Research Unit MPTex, UR17ES33, 5019 Monastir, Tunisia
hajnasr.mouna@enim.u-monastir.tn

[2] University of Monastir, National Engineering School of Monastir, Mechanical Engineering Laboratory (LGM), LR99ES32, 5019 Monastir, Tunisia

Abstract. It is widely recognized that the denim industry, including jeans production, generates significant environmental impacts, and therefore requires a detailed assessment of its production practices. In this context, a life cycle assessment (LCA) was conducted to evaluate the environmental performance of women's jeans manufactured in Tunisia, applying a gate-to-gate approach. To this end, data were collected from a jeans manufacturing plant in Tunisia where the main steps of production were covered, including garment manufacture, industrial washing, and finishing. The findings revealed that the total impacts for each selected impact category were 3.8 kg CO_2 eq for global warming potential (GWP), 7.5×10-3 kg SO2 eq for terrestrial acidification (TA), 1.6×10-3 kg P eq for freshwater eutrophication (FE), and 56×10-3 m3 for water consumption (WC). Results highlighted key environmental issues, including electricity and heat consumption, high water use during processing, along with metal components used in trimmings. Overall, this study emphasizes the importance of LCA as a tool for identifying environmental hot spots with the aim of guiding the development of strategies and initiatives, based on existing literature, to mitigate these impacts.

Keywords: Life Cycle Assessment · Environmental Impact · Tunisian Apparel sector · Women Jeans

1 Introduction

Over the last few years, the global market for denim clothing has grown considerably, from around 100 billion USD in 2018 to around 128 billion USD in 2023 [1]. In 2012, an estimated 1.8 billion pairs of denim pants were sold [2]. By 2022, this number had risen to around 3.1 billion pairs, and predictions indicate that it will exceed 3.46 billion pairs by 2026 [3]. The Tunisian denim industry is a significant player in the global jeans market, thanks to its strategic position close to Europe, which offers a number of competitive benefits, such as shorter delivery times and lower transport costs [4]. As a result, Tunisia plays a key role as a jeans exporter, particularly to European markets. In 2021, it accounted for 8.63% of EU denim imports, alongside major suppliers such as Bangladesh, Turkey and Pakistan [5].

M. Hamdaoui et al. (Eds.): ITDM'3 2025, PTT, pp. 171–179, 2026.
https://doi.org/10.1007/978-981-95-7704-0_16

Jeans are one of the mainstays of the global apparel market, widely recognized and accepted by all age groups. However, their production in developing countries comes at a high environmental cost [2], involving intensive use of cotton, water and chemicals [6], and contributing to significant climate and water challenges [2]. It was stated that a single pair of cotton jeans (340 g) produced 33.4 kg of carbon dioxide equivalent (kg CO_2 eq) emissions and used 3781 l of water [7]. These figures, when multiplied by the annual worldwide production of jeans, generate a considerable impact on the environment.

A holistic approach is essential when assessing the environmental impact of jeans production, covering all stages of the production process from the initial stage to final product [2]. A pair of jeans' lifecycle includes five main stages: the extraction of raw materials, the production of the fabric, the manufacture of the jeans garment, the use phase and lastly the end-of-life [6]. Each of these steps contributes to the environmental impact. As for sourcing raw materials, which mainly involves growing cotton, consumes large quantities of water, energy, land, pesticides and fertilizers [2, 8]. The fabric production phase which encompasses spinning, warping, dyeing, sizing and weaving [9, 10], all of which consume large amounts of energy [11] and toxic chemicals inputs. Garment manufacture phase encompasses cutting, sewing, washing and finishing [12]. The washing phase implies significant water consumption [2] and requires various chemical treatments to achieve the desired look [6]. Depending on fashion trends, jeans washing techniques evolve, and the most common techniques are laser treatment, stone washing, enzymatic washing and hypochlorite bleaching, often referred to as acid washing [2]. The finishing stage of jeans includes the incorporation of various components such as buttons, rivets, leather patch, labels and packaging [6]. The usage phase involves daily action like wearing and maintenance activities such as washing, drying and ironing [7, 13]. The latter can entail substantial water and energy consumption, depending on individual user behavior [2, 14]. Finally, the end-of-life phase can involve reuse, recycling, landfill or incineration, depending on the consumer's choice [6, 10, 13].

Studies quantifying the environmental impact of jeans, particularly in the context of Tunisian manufacturing, remain very limited. The present study aims to fill this gap by providing valuable insights using recent and reliable primary data, which specifically covers the key stages of jeans manufacturing: garment manufacture, industrial washing and finishing. The objective of the study is to evaluate the environmental performance of a pair of women's jeans and to pinpoint key factors contributing to its overall impact.

This study is organized as follows: the research methodology is presented in Section 2, including details regarding the functional unit (FU) of the study, system boundaries, data collection approach and the selected impact categories. Section 3 presents the results of the environmental impact assessment, highlighting the main contributors (hotspots) to each selected impact category and identifying key areas for improvement in order to mitigate the impact of the identified hotspots. The conclusions are presented in Section 4.

2 Methodology

A standard pair of jeans is studied as the reference product, given its status as an indispensable and universally worn item of clothing [2]. This study adheres to the standards outlined in ISO 14044:2006 and ISO 14040:2006 [15, 16], which provide the framework for conducting Life Cycle Assessment (LCA). Data analysis was carried out using the faculty SimaPro® software (version 9.5.0.2). The LCA approach involves defining the goal and scope of the study, compiling a Life Cycle Inventory (LCI), Life Cycle Impact Assessment (LCIA), and interpreting the results. Conducting company-specific LCA is essential for identifying key environmental impact contributors. This study, presents the current environmental performance of one pair of women jeans, based on data from a Tunisian jeans manufacturer. Thus, it enables the identification of areas for improvement and strengthening sustainability in the jeans sector.

2.1 Goal and Scope

This study aims to evaluate the environmental performance associated with the production of a pair of women's jeans in the Tunisian textile industry. It also aims to identify the main hotspots contributing to its environmental impact throughout the manufacturing process. A gate-to-gate LCA approach is adopted in this study. The FU is defined as "the production of one women's pair of jeans, size 28, that weighs 520 grams, manufactured in Tunisia with a mixed wash treatment combining enzymes and pumice stones". This analysis is deliberately limited to the main manufacturing processes and does not cover the upstream phase of producing cotton and fabric or the downstream stages relating to the use and end of life of the product. The system boundaries considered in this study are illustrated in Fig. 1, which are limited to the main production stages, namely garment manufacture, industrial washing and the finishing phase.

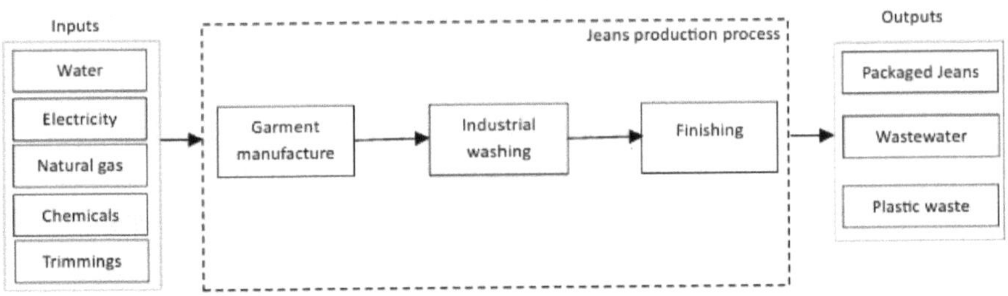

Fig. 1. Assessed pair of jeans - life cycle system boundaries.

2.2 Life Cycle Inventory

Jeans Production Process

The study concentrates on the jeans manufacturing process in Tunisia, which typically begins with the fabric cutting and assembly on site using electricity, heat, water and trimmings such as zippers, labels and threads. This step is followed by mixed industrial washing, which consumes additional utilities, pumice and chemicals, and generates wastewater. Finally, the finishing stage uses a similar range of resources, as well as other trims such as buttons, rivets, leather patch, and packaging.

Data Collection

All data collected is evaluated according to the indicated FU (i.e. a pair of jeans weighing 0.52 kg). There are two types of data used for the study. Primary data was collected in 2024 from a Tunisian jeans manufacturing plant through direct interviews, on-site observations and internal reports (2022-2023). To complete the analysis, secondary data were obtained mainly from Ecoinvent (V.3.9.1) database [17] and relevant literature. Table 1 presents a synopsis of the data used in the study. It includes sub-processes, as well as associated inputs such as water, electricity, heat, chemicals and auxiliary materials, and outputs such as wastewater and plastic waste. This study excludes the production of raw materials, the manufacture of denim fabric, consumer use and end-of-life stage, as well as capital goods and transportation due to the proximity of the facilities (less than 3 km).

Table 1. Jeans production processes and data sources.

Process	Data	Data source
Garment Manufacture	Heat (Natural gas)	[6]
	Electricity	Site-specific
	Water	[6]
	Trims	Site-specific
Industrial washing	Heat (Natural gas)	Site-specific
	Electricity	Site-specific
	Water	Site-specific
	Chemicals	Site-specific
	Wastewater	Site-specific
Finishing	Heat (Natural gas)	[6]
	Electricity	Site-specific
	Water	[6]
	Trims and packaging	Site-specific
	Plastic waste	[6]

2.3 Life Cycle Impact Assessment

The life cycle analysis was carried out using SimaPro software (V.9.5.0.2), following a gate-to-gate approach. A total of four environmental impacts were identified and assessed: Global Warming Potential (GWP), Terrestrial Acidification (TA), Freshwater Eutrophication (FE) and Water Consumption (WC). These environmental impacts were quantified using the ReCiPe 2016 (V.1.1) midpoint (H) LCIA method. These indicators were selected based on those deemed as priorities by F.S. Fidan et al. (2020) due to their relevance to the main environmental challenges within the textile sector [18].

3 Results and Discussion

The LCIA results for the production of a pair of women's jeans in Tunisia, considering a gate-to-gate boundaries, are presented within four impact categories: GWP, TA, FE, and WC. Table 2 depicts the characterized LCIA findings of the production process of the assessed jeans (0.52 kg).

Table 2. Characterized LCIA results of the assessed jeans production

Impact category	Unit	Result
Global Warming Potential	kg CO_2 eq.	3.8
Terrestrial Acidification	kg SO_2 eq.	7.5×10^{-3}
Freshwater Eutrophication	kg P eq.	1.6×10^{-3}
Water Consumption	m^3	56×10^{-3}

3.1 Global Warming Potential

The chosen midpoint characterization factor for climate change is GWP which is a commonly used measure of the integrated infrared radiative forcing increase of a greenhouse gas (GHG) [19]. This study reveals that the producing of a pair of women's jeans (0.52 kg) results in 3.8 kg CO_2 eq. The climate change impact of the examined jeans is largely attributable to the dependence on fossil fuels for generating heat and electricity. Heating is the main contributor, accounting for 55.3% of the total impact, followed by electricity consumption at 22.61%. The findings align with earlier studies, highlighting that the use of fossil fuels for energy production contributes significantly to CO_2 emissions [2, 6].

To our knowledge, previous studies have not specifically evaluated a "mix washed" jeans model, which makes the present case study an original contribution to the existing literature. Nevertheless, few relevant references can be found in the literature. Periyasamy and Duraisamy (2018) assessed the carbon footprint of one model of men's jeans (0.57 kg) processed with a medium stonewash and found that the production step accounted for a 6.9 kg CO_2 eq [20]. Also, Åslund Hedman (2018) evaluated the impact of two pairs of stonewashed men jeans: the first one (0,72 kg) was produced in Tunisia, and had an impact of 7.41 kg CO_2 eq, while the second one (0,6 kg) was produced in Italy, and had a significantly lower impact of 3.8 kg CO_2 eq [6]. This latter result is consistent with the findings of this study.

3.2 Terrestrial Acidification

Terrestrial acidification potential refers to the release of inorganic substances including sulphates, nitrates, and phosphates into the atmosphere, which alter soil acidity. Since most plants require a specific range of acidity to thrive, major changes can be detrimental to many species [21]. TA is estimated at 7.5×10-3 kg SO2 eq. for the analysed jeans, mainly due to the metals used in buttons, zippers, and rivets, such as brass, representing 43.4% of the total impact. This is largely attributed to the copper content of brass, which is supported by the findings of a previous study [22], where the use of copper in jeans garments was reported as a key influencing factor for TA.

Other significant contributors to TA include heating with natural gas (16%) and the use of sodium hypochlorite (8.12%) within the washing process. The results are consistent with previous studies that have shown that acidifying emissions are linked to the combustion of fossil fuels, such as natural gas [6, 23], and to the use of chemicals [23]. In this study, TA is found to be lower than the value reported by Åslund Hedman (2018) for jeans (0.72 kg) produced in Tunisia (42×10-3 kg SO2 eq) and also lower than the jeans (0.6 kg) produced in Italy (16×10-3 kg SO2 eq) [6].

3.3 Freshwater Eutrophication

Eutrophication refers in general to the enrichment of water bodies with nutrients, in particular phosphorus and nitrogen, causing increased growth of organisms such as algae. This process disrupts aquatic ecosystems and can lead to a loss of biodiversity [21]. In freshwater bodies, phosphorus is the main limiting nutrient [6]; therefore, the fate of phosphorus emissions serves as the basis for the midpoint characterization factors related to freshwater eutrophication [19]. In terms of FE, the total impact of the assessed jeans accounted for 1.6×10-3 kg P eq. The findings reveal four main contributors: plastic waste, which accounts for 39% of the total impact, metals such as brass, which contributes 22%, wastewater emissions, which account for 10.2% of the total impact and sodium hypochlorite, which accounts for 4.54%.

This is consistent with the findings of Åslund Hedman (2018), who identified the extraction of metals used for buttons and zippers, as well as substances used in the washing process, as key contributors to FE [6]. Moreover, it aligns with previous research indicating that eutrophication in the textile industry is largely influenced by wastewater releases, notably owing to the emissions of nitrogen and ammonia [24]. In this study, FE is found to be lower than the value reported by Åslund Hedman (2018) for jeans produced in Tunisia (4,1×10-3 kg P eq), whilst it is consistent with the result reported for jeans produced in Italy (2×10-3 kg P eq) [6].

3.4 Water Consumption

Water consumption is a commonly accepted indicator for quantifying increases in water use. The midpoint characterization factor is defined as the volume of water consumed per volume of extracted water [19]. A total impact of 56×10-3 m^3 of is recorded during the production of the women's jeans under study. This is predominantly due to the water used during production, which accounts for 92.3% of the total impact. As anticipated, this reflects the substantial demand for water associated with jeans manufacturing, particularly during industrial washing processes [2]. In comparison, Åslund Hedman's study [6] reports a water consumption of 110×10^{-3} m^3 during jeans production process, which is much higher than the obtained results of the present study.

3.5 Opportunities for Analysis Improvement

The environmental impacts assessed in this study arise from four main factors: electricity and heating consumption, metals like brass used for trimming, and water usage. This allows decision-makers to identify areas for improvement and implement effective eco-design solutions [25]. In fact, LCA enables objective scenario comparisons, serving as a powerful tool for alternative assessments [26].

This section explores literature-based strategies aimed at reducing the environmental impact of the jeans industry, highlighting key innovations and initiatives throughout the production process to address the identified hotspots. Such measures include minimizing evaporation losses, installing effective recycling systems [27], and using technologies such as lasers, ozone, and Jeanologia e-flow, which have proven efficient in reducing water and energy use during industrial washing [8, 28]. Manufacturers are also switching to renewable energy sources like solar and wind power [6], while also recovering heat from machines for drying processes [8]. Another relevant initiative is adopting combined heat and power (CHP) plant. CHP systems are recognized as one of the most energy-efficient solutions in the textile industry, since they significantly reduce CO_2 emissions by using less primary energy sources [18]. Additionally, eco-friendly buttons and accessories, such as "No Impact Buttons" manufactured from natural materials, are increasingly used [28], along with PET zippers, further contributing to sustainability efforts [29].

4 Conclusions

In this study, the environmental performance of women's jeans production in Tunisia is assessed within a gate-to-gate life cycle assessment (LCA). The findings revealed that the production of a single pair of women's jeans (0.52 kg), from garment manufacturing to packaging, is associated with significant environmental impacts. These include 3.8 kg CO_2 eq for climate change, 7.5 × 10-3 kg SO2 eq for terrestrial acidification, 1.6 × 10-3 kg P eq for freshwater eutrophication, and 56 × 10-3 m3 in terms of water consumption. A detailed examination of each impact category identified the main contributors, with the most significant impacts arising from dependence on fossil fuels for electricity and heating (natural gas), the use of metals such as brass, and water-intensive usage.

Within the selected impact categories, the assessed jeans show relatively low environmental impacts compared to other jeans reported in the literature. The main difference between the jeans evaluated in this study and those reported in the literature resides in the washing process, as the present study used a mix washing technique (combination of enzymes and pumice), whereas previous studies relied on conventional pumice washing process.

Overall, this research provides a useful framework for identifying priority areas for the implementation of effective sustainable development strategies. By highlighting key issues and drawing on literature-based insights, this study contributes to promoting more environmentally friendly practices and advancing the jeans industry towards a more sustainable future.

Nevertheless, a major limitation of this study stems from the selected gate-to-gate approach, as it excludes upstream and downstream processes. Future research should adopt a broader scope, such as cradle-to-gate or cradle-to-grave approaches, and include other relevant impact categories (i.e., ecotoxicity, human toxicity, and abiotic depletion). This would provide a fuller assessment of the environmental impacts of jeans production.

References

1. Bruna Rocco (2025) Size of the Jeans Wear Market Worldwide from 2018 to 2023. Available online: https://www.statista.com/statistics/541304/size-of-the-global-jeans-market/. Accessed 28 Mar 2026
2. Muthu, S.S.: Sustainability in denim. Woodhead Publishing: Cambridge, UK (2017). http://www.sciencedirect.com:5070/book/9780081020432/sustainability-in-denim
3. Smith, P.: Units of denim jeans on the global market 2021 to 2026 (2023). Available online: https://www.statista.com/statistics/1368005/global-units-of-denim-jeans-market/. Accessed 28 Mar 2026
4. Montassar, K.: The Framework of Tunisian Textile and Clothing Industry. MPRA Paper (2017). https://ideas.repec.org/p/pra/mprapa/60283.html
5. CETTEX, Analyse de marché de l'Union Européenne (2022). https://ideas.repec.org/p/pra/mprapa/60283.html. Accessed 28 Mar 2026
6. Åslund Hedman, E.: Comparative life cycle assessment of jeans: a case study performed at Nudie Jeans (2018). https://kth.divaportal.org/smash/get/diva2:1221104/FULLTEXT01.pdf
7. Levi Strauss & Co.: The Life Cycle of a jean. Understanding the environmental impact of a pair of Levi's ® 501 ® jeans (2015). https://www.levistrauss.com/wp-content/uploads/2015/03/Full-LCA-Results-Deck-FINAL.pdf
8. Ebbers, L. Towards a Brighter Blue | Creating an Impact Index to Increase Adoption for Sustainable Jeans (2019). https://repository.tudelft.nl/record/uuid:326178cc-051e-4b02-91b1-ce4d90ff3dc6
9. Fidan, F.Ş., Aydoğan, E.K., Uzal, N.: Multi-Dimensional Sustainability Evaluation of Indigo Rope Dyeing with a Life Cycle Approach and Hesitant Fuzzy Analytic Hierarchy Process. J. Clean. Prod. **309**, 127454 (2021). https://doi.org/10.1016/J.JCLEPRO.2021.127454
10. Nellström, M., Saric, M.: A comparative life cycle assessment of nudie jeans' repair and reuse concept (2019). https://hdl.handle.net/20.500.12380/300088
11. Zhao, M. et al.: Virtual Carbon and Water Flows Embodied in Global Fashion Trade-a Case Study of Denim Products. J. Clean. Prod. **303**, 127080 (2021)
12. Rupo, D.: Sustainability and Law. In: Sustainability and Law (2020). https://doi.org/10.1007/978-3-030-42630-9
13. Şener Fidan, F., Kızılkaya Aydoğan, E., Uzal, N.: The Impact of Organic Cotton Use and Consumer Habits in the Sustainability of Jean Production Using the LCA Approach. ESPR. **30**, 8853–8867 (2022). https://doi.org/10.1007/S11356-022-22872-6
14. Luo, Y., Wu, X., Ding, X.: Carbon and Water Footprints Assessment of Cotton Jeans Using the Method Based on Modularity : A Full Life Cycle Perspective. J. Clean. Prod. (2021). https://doi.org/10.1016/j.jclepro.2021.130042
15. The International Organization for Standardization (2006b) ISO 14044:2006 - Environmental management — Life cycle assessment — Requirements and guidelines. https://www.iso.org/standard/38498.html. Accessed 28 Mar 2026
16. The International Organization for Standardization (2006a) ISO 14040:2006, Environmental management — Life cycle assessment — Principles and framework. https://www.iso.org/obp/ui#iso:std:iso:14040:ed-2:v1:en. Accessed 28 Mar 2026
17. Ecoinvent, "Ecoinvent," 2025. https://ecoinvent.org/database/. Accessed 28 Mar 2026

18. Şener Fidan, F., Kızılkaya Aydoğan, E., Uzal, N.: An Integrated Life Cycle Assessment Approach for Denim Fabric Production Using Recycled Cotton Fibers and Combined Heat and Power Plant. J. Clean. Prod. **287**, 125439 (2020). https://doi.org/10.1016/J.JCLEPRO.2020.125439
19. Huijbregts, M.A.J. et al.: ReCiPe2016: A Harmonised Life Cycle Impact Assessment Method at Midpoint and Endpoint Level. Int. J. Life Cycle Assess. **22**, 138–147 (2017). https://doi.org/10.1007/S11367-016-1246-Y
20. Periyasamy, A.P., Duraisamy, G.: Carbon Footprint on Denim Manufacturing, pp. 1–18. Springer International Publishing, Handbook of Ecomaterials (2018). https://doi.org/10.1007/978-3-319-48281-1_112-1
21. Huijbregts, M.A.J. et al.: ReCiPe 2016 v1.1: A Harmonized Life Cycle Impact Assessment Method at Midpoint and Endpoint Level. *RIVM Report 2016-0104*, p. 201 (2017)
22. Bongiovanni, R.G., Tuninetti, L.: Análisis Del Ciclo de Vida de Un Jean Producido En Argentina. R. Latino-amer. em Aval. do Ciclo de Vida. **2**(1), 9–34 (2018). https://doi.org/10.18225/LALCA.V2I1.3942
23. Kazan, H., Akgul, D., Kerc, A.: Life Cycle Assessment of Cotton Woven Shirts and Alternative Manufacturing Techniques. Clean Techn. Environ. Policy. **22**(4), 849–864 (2020). https://doi.org/10.1007/S10098-020-01826-X
24. Cotton Incorporated. LCA update of cotton fiber and fabric life cycle inventory. In: Cotton Council International (2017). http://resource.cottoninc.com/LCA/2016-LCA-Full-Report-Update.pdf
25. Fonseca, A., Ramalho, E., Gouveia, A., Henriques, R., Figueiredo, F., Nunes, J.: Systematic insights into a textile industry: reviewing life cycle assessment and eco-design. Sustainability (Switzerland). **15**, 15267 (2023). https://doi.org/10.3390/SU152115267/S1
26. Morita, A.M.; Moore, C.C.S.; Nogueira, A.R.; Kulay, L.; Ravagnani, M.A. da S.S. Assessment of Potential Alternatives for Improving Environmental Trouser Jeans Manufacturing Performance in Brazil. J. Clean. Prod. 2020, 247, 119156, doi:https://doi.org/10.1016/J.JCLEPRO.2019.119156.
27. Hoekstra, A.Y.: The water footprint of industry. In: Assessing and Measuring Environmental Impact and Sustainability. Elsevier Inc (2015). https://doi.org/10.1016/B978-0-12-799968-5.00007-5
28. Muthu, S.S.: Textiles and Clothing Sustainability Recycled and Upcycled Textiles and Fashion. In: Textile Science and Clothing Technology (2017b). http://www.springer.com/series/13111
29. Zhang, Y., Wu, X.Y., Wang, L.L., Ding, X.M.: The Industrial Water Footprint of Zippers. Water Sci. Technol. **70**, 1025–1031 (2014). https://doi.org/10.2166/wst.2014.323

Sustainability in the Textile and Clothing Industry: Challenges and Opportunities

Ross Alexander(✉)

Green Threads, DPP Ltd, Southampton, England
ross@greenthreadsdpp.com

1 Introduction

The textile and clothing industry, a cornerstone of global economies valued at over $1.7 trillion, faces mounting pressure to adopt sustainable practices amid escalating environmental concerns. As consumers become increasingly aware of the ecological footprint of their wardrobes, regulatory frameworks are evolving to foster transparency and accountability. Digital product passports (DPPs) are emerging as a pivotal tool providing detailed information on a garment's lifecycle from raw materials to manufacturing processes.

These digital records, often accessible via QR codes, enable consumers to trace the origins and ethical standards of their clothing, empowering informed decisions that align with personal values. Similarly, end-of-life considerations, such as mandatory recycling programs and extended producer responsibility schemes, are compelling brands to design for durability and recyclability. Taxes on non-sustainable materials or waste exports further incentivize shifts in behavior, penalizing high-impact practices while rewarding eco-friendly alternatives.

Together, these mechanisms not only demystify the opaque supply chains of fashion but also encourage consumers to rethink their purchasing habits, viewing each buy as an investment in a sustainable future.

Historical precedents illustrate how transparency can profoundly alter consumer behavior. The revelation of "blood diamonds" in the 1990s, linked to funding civil wars in Africa, sparked global outrage and led to the Kimberley Process Certification Scheme in 2003. This initiative drastically reduced trade in conflict diamonds by 99%, as consumers boycotted unethical sources and demanded certified, traceable gems. Ethical consumerism surged, with buyers shifting toward lab-grown or certified diamonds, demonstrating how visibility into sourcing can drive market transformations.

A parallel example is cobalt, essential for batteries in electronics and electric vehicles, where exposés on child labor and hazardous mining in the Democratic Republic of Congo prompted demands for responsible sourcing. Initiatives like the Responsible Cobalt Initiative have led consumers to favor brands with transparent supply chains, influencing companies to audit suppliers and invest in ethical alternatives, thereby reshaping purchasing regions and behaviors. These cases underscore a key principle: when consumers gain insight into production ethics, they wield their buying power to demand change, dictating what manufacturers produce and how.

M. Hamdaoui et al. (Eds.): ITDM'3 2025, PTT, pp. 180–185, 2026.
https://doi.org/10.1007/978-981-95-7704-0_17

In this context, consumers emerge as central decision-makers in the fashion ecosystem. Every purchase is an investment in the future, signaling to producers the demand for sustainable options. As transparency tools like DPPs proliferate, shoppers can prioritize items that align with their core values. This consumer-driven shift is crucial, as it addresses systemic issues from resource extraction to waste management.

Hemp fiber, for example, stands out as a promising material innovation, offering high durability, low environmental demands, and ease of recycling or biodegradation—qualities that transparency can highlight to guide better choices. By weaving transparency throughout the supply chain, the industry can transition from exploitative fast fashion to resilient, ethical models, with consumers at the helm.

2 Core Challenges in the Textile and Clothing Industry

The fashion sector grapples with profound sustainability challenges, many of which consumers can influence through informed choices. Transparency in production plays a vital role here, as it reveals the true costs of materials and processes, enabling buyers to favor ethical options that address multiple pain points from durability to end-of-life management.

2.1 Raw Material Selection

The choice of raw materials is foundational to the industry's environmental impact, with traceability often lacking, obscuring the origins and effects of production. Natural fibers, such as cotton and wool, are derived from plants or animals and offer benefits like breathability and biodegradability, but they can be resource-intensive—cotton farming alone consumes vast amounts of water and pesticides. Synthetic fibers, like polyester and nylon, are petroleum-based, providing durability and low cost but contributing to microplastic pollution and non-biodegradable waste. The tradeoffs are stark: naturals promote renewability but may involve higher land use, while synthetics excel in versatility yet exacerbate fossil fuel dependency.

Low-durability varieties, such as upland cotton (Gossypium hirsutum), with its short staples, have fueled the fast fashion boom by enabling cheap, mass-produced garments. Paired with inexpensive polyester, this has created a market of disposable clothing, prioritizing volume over quality.

However, high-durability options, like Pima or Egyptian cotton with longer fibers, offer superior strength, longevity, and comfort, leading to longer garment lifespans and potentially higher incomes for producers through premium pricing. Transparency tools, such as DPPs, can showcase these differences, allowing consumers to opt for high-quality materials.

Imperium Hemp fiber from companies like Heartland exemplifies a superior alternative: requiring minimal water and no pesticides, it provides exceptional durability while being recyclable or compostable, mitigating the pitfalls of both natural and synthetic staples. By demanding transparency, consumers can steer the industry toward such innovations, reducing reliance on resource-heavy or polluting materials.

2.2 Durability and Lifetime Use

Fast fashion's set-in stone a "disposable" mentality amongst consumers, where garments are designed for short-term wear. This has exacerbated waste generation, with consumers discarding items after mere handfuls of uses. This culture thrives on trends and low prices, but it ignores the long-term environmental toll, including landfill overflow and resource depletion.

The shift away from this model is accelerated by global policy changes, as developing countries increasingly reject waste imports from wealthier nations. China's 2018 ban on 24 types of solid waste, including textiles, has disrupted the flow of discarded clothing, forcing first-world countries to confront their overconsumption. Similar restrictions in Rwanda and other East African nations aim to protect local industries from secondhand influxes, highlighting how the disposable mindset is no longer viable.

Consumers hold the power to demand durability through purchases that favor long-lasting items. Transparency reveals production quality, such as fiber strength and manufacturing ethics, influencing decisions toward resilient materials.

Hemp fiber again shines here, with its natural robustness extending garment lifetimes and reducing replacement frequency, all while being traceable to ensure ethical sourcing. As bans on waste exports tighten, this consumer-led push for quality over quantity can transform the industry, promoting a circular economy where durability is paramount.

2.3 End-of-Life Management

At the end of a garment's life, collection and recyclability pose significant hurdles, with most textiles ending up in landfills or incinerators. Natural fibers like cotton are more recyclable mechanically, often shredded into new yarns, but blends with synthetics complicate processes, lowering quality and feasibility.

Synthetics, while potentially chemically recyclable, release microplastics and require energy-intensive methods. Globally, only about 14.7% of textiles are recycled, underscoring the need for better systems.

Compounding this, countries like Malaysia and Uganda are imposing restrictions or bans on textile waste imports, rejecting the role of "dumping grounds" for first-world excess. The EU's proposed bans on textile destruction and waste exports further signal a global reckoning.

Transparency is key to resolving this, as it informs consumers about recyclability, encouraging support for designs that use all-natural fibers like cotton and hemp rather than blends of natural and synthetic. Hemp fiber facilitates end-of-life solutions, being fully biodegradable in landfills or compost and easily recyclable due to its purity and strength. By prioritizing transparent, recyclable options, consumers can drive innovations that close the loop on textile waste.

3 Solutions to Resolve Consumer Behaviors

Addressing these consumer behavior challenges requires innovative solutions that leverage transparency and advanced materials. The goal is to empower consumers to make value-aligned choices, fostering a sustainable industry.

3.1 Transparency & Digital Passports

Green Threads' digital passport technology exemplifies how transparency can revolutionize fashion. By creating digital twins of garments, it allows consumers and regulators to access comprehensive data on sourcing, labor, and environmental impact via simple scans.

These combats greenwashing and aligns purchases with ethics, much like the shifts seen in diamonds and cobalt. Integrated with EU DPP mandates under the Ecodesign for Sustainable Products Regulation (ESPR), which became effective in June 2024 and prioritizes textiles for implementation by 2027, it dictates behaviors by highlighting sustainable practices, such as using traceable hemp, ultimately reshaping manufacturer priorities.

DPPs are fundamentally altering consumer purchasing mentalities by fostering a culture of informed, conscientious decision-making. In an era where 78% of consumers report changing habits to reduce environmental impact, DPPs provide verifiable data on a product's origins, materials, and footprint, enabling buyers to evaluate sustainability claims directly. This transparency shifts the mindset from viewing clothing as disposable commodities to long-term investments, as shoppers can assess durability, repairability, and resale potential—potentially doubling a product's lifetime value through enhanced circularity and secondhand markets.

For instance, by scanning a QR code linked to a DPP, consumers gain insights into ethical labor practices and carbon emissions, building trust and encouraging loyalty to brands that demonstrate accountability. This educational aspect combats impulse buying, as evidenced by surveys showing that access to lifecycle information prompts 65% of fashion consumers to opt for eco-friendly options, mirroring behavioral changes in other sectors where transparency drove ethical sourcing.

As EU regulations mandate DPPs for textiles by 2027, with a working plan announced in April 2025, this tool not only enforces compliance but also empowers consumers to demand higher standards, pressuring manufacturers to innovate in traceable, durable materials like hemp. Green Threads, focused on apparel sectors like workwear and outdoor clothing, simplifies this process by offering user-friendly platforms that integrate seamlessly with supply chains, turning regulatory requirements into opportunities for consumer engagement and brand differentiation.

3.2 New Material Selection

Heartland's Imperium hemp fiber offers a multifaceted solution, using minimal water and no pesticides while yielding low-cost materials that enhance product strength and sustainability. Its high durability counters fast fashion's disposability, and its biodegradability simplifies end-of-life, making it recyclable or compostable. When paired with transparency tools, hemp becomes a consumer-favored option, solving raw material tradeoffs and promoting longer lifetimes.

Imperium hemp fiber from Heartland Industries stands out as a premier material solution, directly addressing durability challenges and driving extended garment use in a market plagued by short-lived products. Engineered through proprietary processes,

Heartland's hemp fibers are exceptionally strong providing superior tensile strength that resists wear and tear, thereby extending the functional lifespan of clothing items.

This durability combats the fast fashion model's emphasis on low-cost, low-quality materials, where garments often fail after just 7–10 wears; in contrast, hemp-reinforced textiles can endure many more cycles, reducing the need for frequent replacements and curbing overconsumption.

Heartland's approach leverages hemp's natural properties—requiring 95% less water than cotton to create low-cost sustainable fiber that not only bolsters mechanical performance but also lower the overall environmental footprint reducing emissions by 70% compared to Cotton.

Transparency amplifies hemp's advantages, as DPPs can detail these attributes—such as traceability from farm to fabric and quantifiable durability metrics—allowing consumers to visualize the long-term value of their purchases. This visibility shifts buying behaviors toward items that promise extended use, fostering a mentality where quality trumps quantity. For example, hemp's antimicrobial and breathable qualities, combined with its biodegradability, ensure garments remain functional longer and degrade harmlessly at end-of-life, aligning with circular economic principles.

Sustainability in textiles hinges on consumer empowerment through transparency and innovative materials like hemp. By addressing challenges in materials, durability, and end-of-life solutions like Green Threads' DPPs and Heartland's fibers, the industry can evolve. As regulatory pressures mount, including the EU's phased DPP rollout through 2033, consumers' choices will dictate a greener future, echoing the transformative impacts seen in other sectors.

3.3 Legislation and Regulatory Frameworks

Based on the truly global nature of supply chains within the industry, while the implementation of regulation is being driven by the EU the implications are far more far reaching. Whilst it will be necessary to implement new protocols and processes to achieve compliance for products being sold within the EU, it is highly likely that Brands and Producers will simply adopt new practices across their ranges and as such consumers globally will benefit from the increased focus on Transparency and the communication of data at point of sale.

By introducing a variety of different regulations which are tailored to achieve specific outcomes in the EU, Clothing suppliers will be increasingly motivated to implement eco-design principles in order to minimise potential costs and financial penalties.

One such regulation is EPR (Extended Producer Responsibility) EPR requires **brands, manufacturers, and importers** to take responsibility for the **entire lifecycle** of their products including post-consumer waste. This includes:

- **Collection** of discarded textiles
- **Sorting and recycling**
- **Funding infrastructure** for waste management
- **Reporting and compliance** with national regulations

The calculation of the costs associated with EPR will be facilitated by Eco-modulation

This is a mechanism within Extended Producer Responsibility (EPR) schemes that adjusts the fees producers pay based on the **environmental performance** of their products. It's a strategic lever to **incentivize sustainable design** and discourage wasteful practices.

Based on the premise that improving the environmental characteristics of your products will result in lower EPR costs there will be significant financial incentives for the adoption of low impact materials such as Hemp within supply chains.

Dyeing and Finishing

Comparative Study of Extracting Dyes Methods from Mitragyna Inermis Bark

Pulchérie Marie Laure Kantiono[1,2](✉), Hassen Hedfi[3], Yssouf Karanga[1], Mohamed Hamdaoui[2], and Moussa Bougouma[1]

[1] Laboratoire de Chimie Analytique, de Physique Spatiale Et Energétique (L@CAPSE), Université Norbert ZONGO, BP 376 Koudougou, Burkina Faso
kantiono.pulcherie@gmail.com

[2] University of Monastir, National Engineering School of Monastir, Textile Materials and Processes Research Unit MPTex, UR17ES33, 5019 Monastir, Tunisia

[3] University of Monastir, National Engineering School of Monastir, Mechanical Engineering Laboratory (LGM), LR99ES32, 5019 Monastir, Tunisia

Abstract. For centuries, people have been interested in natural dyes for dyeing applications. Natural dyes are known for their phytochemical properties, which are highly sought-after in many fields, including dyeing. The extraction of compounds from plant matter is the key to obtaining the dyestuff. A number of research projects have focused on finding and developing better extraction methods. Modern environmentally-friendly extraction methods, such as microwaves, accelerated solvents and ultrasound, are in the ascendancy. The aim of this study was to compare three methods of extracting natural dyes from the bark of Mitragyna Inermis, a plant belonging to the Rubiaceae family, for use in textiles. The methods evaluated were maceration, digestion and ultrasound- assisted extraction. The three extraction methods were optimized according to the parameter's extraction temperature and extraction time. The effectiveness of each method was assessed by calculating the extraction yield after optimization and measuring the colour strength (K/S) after application to fabric.

Keywords: Natural dyes · Extraction methods · Colour strength

1 Introduction

Natural dyes derived from plants [1] are increasingly in demand for textile dyeing, to the detriment of chemical dyes which pose a threat to humans and the environment, not only because they are synthesized but also because their application to textiles requires intermediate products, a source of environmental pollution [2].

Natural dyes are in great demand because of their biodegradability and low environmental impact. Several studies have been carried out on the phytochemicals in Mitragyna Inermis. This plant is widely used in traditional medicine for its bioactive compounds such as flavonoids and tannins. It is recognized for its many therapeutic properties [3, 4] and is much in demand in several African countries. In Mali, the leaves of Mitragyna

M. Hamdaoui et al. (Eds.): ITDM'3 2025, PTT, pp. 189–201, 2026.
https://doi.org/10.1007/978-981-95-7704-0_18

Inermis are widely used in traditional pharmacopoeia in association with other plants to treat illnesses such as jaundice, syphilis and arthritis. In Burkina Faso, studies have shown that a decoction of Mitragyna Inermis is used to treat neuropsychiatric disorders [5].

In Côte d'Ivoire, Senegal and Benin, the bark is prescribed for stomach pains, dysentery and schistosomiasis. A decoction of the roots is used to treat malaria [6]. A decoction of the leaves and bark is used as an antidiabetic and antipyretic, as well as a treatment for hypertension and tuberculosis [7, 8].

Studies have shown that Mitragyna Inermis extract has antibacterial properties against the pathogenic bacteria Staphylococcus aureus and Klebsiella pneumoniae, as well as Escherichia coli [9]. These bacteria, which survive in textiles, can transmit infections, especially in hospitals. Mitragyna Inermis extract used as a dye could therefore be of interest in the medical field. The dyed fabric could be used as a bactericidal blanket or bed sheet in hospitals.

The extraction of natural colorants is based on a variety of methods adapted to the nature of the raw materials and the compounds sought. The most common techniques include solid-liquid extraction, which includes maceration, infusion, digestion and decoction, releasing soluble pigments into an often aqueous or organic solvent. These methods are simple to apply and relatively inexpensive. Solvent extraction, using liquids such as ethanol, hexane or aqueous methanol, is also widely employed, notably to isolate specific pigments such as carotenoids or anthocyanins [10], although this method can have drawbacks linked to solvent consumption and potential toxicity. Other advanced techniques, such as adsorption on resins, ultrasound-assisted extraction or pulsed electric field treatments, can improve the yield and purity of extracted colorants while reducing environmental impact.

The complexity of the techniques and the high cost of the equipment make these techniques less accessible to craft structures. Disadvantages of conventional extraction techniques include long extraction times, the use of substantial quantities of solvents and, sometimes, numerous extraction steps. In addition, significant quantities of thermolabile phytochemicals are decomposed or degraded during heating when extracting by decoction, infusion or digestion. However, these extraction techniques are still used to extract natural dyes for textile use. With a view to making dyes available for dyeing, a study was carried out into the yield of three extraction methods.

Three extraction methods were used in our study. These were extraction by maceration, extraction by digestion and extraction assisted by ultrasound.

Maceration

The maceration extraction method consists of leaving the plant material (fresh or dry) in a solvent at room temperature for a sufficiently long time with intermittent stirring [11]. At the end of the extraction, the solution is filtered and then stored in a closed container to avoid possible contamination. The liquid extract obtained is cleaned by filtration or decantation. To obtain the dye in powder form, the liquid extract is condensed by evaporation to remove the solvent. The condensate is then placed in a freeze-dryer to obtain the dye powder. The advantage of this method is that it preserves the integrity of the molecules. However, the procedure is lengthy and the extraction yield is low compared with other methods.

Digestion

Digestion is an extraction method which uses gentle heating during extraction, accompanied by stirring at regular intervals [11]. However, care must be taken to ensure that the temperature does not alter the phytochemical compounds in the plant material. Most of the time, temperatures are maintained between 27 °C and 40 °C, but they can reach 50 °C for materials such as bark. During extraction, the plant material is placed in a container containing the appropriate solvent preheated to the temperatures indicated. The optimum temperature is maintained for a period ranging from half an hour to 24 h [11]. This is an environmentally friendly method as it does not require prolonged heating. However, it does require the use of equipment to ensure continuous agitation.

Extraction Assisted by Ultrasound

Ultrasound is an electromagnetic wave with frequencies ranging from 20 kHz to 2000 kHz. They propagate through the medium causing expansion and contraction, in accordance with the wave nature of the medium. The mechanical effect of the acoustic cavitation generated by ultrasound increases the contact surface between the solvent and the plant matrix, as well as the permeability of the cell walls. Ultrasound accelerates the release of phytocompounds.

To improve and optimize extraction efficiency, it is essential to study ultrasound parameters in parallel. Studies have shown that the frequency used can modify and favorably influence the extraction of compounds. Indeed, yields of phenolic compounds are higher at a lower frequency of 40 kHz than at 120 kHz [12]. The use of ultrasound reduced the extraction time by 30 min, with better yields compared to the maceration method, which required 120 min. The choice of solvent also influences extraction yields. One study showed that extracting phenols from rhizomes with 75.3% ethyl alcohol over an extraction time of 40 min gave higher yields than those obtained with the solvent alone [13].

To compare the different extraction methods, chemical screening was carried out to ensure the composition of each extract. The extraction yield was then calculated. Before applying the extracts to fabric, their stability under storage conditions was studied. Finally, color yield was measured on samples dyed with the extracts obtained for each method.

2 Material and Methods

2.1 Presentation of the Plant

The plant we studied was Mitragyna Inermis (Willd) Kuntze. It is a tree (see Fig. 1) belonging to the Rubiaceae family and varies in height from 8 to 10 m with a straight trunk up to 60 cm in diameter. The leaves are elliptical, 7 cm long and 4 cm wide. The flowers are white and fragrant and the fruit contains several spherical brown seeds [14]. For our study, Mitragyna Inermis bark was collected in June 2024 in Ouagadougou,

Burkina Faso. The bark was cleaned and then air-dried in the dark. After drying, the bark was crushed and sieved using a 0.5 mm diameter sieve (see Fig. 2).

(a) (b)

Fig. 1. Mitragyna Inermis: (a) Tree and (b) Leaves

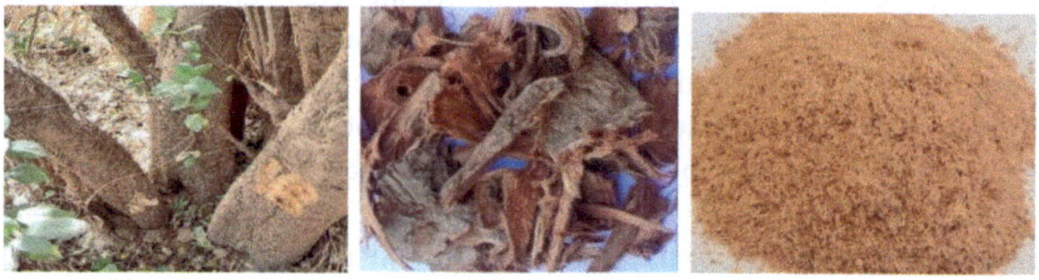

(a) (b) (c)

Fig. 2. Mitragyna Inermis bark: (a) Bark, (b) Barks dried indoors and (c) crushed and sieved bark

2.2 Extraction

For ecological reasons, we used an aqueous extraction with water as the solvent. The best yield for each method was considered after optimization of the extraction process. The optimization parameters considered were extraction temperature (T) and extraction time (t). To this end, 50 g of plant material were used for the three methods.

For maceration, the plant material was placed in 1 L of water at different times (two levels: 24 h and 48 h) at room temperature (25 °C to 27 °C). The extraction was carried out in a closed bottle to avoid contamination.

For extraction by digestion, the plant material was placed in 1 L of water. The temperature was varied between 27 °C and 45 °C and the extraction time was 90 min, 120 min and 240 min. The mixture was kept under constant stirring using a MAGNETIC STIRRER SH-2 type heated stirrer for stirring and heating, and a magnet bar, which was carefully cleaned and then placed in the container.

Ultrasonic extraction was carried out using a GRANT XUB5 type machine. The temperature parameter was varied between 27 °C and 45 °C for extraction times of 30 min, 45 min and 60 min.

The extracts were frozen and dried in an ALPHA 1–2 LD plus freeze-dryer. The dry dyes obtained were weighed and the yields determined by Eq. (1):

$$Yield\,(\%) = \frac{mass\ of\ dry\ dye}{mass\ of\ plant\ material} \times 100 \tag{1}$$

2.3 Chemical Screening

Phytochemical screening was carried out on Mitragyna Inermis bark extract using the method employed by Toklo et al. [15]. This is a qualitative chemical analysis aimed at identifying the main families of compounds, based on differential precipitation and staining reactions.

2.4 Extract Stability

It is essential to study the stability of extracts to ensure that bioactive compounds remain intact and effective at the time of use. Indeed, secondary metabolites present in extracts can be degraded or transformed by environmental factors such as storage, temperature and pH, which could alter their tinctorial, pharmacological or biological properties. The stability of the extracts under storage conditions [16] was studied. 1 L of extract was dispensed into vials and stored in a room conditioned to ISO 139. Absorbance was measured every 1 h for 7h and then every 12 h for five (05) days using a HACH DR 3900 UV-visible spectrophotometer.

Extracts were tested for pH and temperature stability. For pH, solutions were prepared at different pH values (5, 7 and 9) and absorbance was measured for each pH value. To study the effect of temperature, extracts were placed in an oven at temperatures ranging from 50 °C to 90 °C. Extracts were diluted 1:10 and absorbance measured using a UV-visible spectrophotometer.

2.5 Dyeing Process

The dyes were applied to a 100% debouched and bleached cotton fabric (see Table 1). Dyeing was carried out using an AHIBA NUANCE laboratory-scale dyeing machine.

Table 1. Fabric characteristics

Characteristics		Standards used
Composition	Cotton	ISO TR 11827 [17]
Structure	Plain	NF G07-154 [18]
Thickness	0.72 mm	ISO 5084 [19]
Weight per unit area	188 g/m^2	NF EN 12127 [20]
Warp count	40/2	ISO 7211-5 [21]
Weft count	40/2	ISO 7211-5 [21]
Warping (yarn/cm)	16	ISO 7211-2 [22]
Wefting (weft/cm)	12	ISO 7211-2 [22]

The bath ratio was 1/40 with a dye quantity of 1 g/L. Dyeing was carried out at neutral pH and the dyeing temperature and time were set at 90 °C and 45 min respectively. After dyeing, the samples were rinsed and dried at room temperature.

The amounts of color on the fabrics, estimated by the color intensity values (K/S), were measured using a DATACOLOR 500 spectrophotometer (Illuminant D65/10° observer (with Color tools software (Data color International, USA).

3 Results and Discussion

3.1 Chemical Screening

The results of the chemical screening of Mitragyna Inermis bark extract for the three extraction methods are summarized in Table 2.

Table 2. Chemical screening of the extract from the bark of Mitragyna inermis.

Chemical group	Maceration	Digestion	Ultrasound
Tannins	+	+	+
Flavonoids	+	+	+
Alkaloids	+	+	+
Terpenes and sterols	+	+	+
Quinones	+	+	+
Saponins	+	+	+

(+): Presence (-): absence

Chemical screening of Mitragyna Inermis bark extract reveals the presence of a number of major compounds such as tannins, alkaloids, flavonoids, quinone derivatives, terpenoids, sterols and saponins. Table 2 shows that these compounds are all extracted efficiently by maceration, digestion and ultrasound, so that all three methods are equally effective in obtaining extracts rich in secondary metabolites.

3.2 Extract Stability

The figure (see Fig. 3-a) shows the evolution of absorbance, which reflects dye concentration, as a function of storage time for the three extraction methods compared. There is a progressive decrease in absorbance for all three methods, indicating degradation of the extract over time. All extracts show good short-term stability (6H). Ultrasound gives a higher initial concentration, but less long-term stability. Maceration offers the best extract stability under the storage conditions tested, as degradation is slower.

The results of the study of the extracts thermal stability (see Fig. 3-b) show that the compounds responsible for color do not degrade significantly between 50 °C and 90 °C.

On the contrary, their apparent concentration increases, which is often linked to better extraction.

The similarity of the curves at acidic and neutral pH indicates a greater stability of colored compounds under acidic to neutral conditions (see Fig. 3-c). The increase in absorbance at basic pH suggests a structural modification of the compounds present in the extract, probably linked to ionization or deprotonation of certain functional groups (such as phenols or flavonoids), which modifies their absorption spectrum.

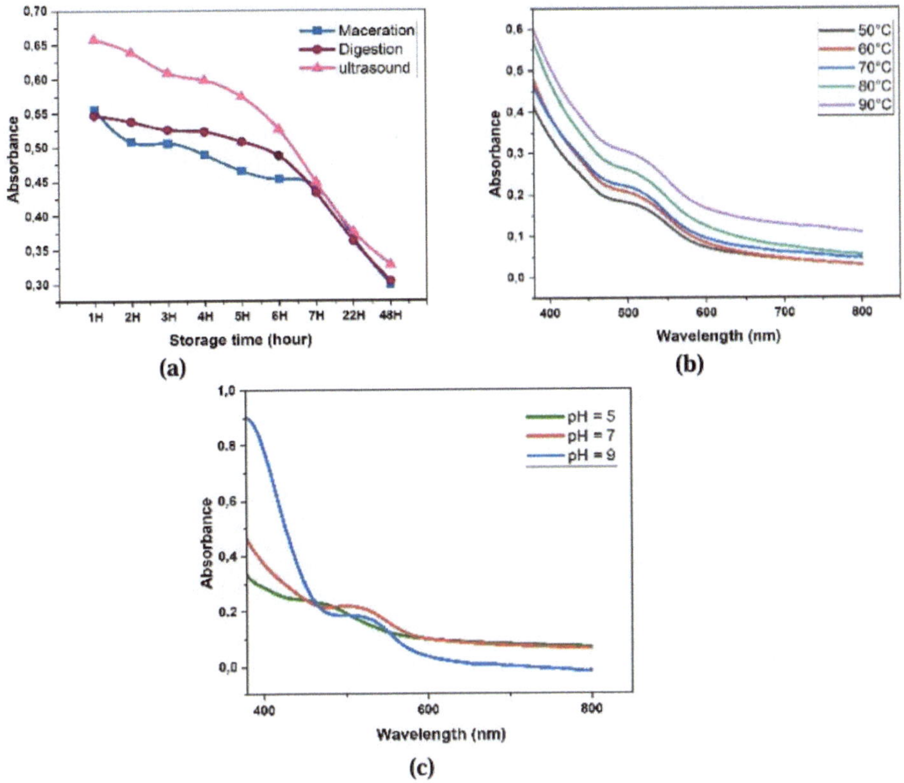

Fig. 3. Extract stability-Effect of: (a) Storage, (b) Temperature and (c) pH

3.3 Extraction Yield

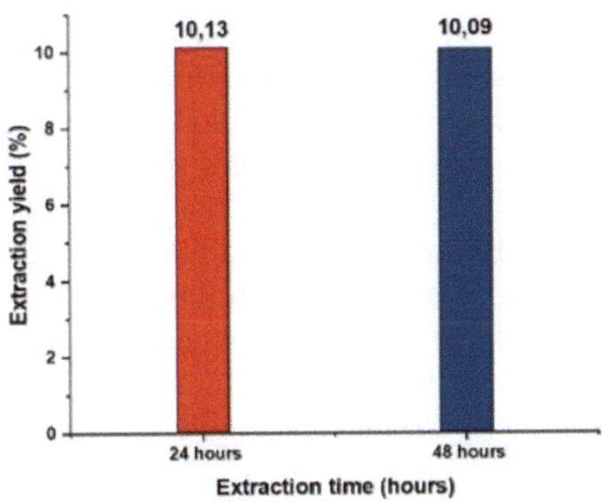

Fig. 4. Extraction yield of dyes obtained after maceration

Graphical analysis of the results of calculating extraction yield and measuring colour strength shows that for each extraction method there is an optimum temperature and an optimum extraction time beyond which extraction efficiency decreases.

Indeed, the variation in the duration of extraction by maceration shows that the yield (see Fig. 4) is slightly higher after an extraction time of 24 h.

The best colour yield was obtained after maceration for 24 h (see Fig. 5), suggesting that extending the extraction time beyond 24 h does not improve extraction. It is true that extraction from the bark requires more time to release as many molecules as possible [11], but other factors such as the state of the coarsely or finely chopped, ground plant material play a key role in the speed of extraction.

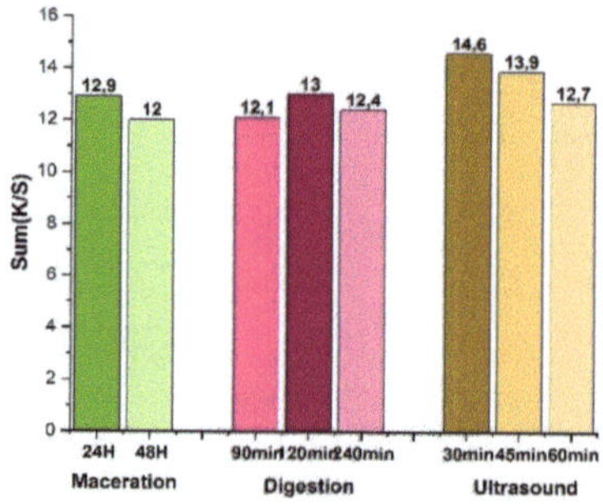

Fig. 5. K/S color strength of the three extraction methods

In our case, the plant material was ground to a powder, which increases the contact angle between the plant matrix and the solvent (which is water), and accelerates the release of the colour molecules. In addition, most of the colour compounds, such as anthocyanins and polyphenols, are rapidly extracted from the start of the maceration process [23]. The degradation of the colour due to prolonged maceration makes the extract less colourful, which reduces the K/S.

For ultrasound-assisted extraction, the best extraction yield and the best K/S were obtained after a short extraction time (30 min) (see Figs. 7 and 5). Yields decreased considerably after this time. This shows that extraction times that are too long have a negative effect on extraction yields.

Indeed, long durations can generate excessive cavitation, leading to the degradation of compounds such as polyphenols [24]. In addition, re-adsorption and precipitation phenomena lead to a decrease in the concentration of the dye in the extract, which reduces the colour yield after dyeing for samples dyed with the extract obtained by maceration and ultrasound.

In contrast to the maceration and ultrasound methods, the extraction yield using the digestion method (see Fig. 6) increases with time up to 120 min and stabilizes at 240 min. This suggests that most of the extraction occurs in the 0 to 120 min phase and that beyond 120 min there will be no significant improvement in yield. In addition, the colour yield increases from 90 to 120 min and then decreases from 240 min.

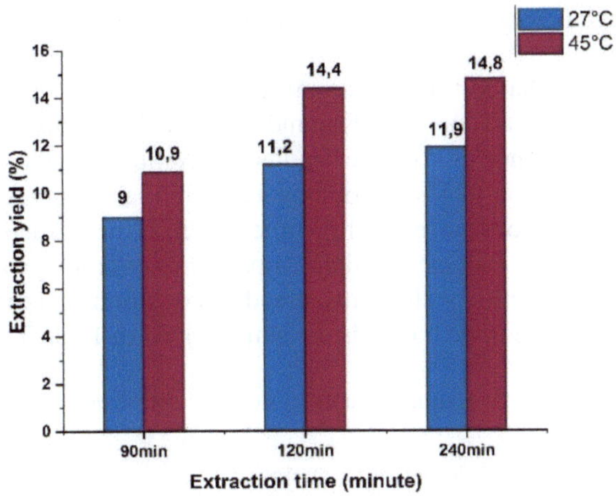

Fig. 6. Extraction yield of dyes obtained after extraction digestion

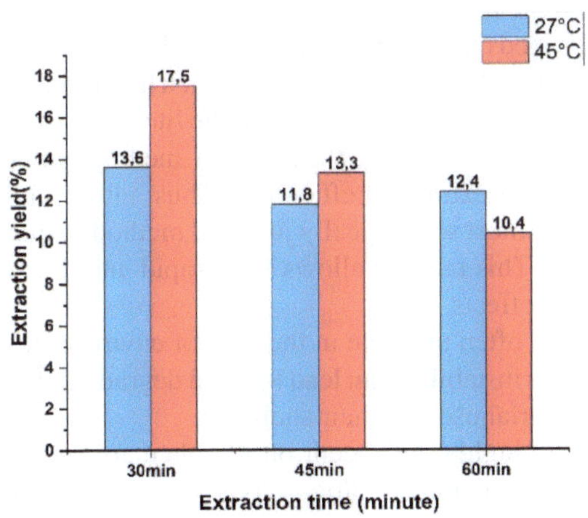

Fig. 7. Extraction yield of dyes obtained after extraction ultrasound

This indicates that there is an optimum extraction time beyond which the dye degrades, resulting in a drop in colour rendering.

The results of the study of the effect of extraction temperature for the two methods (digestion and ultrasound) (see Figs. 6 and 7) show that the yield is high at a temperature of 45 °C.

This shows that increasing the temperature favors extraction. Indeed, when the temperature is increased, the diffusivity of the solid matrix increases, which accelerates the mass transfer of the compounds from the plant material to the water. As a result, solubility increases, allowing maximum pigment release. These results enable us to confirm that Mitragyna Inermis bark contains dye molecules capable of binding to 100% cotton fabric.

Studies have shown that Mitragyna Inermis leaves contain flavonoids, anthocyanins, quinones and tannins [24], all compounds found in textile dyes. This confirms our results after dyeing. Several authors [25] have studied the influence of different extraction conditions on the extraction yields of compounds from plant sources, highlighting the necessity to determine the optimal extraction time as well as the extraction temperature to avoid dye losses caused by excessively high temperatures or prolonged extraction times, and to maximize the extraction yield. Thus, the ideal conditions in terms of temperature and extraction time for the three methods are summarized in Table 3.

Table 3. Extraction conditions after optimization

Methods of extraction	Conditions of extraction	Extraction yield (%)	K/S
Maceration	27 °C ± 4 °C, 24h	10.1	12.9
Digestion	45 °C, 120 min	14.8	13.0
Utrasound	45 °C, 30 min	17.5	14.6

The ultrasound-assisted extraction method shows the highest extraction yield, with a yield of 17.5%, followed by digestion extraction, which has a yield of 14.8%. Maceration extraction presents the lowest extraction yield. In the literature, authors have shown [26] that ultrasound-assisted and maceration extraction methods indicate that maceration techniques yield the lowest extraction efficiency. Thus, ultrasound- assisted extraction at 45 °C for 30 min is the most scientifically justified method to maximize the extraction of coloring compounds. This method allows for a rapid and maximal extraction while minimizing degradation effects.

Natural colorants are often unstable in the face of environmental variations such as temperature and pH. This instability can lead to rapid degradation of extracted pigments, limiting both their industrial use and their shelf life.

Extraction yields are highly dependent on the plant source used, the nature of the pigment, and above all extraction conditions (type of solvent, extraction method, temperature, time). The solvent used plays a crucial role in extraction efficiency. In our study, the use of water as an extraction solvent may justify the relatively low extraction yields obtained. Studies have shown that extraction yields are better in one solvent (water/ethanol) [27, 28]. In addition, extracts obtained by conventional methods (maceration, digestion) show good storage stability compared with extracts obtained by ultrasound-assisted extraction. Indeed, the reduced stability of extracts obtained by ultrasound is mainly due to physical effects such as intense cavitation, which promotes rapid release of compounds, but can also lead to partial degradation of sensitive molecules, particularly natural colorants, thus reducing their stability during storage, due to the local increase in temperature and mechanical stress. Finally, high ultrasonic power and too long an extraction time increase the risk of degradation of sensitive compounds, affecting the final stability of the extract [29]. Ultrasonic power should therefore be considered an important parameter in ultrasonic-assisted extraction.

The K/S coefficient is an optical indicator linked to the concentration and absorption of the dye on the textile. It therefore depends on the chemical structure of the extracted molecules. Dyes are made up of chromophore groups (responsible for color) and auxochromes, which determine their capacity to absorb and interact with the fiber. These chemical groups must be compatible with cellulose to ensure good fixation and dye fastness.

The molecular structure of compounds such as polyphenols, tannins and quinones present in the extract affects color, brightness, stability and lightfastness. The antibacterial activity of the extract is confirmed by the presence of alkaloid compounds. Further chromatographic analysis will confirm whether the same chemical structure is extracted in all three extraction methods.

4 Conclusion

The objective of this study was to compare three extraction methods of dyes from a dye plant, Mitragyna inermis. The three extractions methods maceration, digestion, and ultrasound-assisted extraction were optimized. Optimal conditions were obtained for each method: Maceration at 27 °C for 24 h, Digestion at 45 °C for 120 min, and Ultrasound at 45 °C for 30 min. The color strength was also measured to first ensure that the extracts can be fixed onto a textile substrate and to confirm the optimization results.

A comparative study of extraction methods for Mitragyna inermis bark highlighted the decisive influence of temperature and duration parameters on extraction efficiency, as well as on the quality of extracts intended for textile application.

The results show that increasing the temperature to 45 °C significantly enhances extraction efficiency compared to a lower temperature (27 °C), while the optimum extraction time is around 120 min, beyond which efficiency does not improve and may even decline.

These observations suggest that for a textile application, where the quality and stability of natural dyes are essential, it is preferable to adopt an extraction method combining moderate temperature and controlled extraction time. This approach guarantees a good compromise between yields and extract quality, while limiting the risk of degradation of active compounds.

Optimizing extraction conditions for Mitragyna inermis bark is a key step towards valorizing this natural resource in the textile industry, by offering an ecological and sustainable alternative to synthetic dyes.

In future studies, we will be able to investigate the impact of extraction parameters on the physico-chemical properties of dyes for the development of an antibacterial fabric based on Mitragyna Inermis extracts.

References

1. Nacro, M., Millogo-Rasolodimbi, J.: Plantes tinctoriales et plantes à tanins du Burkina Faso. Scientifika (1993)
2. Hedi Ben, M., Oualid, B., Dorra, D., Daniel, B., Leila, C.G., Ridha, M.: Les colorants textiles sources de contamination de l'eau: Criblage de la toxicité et des méthodes de traitement. Revue des sciences de l'eau 209–238 (2011)

3. Abda, B., et al.: Antibacterial and anti-coronavirus investigation of selected Senegalese plant species according to an ethnobotanical survey. J. Ethnopharmacol. (328), 118070 (2024)
4. Placide, M.T., et al.: Phytochemistry and pharmacological review of Mitragyna inermis (Willd.) Kuntze (Rubiaceae). J. Pharmacognosy Phytochem. **9**(2), 22–30 (2020)
5. Kinda, P.T., Zerbo, P., Guenné, S., Compaoré, M., Ciobica, A., Kiendrebeogo, M.: Medicinal plants used for neuropsychiatric disorders treatment in the hauts bassins region of Burkina Faso. J. Med. 1–21 (2017)
6. Kerharo, J., Adam, J.G.: La pharmacopée sénégalaise traditionnelle. In Éditions Vigot frères Paris (1974)
7. Konkon, N.G., Ouatara, D., Kpan, W.B., Kouakou, T.H.: Medicinal plants used for treatment of diabetes by traditional practitioners in the markets of Abidjan district in Côte d'Ivoire. J. Med. Plants Stud. **5**(2), 39–48 (2017)
8. Dassou, H.G., et al.: Diversité, usages vétérinaires et vulnérabilité des plantes médicinales au Nord-Bénin. Int. J. Biol. Chem. Sci. **8**(1), 189–210 (2014)
9. Alowanou, G.G., Olounlade, A.P., Azando, E.V.B., Dedehou, V., Daga, F.D., Hounzangbead-ote, S.M.: A review of bridelia ferruginea: combretum glutinosum and mitragina inermis plants used in zootherapeutic remedies in West Africa: historical origins, current uses and implications for conservation. J. Appl. Biosci. (87) 8003–8014 (2015)
10. Jeyanthi, R.L., Sharmila, S., Merina, P.D., Candace, S.: Extraction and purification of carotenoids from vegetables. J. Chem. Pharm. Res. **6**(4), 594–598 (2014)
11. Handa, S.S., Khanuja, S.P.S, Longo, G., Rakesh, D.D.: Extraction technologies for medicinal and aromatic plants. Italy: United Nations Industrial Development Organization and the International Centre for Science and High Technology, 1st edn. no. 66 (2008)
12. González-Centeno, M.R., Knoerzer, K., Sabarez, H., Simal, S., Rosselló, C., Femenia, A.: Effect of acoustic frequency and power density on the aqueous ultrasonic-assisted extraction of grape pomace (vitis vinifera L.) - a response surface approach, ultrason. Sonochem **21**(6) 2176–2184 (2014)
13. Wang, X., Wu, Y., Chen, G., Yue, W., Liang, Q., Wu, Q.: Optimisation of ultrasound assisted extraction of phenolic compounds from Sparganii rhizoma with response surface methodology, ultrason. Sonochem. **20**(3), 846–854 (2013)
14. Adjanohoun, E., et al.: Contribution aux études ethnobotaniques et floristiques en République du Bénin. Médecine traditionnelle et pharmacopée. In: Agence de coopération culturelle et technique Paris, p.895 (1989)
15. Toklo, P.M., Yayi, E., Sakirigui, A., Alowanou, G.G., Assogba, M.F., Kakpo, B., Hounzangb, S., Gbenou, J.D.: Phytochemical Screening and in vitro effect of watery triturated extract of Mitragyna inermis (Willd.) Kuntz (Rubiaceae) on Haemonchus contortus. Int. J. Adv. Res. 7 (12) 762–767(2019)
16. Younsook, S., Yoo, D.: Storage stability and color reproducibility of yellow and red dyes extracted from carthamus tinctorius L. Text. Color. Finish. **24**(3), 165–172(2012)
17. ISO: Textile fibres — Identification of natural and man-made fibres — Part 1: General principles (ISO/TR 11827:2012). International Organization for Standardization (2012)
18. AFNOR: Methods for analyzing certain fabric construction characteristics - Determining the weave of a fabric. NF G07–154, Paris: AFNOR (1984)
19. International Organization for Standardization. (année). ISO 5084: Textile —Determination of thickness of textiles and textile products (édition). Genève, Suisse: ISO
20. AFNOR: NF EN 12127: Textiles - Fabrics - Determination of basis weight on small samples. Paris: AFNOR, mars, 9 (1998).
21. International Organization for Standardization: Textiles — Methods for analysis of woven fabrics construction — Part 5: Determination of linear density of yarn removed from fabric (ISO 7211–5:2020). ISO (2020)

22. ISO 7211–2:2024: Textiles - Methods of analysis of fabric construction - Part 2: Determination of yarn count per unit length. International Organization for Standardization, Genève, Suisse (2024)
23. Vitisphere. https://www.vitisphere.com/actualite-76089-maceration.html. Accessed 18 Apr 2025
24. Vardanega, R., Santos, D.T., Meireles, M.A.A.: Intensification of bioactive compounds extraction from medicinal plants using ultrasonic irradiation. Pharmacognosy Rev. **8**(16), 88–95 (2014)
25. Sarkar, S., Alvarez, V.H., Saldaña, M.D.A.: Relevance of ions in pressurized fluid extraction of carbohydrates and phenolics from barley hull. J. Supercrit. Fluids (93) 27–37 (2014)
26. Jin, S., et al.: Microwave-assisted extraction of flavonoids from *Cajanus cajan* leaves. Chin. Tradit. Herbal Drugs **42**(11), 2235–2239 (2011)
27. Benmeziane, F., Djamai, R., Cadot, Y., Seridi, R.: Optimization of extraction parameters of phenolic compounds from Algerian fresh table grapes, (Vitis Vinifera). Int. Food Res. J. **21**(3), 1061–1065 (2014)
28. Jokic, S., Velic, D., Bilic, M., Bucic-Kojic, A., Plan Inic, M., Tomas, S.: Modelling of the process of solid-liquid extraction of total polyphenols from soybeans. J. Food Sci. **28**, 206–212 (2010)
29. Jana, ŠŽ, et al.: Ultrasound as a promising tool for the green extraction of specialized metabolites from some culinary spices. Molecules **26**, 1866 (2021)

Benefit Study of Ecological compounds GALACID XT88 Based on Lactic Acid and GALASOLV NF62 Based on Ethyl Lactate and their impact on the surface and Mecanical properties of cotton Fabrics

Nasr Litim[1(✉)], Ayda Baffoun[2], Dorra Ben Kahla[1], and Saber Ben Abdessalem[2]

[1] University of Monastir, National Engineering School of Monastir, 5019 Monastir, Tunisia
nasr.litim@gmail.com, litnasr07@gmail.com

[2] University of Monastir, National Engineering School of Monastir, Textile Materials and Processes Research Unit MPTex, UR17ES33, 5019 Monastir, Tunisia

Abstract. Two compounds "GALACID XT 88" based on lactic acid and "GALASOLV NF 62" based on ethyl lactate as being a green solvent, which are biodegradable and renewable compounds. In this study, we attempted these compounds to use in finishing treatment, especially applied on cotton denim fabric in order to get 3D effects like fashion require. The mechanical, morphological, microscopic characterizations as well as the evaluations obtained were elaborated on the treated fabrics in two steps; before wash BW and after wash AW. In two steps of finishing, both compounds give acceptable Dry crease recovery angle, but it seems that compound based on ethyl lactate is better in DCRA quality compared to compound based on lactic acid after wash. We finish that the ecological compounds gives a satisfy quality of 3 D effects. Consequently, these compounds preserve the mechanicals properties during two steps (GALACID XT 88 causes loss of 25% max. and GALASOLV NF 62 cause loss of 8% max.). These results are proved using Kawabata evaluation system and Surface morphology by SEM. Indeed, these compounds improve the quality of treated fabric.

Keywords: Ecological compounds · Surface morphology · Mechanical properties · KES · Finishing

1 Introduction

A lot of finishes can create pollutant byproducts in the water stream. The use of oils, resins or other chemicals in finishing treatments is so varied and so extensive that it is impossible to consider them all in a brief survey of this kind; however, the application of flame-retardant, softening, durable-press, antistatic, soil-release, stain resistant, waterproofing or oil-repellent finishes invariably uses materials which are injurious to the environment if discarded [1]. Ecofriendly textile processing is presented as a global challenge, as ecological criteria are more and more accepted in all parts of the world in selecting

M. Hamdaoui et al. (Eds.): ITDM'3 2025, PTT, pp. 202–217, 2026.
https://doi.org/10.1007/978-981-95-7704-0_19

consumer goods and 'green' products can command a higher price. Two aspects of textile production, the restraint of harmful products and the decrease in air or water pollution, must be tackled in particular [2]. Various textile finishing agents and acid dying of food require low cost. For this, they use lactic acid to compete with cheaper inorganic acid. In fact, lactic acid is being used in many little scale applications like pH adjustment hardening baths for cellophanes used in food packaging, finishing agent for phenol formaldehyde resins, alkyd resin modifier, solder flux, lithographic and textile printing developers, adhesive formulations, electroplating and electro polishing baths, detergent builders. Lactic acid esters like ethyl/butyl lactate can be used as green solvents. They are high boiling, non-toxic and degradable components [3]. Environmentally friendly, 'green' solvents are another potential growth area for lactic acid derivatives, particularly lactate esters of low molecular weight alcohols such as ethyl, propyl and butyl lactate [4, 5]. The US Environmental Protection Agency (EPA) recently classified the lactate esters ethyl and butyl lactate as Class 4A inert ingredient for use in the formulation of pesticides and other bioactive resins. The Class 4A designation is given to compounds that have demonstrated negligible toxicity and an excellent environmental profile. Thus, a good range of specialty applications and commercial uses could be developed with these nontoxic, environmentally friendly lactate ester solvents with other biologically derived solvents [6]. This work will therefore be devoted to the study of the behavior of an ecological compound class provided by the Belgian firm "GALACTIC".

In this paper, we study the effect of two types of ecologic compounds "GALACID XT 88 "and "GALASOLV NF 62", typically they are used to permanent wrinkle treatments and 3D effects of denim fabrics. To establish the compound effects during the finishing process Pad Dry Cure "PDC" with the two selected compounds, we characterize the samples treated at two phases of the finishing process; before washing BW and after washing AW. Mainly, the treatment was affected on support textile, which is the denim cotton fabric. Finishing compounds have been used to impart crease recovery properties in a process. Various factors, such as the nature of compound, the concentration and curing temperature were considered. These factors should have a significant effect on surface morphology of treated fabrics, on treated fabric thickness, on the air permeability, on 3D effects fashion, on the geometrical roughness SMD, properties of treated fabric with the four KAWABATA Evaluation systems and particularly on the mechanical properties (strength loss). This study provides main idea about the link between the characteristics of denim fabrics and ecological compound effect in finishing treatment.

2 Experimental

2.1 Materials and Methods

The aim of this paper is to inspect the influence of ecological compound characteristic and finishing conditions on mechanical and surface properties of cotton Denim fabrics. For this purpose, surface morphology of treated fabrics with ecological compounds at various concentrations was examined by Scanning Electronic Microscopy. The micrographs were taken with a SEM "Hitachi SU 3500". We used a sputter coater to pre-coat conductive gold onto the surface before observing the microstructure at 22Kv. All the mechanicals' modulus (Shear, compression, Flexural and roughness SMD) were determinate by Kawabata Evaluation System. These characteristics are closely associated to the hand feeling of fabrics. The detection of the surface roughness is very sensitive, and the precision is about 0.5 μm. The thickness of the treated fabric was evaluated with the Kroeplin 67 device (for a precision of 0.5 μm). Ten measures were taken for every sample. The air permeability for the treated fabrics is evaluated with TESTEX brand textile fabric air permeability tester TF164 which conforms to GB/T 5453, ISO 9237 and ISO 9073:15–2007. The crease recovery properties were determined for dried finished fabric according to the standard norm ISO 2313 NFG 07–110 by means of a crease recovery tester model M003A SDL Atlas. The used load was 0.5 kg for 5 min, at temperature room. The samples for crease recovery measurements were typically cut according to a special standard rectangular shape. The recorded value of crease recovery angle is the typical of five measurements. The visual assessment of the degree of 3D effects; we referred to the ISO 9867 standard. The mechanical properties of treated samples are evaluated in warp direction with the MTS Criterion™ Universal Testing Systems after conditioning during 24 h in the relaxed state (22 °C, 60% HR) according to the norm ISO 13934-1, NFGS 07–129-1.

2.2 Treatment Application

The products are synthesized by natural fermentation, having the advantage of being biodegradable and increasingly regulated to limit the impact on the environment and health. They are not hazardous to handle, store or use under normal conditions [9]. Two types of ecologic compounds used recurrently in industry are the subject of this study. They are "GALACID XT 88 "based on lactic acid and "GALASOLV NF 62" based on based on ethyl lactate (bio solvent). More details about chemical and physical properties are briefly presented in Table 1.

Table 1. Chemical and physical properties of ecological compounds

	GALASOLV TM NF-62	GALACID XT 88 (Lactic Acid 88% Indus.)
Raw Formula	$C_5H_{10}O_3$	$C_3H_6O_3$
Molar weight (g .mol^{-1})	118.7	90
Density	1.030	3.86
Viscosity (mPa.s) à 25 °C	22	–
Melting tempearture (°C)	53	53
boiling tempearture (°C)	154	122
Water Miscibility	Miscible with water	Miscible with water, alcohol, glycerol and furfural, an alcohol-ether solution. Slightly soluble in ether. Insoluble in chloroform, petroleum ether, carbon disulfide.
Solubility	Soluble in water and most organic solvents	Soluble in water and other organic solvents
Flash point (°C)	61.5	110

Table 2. Mechanical and physical characteristics of raw fabrics

Characteristics of raw fabrics		value
Thickness (mm)		0.8
Air Permeability ($cm^3/cm^2/s$)		2.86
DCRA	Warp/Weft (°)	180
Tensile strength	Maximum Strength (N)	548.64
	Maximum Elongation (%)	12.73
Shear modulus KES-FB1	G (cN/cm.degré)	0.62
	2HG (cN/cm)	1.84
	2HG5(cN/cm)	2.53
Flexural modulus KES-FB2	B ($cN/cm^2/cm$)	0.179
	2HB (cN.cm/cm)	0.169
Compression modulus KES-FB3	WC ($cN.cm/cm^2$)	0.2
	LC	0.38
	RC (%)	29.07
Area modulus KES-FB4	SMD (μm)	6.475
	MIU	0.175
	MMD	0.028

To study the influence of the finishing compound in terms of nature and concentration and curing temperature, we used denim fabric as a textile support. This fabric having as composition (95% cotton, 5% Lycra), having as weight 326 (g/m^2), more characteristics of raw fabrics are summarized in Table 2.

In the experimental application, "GALACID XT 88 "based on lactic acid and "GALASOLV NF 62" based on ethyl lactate "LE", present two ecological compounds, they have undergone the same method of impregnation application. The compounds are dissolved into water at various concentrations.

More details for the finishing conditions process, such as curing temperature, drying time explain the finishing treatment. More details are summarized in Table 3.

The idea of the 3D effects: "whiskers", "honeycombs" and "stacks" in the case of trousers which is the focus of this study, is taken from the worn-out lines and impression patterns generated by natural wearing [7, 8]. The compound treatment is generally applied specially to obtain 3D wrinkle (or 3D effects). After finishing application, the treated sample is fixed on a special bracket (model), where created 3D effects on special place in fabric. The treated sample underwent a predrying oven at 100 ° C for 20 min in a device called 'Margherita', before compound treatment that's considered important steps. After treatment, the fabrics samples were desized with amylase "Ecoprep" 2 g/L and softened with "CHTTACC" 1 g/L for 10 min at 50 °C, washed with enzyme "Novasi ultra-MC/M" 2 g/L for 20 min at 50 °C, Fabric to liquor ratio used was 1:20. Finally, they are dried for 20 min at 90 °C and conditioned.

Table 3. Treatment Ecological compound conditions

	Ecological compound conditions					
	GALACID 88 XT			GALASOLV 62NF		
Compound concentration (g.l^{-1})	60	100	150	60	100	150
Soaking Time (min.)	10			10		
Dry temperature (°C)	90			90		
Dry time (min.)	15			15		
Curing temperature (°C)	100–140			100–140		
Curing Time (min.)	30			30		

3 Results and Discussion

3.1 Effect of Compound's Characteristic and Finishing Conditions on Surface Morphology of Fabric

SEM micrographs show the morphological surface of cotton fabrics treated with "GALACID XT 88" and "GALASOLV NF 62" ecological compounds with various concentrations (60, 100 and 150 g/L) at 100 °C and 140 °C respectively.

According to the SEM graphs, it can be seen that this ecological compound is deposited on the fibrous structure in the shape of a film and creating inter-fiber bridging whose intensity varies according to the concentration, except without modifying the morphology of the surface of the fibers. The intensity of these inter-fiber bridging proves to be greater in the BW state. It's clear for ecological compound GALACID XT 88 based on lactic acid, when we applied a high curing temperature 140 °C and high compound concentration (100 g/L), generates a small cracking in the treated fiber. On the other hand, and under the same conditions, the surface of the treated fiber is more covered with viscous product which coats it with the "GALASOLV NF 62" based on ethyl lactate "LE" and there is no damage to the surface.

3.2 Effect of Ecological Compound's Characteristic on the Air Permeability

Figure 2 shows Air permeability evaluation at curing temperature of 100 °C and 140 °C and with various compound concentrations, in two steps before BW and after washing AW. It is clear that after washing AW, the value of the air permeability of the samples treated with ecological compounds are greater than those before washing BW. This may be due to the fact that a certain amount of the compound is removed from the surface of the treated fiber which makes the passage of air through the fabric surface easier and consequently high air permeability.

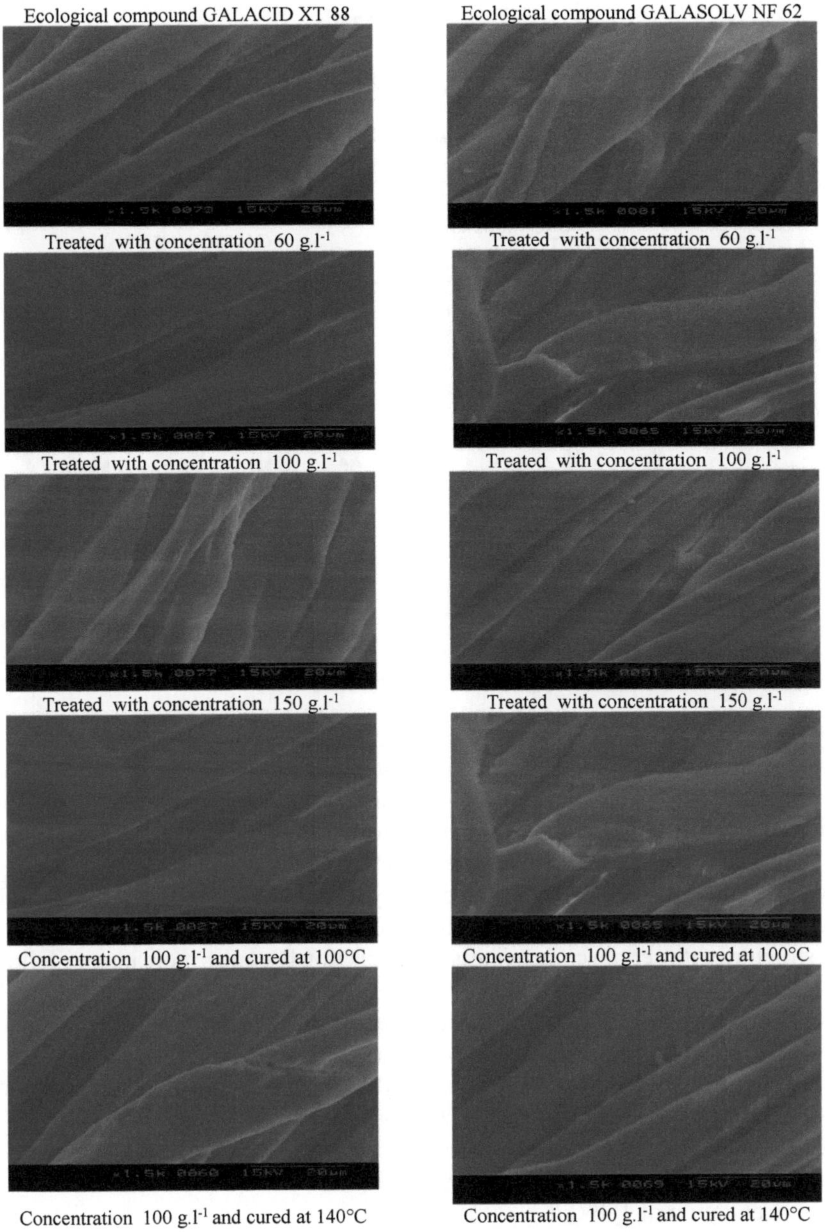

Fig. 1. Surface Morphology SEM of treated cotton fiber with ecological compound GALACID XT 88 and GALASOLV NF 62. All samples have 1500 times magnification.

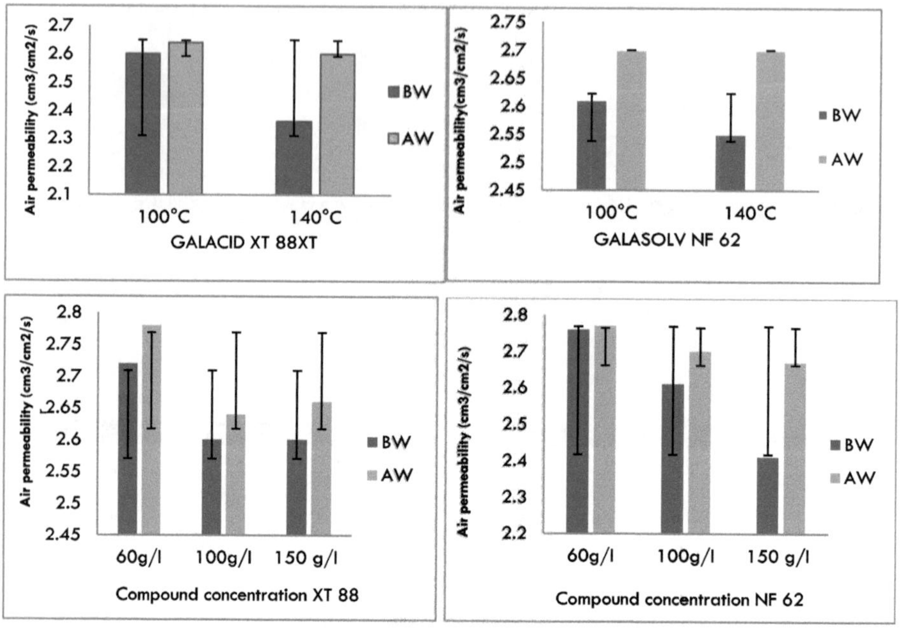

Fig. 2. Air permeability evaluation at curing temperature of 100 °C and 140 °C and with various compound concentrations in two steps before BW and after washing AW

Figure 2 appear that when curing temperature increase the treated specimens present a slight fall in air permeability compared to the raw state. This fall continues to have effect but with little intensity when the compound concentration increase. The textile support, being subjected to the action of the compound which coat the structure and is deposited at the level of the inter-wire spaces, block the empty spaces and immobilizes the amorphous and flexible region by creating cross-linking bridges, which makes it possible to develop a rigid structure with certain compactness. This reaction becomes more intense following an increasing of the curing temperature, which favors the reactivity of the compound and makes the fibrous structure more compact and united, as a result, it favors the reduction of the interstices and strengthens the opposition of the air way.

The effect of two compounds is similar, wherever the permeability marks a slight increase; this is mainly due to the plasticizing behavior of the compound based on lactic acid which ensures at the same time a good closing of the structure, although in addition has a good aeration property. This finding is reflected by tolerable air permeability even with a concentration of 150 (g/L).

3.3 Effect of Ecological Compound's Characteristic on Thikness of Treated Fabric

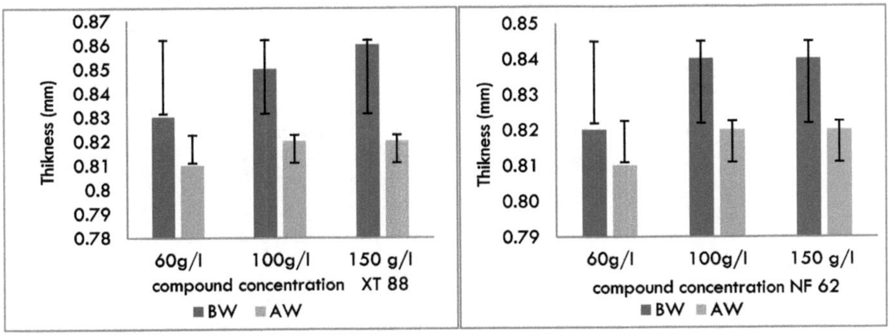

Fig. 3. Thikness of treated fabric at various compound concentrations in two steps before BW and after washing AW

Figure 3 show the thickness of treated fabric at various compound concentrations in two steps before BW and after washing AW. It is distinguished that the thickness of the treated specimens increases before washing BW according to the increase of the compound introduced, this could be justified that the increasing quantity of compound introduced, developed new inter-chain bonds and allows the formation of three-dimensional polymers, which will occupy the empty spaces of the structure, which weakly contributes to increase the thickness of the structures. The same results we have shown in previous studies, except using conventional resins based on glyoxal and acrylic [10–12]. The treatment in an aqueous medium and the high temperature causing the dissociation of the hydrogen bonds as well as the formation of other hydrogen bonds ensure the swelling of the structure (cellulose), this swelling being limited under the action of the cross-linking agent, which blocks the microstructure of the fiber. Mutually compounds show the same thickness variation behavior, with increasing concentration. The compound based on lactic acid, which could be brought back in the form of a film following cross-linking and which has a good ability to preserve an exact dimension, are the main causes of the increase in the thickness of the treated structures, its plasticizing character's which is manifested by softening during treatment and swelling of the treated structure.

3.4 Evaluation of Ecological Compound's Characteristic on 3D Effects of Treated Fabric

Figure 4 show the 3D effects of treated fabric with GALACID XT 88 and GALASOLV NF 62 at curing temperature of 100 and 140 °C.It's clear that, at high curing temperature of 140 °C, we noted that treated specimens give moderately appreciable 3D effects and which become slightly developed at curing temperature of 100 °C. The treated specimens always keep their basic nuances and present a soft and satisfying touch even following a treatment at 140 °C.This comment could be due to the plasticizing behavior

GALACID XT 88 with curing temperature 100 °C

GALACID XT 88 with curing temperature 140°C

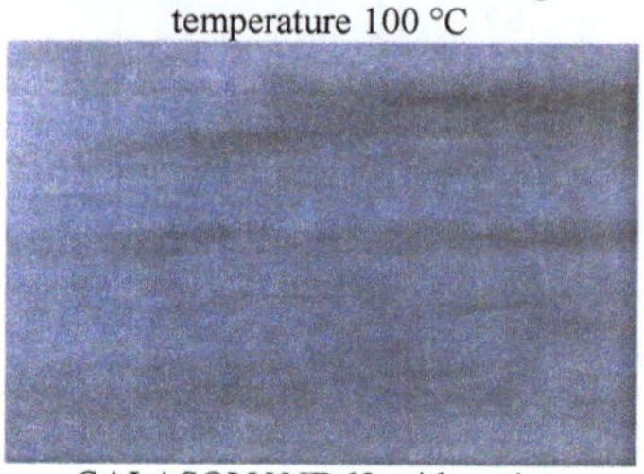

GALASOLV NF 62 with curing temperature 100 °C

GALASOLV NF 62 with curing temperature 140 °C

Fig. 4. 3D effects of fabric treated with GALACID XT 88 and GALASOLV NF 62 at curing temperature of 100 and 140 °C.

of the GALACID XT 88 compoundbased on (lactic acid), which ensures an excellent desirable touch for this application, covering the structure with a thin film which gives it a certain softness and extensibility.

From Figure 5, it is clear that the 3D effect improves by increasing the concentration of the introduced compound. There are appreciable wrinkles appear with a concentration of 60 (g/L) and which become progressively developed with a concentration of 150 (g/L).

The specimens treated with the compound "GALASOLV NF 62" based on ethyl lactate "LE", show a little developed 3D effects which persist moderately following a washing treatment and develop slightly following the increase of the curing temperature, while noting in particular the appearance shiny and soft of the structure. We noted that the treated specimens showed no tendency to yellow following after the application of these compounds, even under the action of a high curing pair (temperature/time).

3.5 Effect of Ecological Compound's Characteristic on the Mechanical Behaviors of Fabric

Figure 6 shows the evaluation of mechanical property loss with various compound concentrations in two steps before BW and after washing AW. Indeed, the main cause of the strength loss is the degree of cross-linking and the blocking of the structure which increases according to the quantity of compound introduced. The compound causes the immobilization of the amorphous and flexible zones of the textile fibers by creating bridges with the cellulose of the cotton, which prevents the arrangement of the molecular chains in the extension zone, the transmission of the tension to a number of weak macromolecular chains and subsequently the inability of the structural elements of the fiber to arrange themselves so as to resist to the mechanical stresses in the same way. The effect

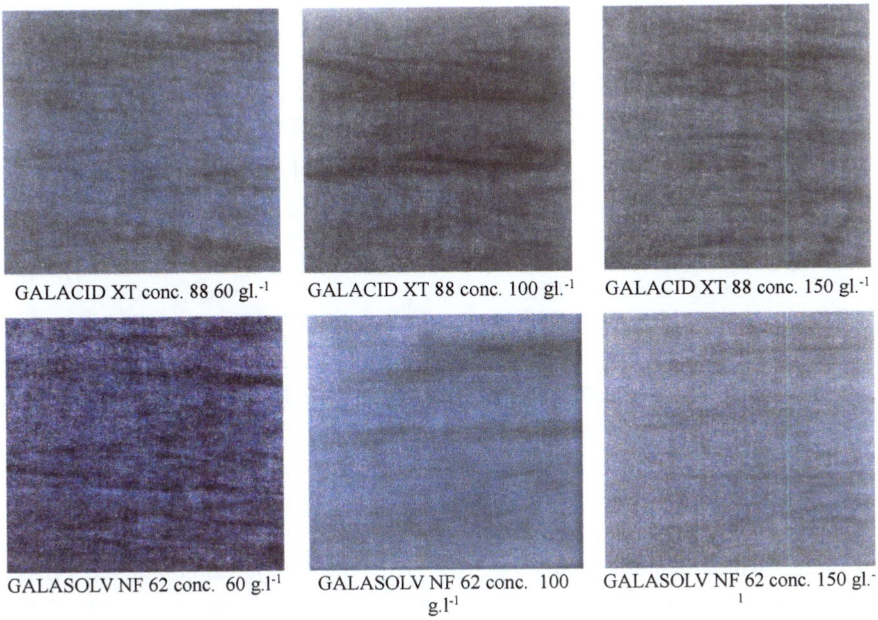

Fig. 5. 3D effects of fabric treated with GALACID XT 88 and GALASOLV NF 62 at various concentrations

becomes progressively accentuated by increasing the concentration of the compound. The mechanical properties losses for these two compounds are evaluated as not considerable as long as they do not exceed of the order of 30%. Indeed, lactic acid being a weak acid, endowed with good mechanical properties, high crystallinity and rigidity which form a film on the treated structure following its plasticizing action, gives the structure a certain cohesion and compactness which is reflected by slight losses. The combined action of the GALACID XT 88 compound based on lactic acid in addition to the curing temperature and curing time are the main factors that cause the fall in mechanical properties, which reached 25% with a compound concentration of 150 (g/L). GALASOLV NF 62 preserved the tensile strength retention, limiting strength loss to a maximum of 8%. The lower concentration impact compared to GALACID XT 88, suggesting less fiber immobilization. After washing, stabilization further confirms improved durability, making NF 62 suitable for long-term textile applications. Other researchers, also state acrylic resin treatments improved tensile and reduced fabric extensibility after washing [10, 15]. Alternatively, this finishing resin, such as DMDHEU or others resins with little amount of formaldehyde imparts a negative effect on the mechanical properties of the finished fabric, such as tensile strength confirmed by many researchers [14, 16, 17].

Certainly, the exhibition of treated fabric with compound based on lactic acid, for a prolonged period of time and under the application of high curing temperature, causes the loss of mechanical properties. This observation was confirmed, previously, by the micro-cracks already seen on the surface fiber treated with this compound (SEM micrographs in Fig. 1). These results proved the sensitivity of this ecological compound to high

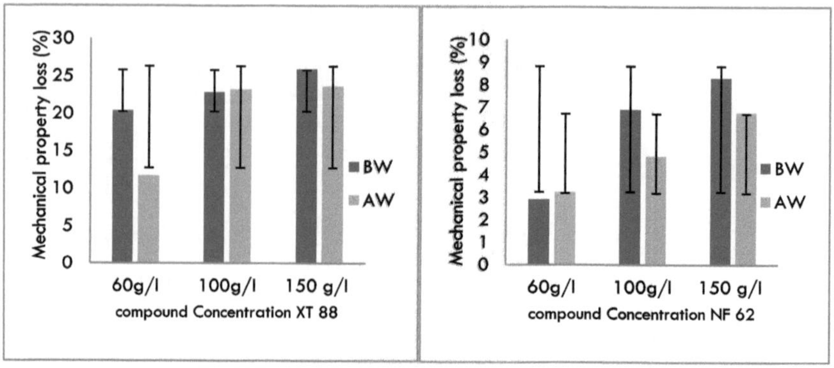

Fig. 6. Evaluation of mechanical property loss with various compound concentrations in two steps before BW and after washing AW

temperatures and which is decisive for this application process. In addition, we add to all this factors, the sensitivity of the cellulosic fiber mixed with elastomeric "Lycra" to high temperatures which degrade their properties [13]. The ecological compound "GALA-SOLV NF 62" based on ethyl lactate "LE" has an important characteristic, it evaporates very slowly which proves that its exposure to this curing temperature, the selected time interval allowed its vaporization without attacking and altering the structure, this makes clear the results of little mechanical loss when concentration of compound increase.

3.6 Evaluation of Ecological Compound on the Mechanical Properties of Fabric with KAWABATA Modules

In fact, it's very important to estimate the finishing fabric treated with chemical compounds, other researchers are investigated the finished support to clarify the textile mechanical properties using KAWABATA Evaluation System [3, 10, 16].

Figure 7 shows the evaluation of ecological compound on the mechanical properties of fabric by the four KAWABATA modules of treated fabrics, in two steps BW and AW. We know that Shear stiffness consists of a parallel deformation with respect to two opposite sides, is an indicator of the strength fabric following low rotational stresses, the lower it is, and the more it describes a flexible and more suitable character to the drape.

Refer to Fig. 7; it's clear that treated specimens with the two compounds have shear strength close to the threshold values facilitating the handling of the fabric as well as comfort in wearing. This magnitude, measured in the weft direction, increases as a function of the concentration of the two compounds. Actually, the greater the concentration, the more intense its action on the immobilization of the amorphous zones of the fiber by bond bridging, which makes the degree of greater cross-linking, the more its action on the interactions between the warp and weft elements of the structure increases, the intensity of which increases during shear stresses to cope with them. Lactic acid, known for its significant plasticizing action, which is manifested by an increase in the extensibility of the textile support, gives it a good attitude to shear. In addition, the presence of the elastomeric fiber in the weft element gives the structure good elasticity properties. These two compounds exhibit good cross-linking stability, as a quality factor. The shear strength shows a slight drop following the washing treatment, which shows good stability of the cross-linking of these compounds after the washing test.

The flex test is a very important test for evaluating the quality of a fabric. Flexion describes the capacity of resistance to deformation under stress normal to the surface of the fabric. Indeed, the bending stiffness reflects the flexibility of the fabric. Otherwise, the higher it is, it indicates a significant resistance to bending. The specimens treated with the two compounds based on lactic acid and "LE" shows a moderately significant increase in the resistance to bending B, following their increase in concentration, where their action proves to be greater in blocking the areas amorphous and flexible of the fibers thus limiting by bridging of connections and consequently the movements of the microstructure. This observation translates into an increase in the resistance to bending to oppose normal stresses on the surface of the fabric while remaining with a value close to the raw state. This magnitude marks stability for the compound based on lactic acid and a slight drop in the washed state using the compound based on "LE" because the washing treatment decreases the rigidity of the structure and confers certain flexibility by dint of the used products.

One of the essential characteristics to be studied for fabrics treated with compounds is the behavior in compression which would be decisive for the use. The behavior of compression following the application of the two compounds is visualized in the graphs (Fig. 7) as a function of the concentration of compound introduced. Indeed, the compression energy WC, which provides information on the feeling of touch, indicates that the more the material dissipates energy, the more it has a soft touch and will contribute to the feeling of comfort when sitting. The treated specimens with these two compounds show good compression energy values at medium concentrations 100 (g./L) and which gradually fall as the concentration increases. The compound based on lactic acid, covers the structure with a thin and elastic film which gives it a soft touch. In the process of compressing the fabric, the inter-fiber space is continuously decreasing, the force required to compress it must be greater than the internal stresses of the yarns and the inter-fiber friction strength. The compound introduced generates the cross-linking of the cellulose and the formation, thanks to numerous bridging of bonds, of a rigid macromolecular network, the greater the compound present, the more it is responsible for the cohesion and adhesion between fibers and by the inter-son sequence. This analysis explains the decrease in WC and the increase in compressive strength following the increase in compound concentration. Indeed, when increasing the concentration, the inter-thread space decreases which intensifies the internal stresses of the threads and the frictional strength, consequently the WC decreases, indicating a softer touch at lower concentrations.

The SMD roughness reflects the surface roughness of the fabric, the lower it is, the more uniform the surface of the fabric and less rough. SMD graphs visualize (Fig. 7), its behavior following a variation of the compound, which always depends on the fabric on which it will be applied. The application of the compound based on lactic acid and based on "LE" reduces the roughness of the treated textile structure compared to the raw state. The increase in the concentration of the compound "GALACID XT 88" makes the SMD value stable, which indicates that an increase in the concentration of this compound does not affect the microstructure of the fiber as well as its morphology and subsequently the characteristics of the textile support. Indeed, a higher compound concentration does not

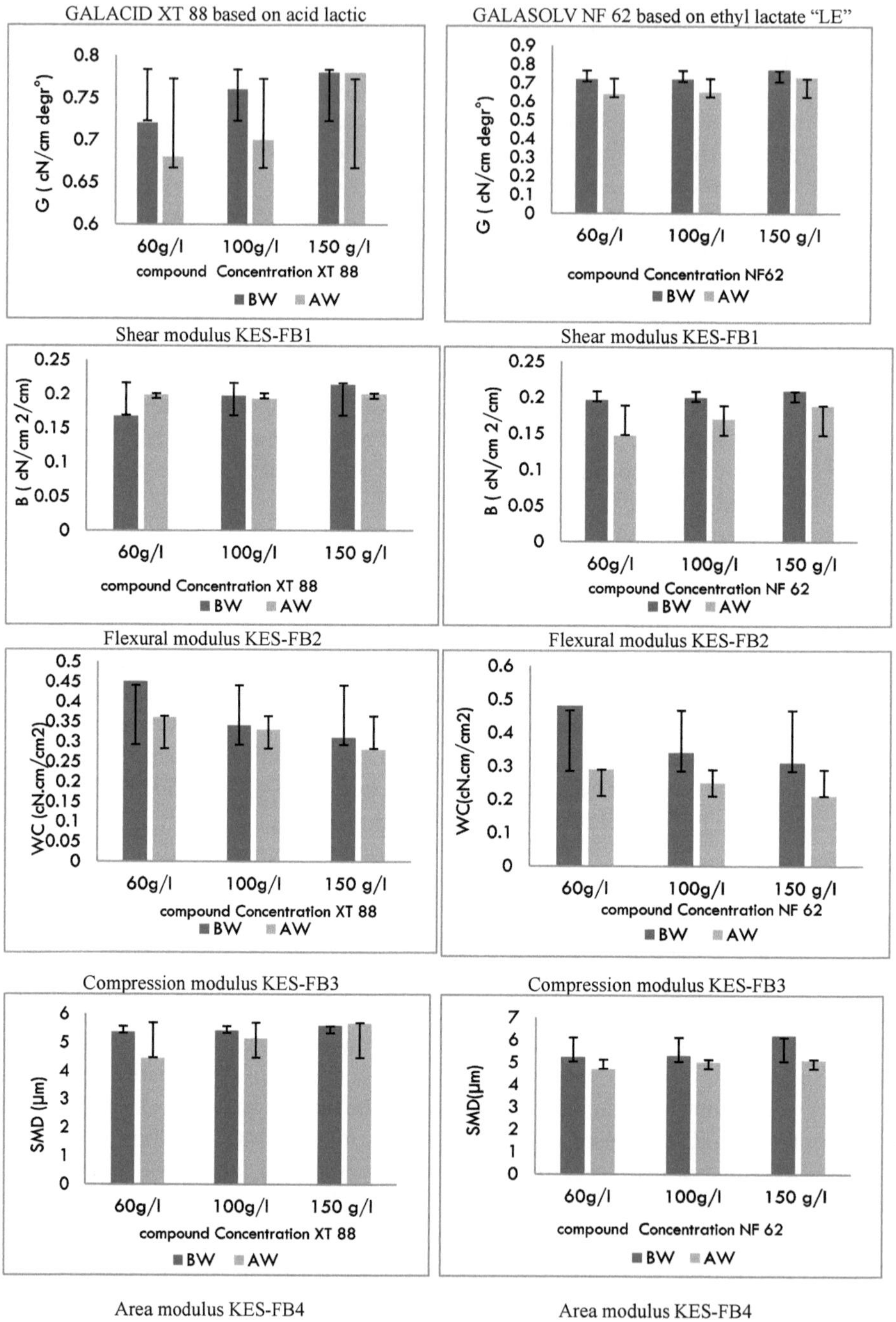

Fig. 7. Evaluation of ecological compound on the mechanical properties of fabric by the four KAWABATA modules of treated fabrics in two steps BW and AW.

cause an increase in SMD, which indicates that the surface has not become rough due to the hardening of the compound. According to the (SEM graphs Fig. 1), it can be seen that the lactic acid covers the fibers of the structure with a film which covers the cellulosic fiber and fills the inter-fiber spaces with compound, which are more visible in the unwashed BWstate, while in the washed state AW, without modifying the morphology of the fiber. The most important aspect to note, that this compound does not cause an alteration of the structure, which is generally manifested by the appearance of bigger crack in the fibrous surface which causes the roughness of the structure, this aspect persists even at high concentrations.

The behavior of the structure by application of the "GALASOLV NF62" compound leads in particular to stabilization at the level of roughness, which allows us to conclude that the introduction of significant proportions of this compound does not cause the alteration of the structure. According to the SEM graphs, it can be seen that this compound is deposited on the fibrous structure in the form of a film and by creating inter-fiber bridging whose intensity varies according to the concentration, but without modifying the morphology of the surface fibers. The intensity of these inter-fiber bridging is greater in the before wash state BW.

3.7 Effect of Compound's Characteristic on DCRA Fabrics

The Fig. 8 shows the evaluation of ecological compound on DCRA of treated fabric (warp/weft) at various concentrations in two steps BW and AW. It is clear that at the two steps of finishing processes: before and after washing, the Dry Crease recovery angle DCRA values enhance after compound treatment (compared to untreated fabrics). Also, the increase in the compound concentration induces a proportional raise of the DCRA values. This consideration confirmed by another researcher [10, 18].

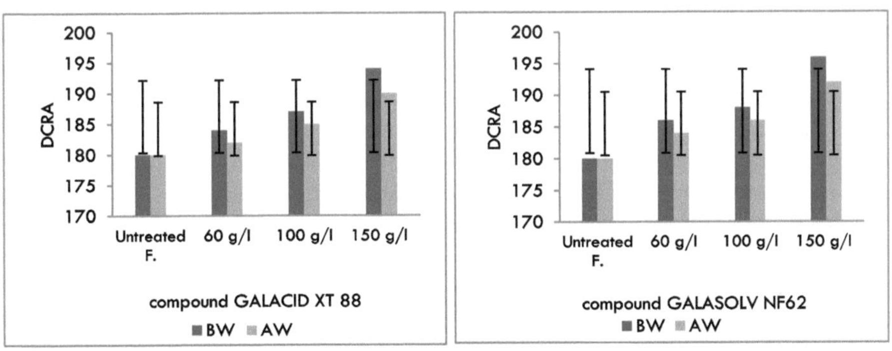

Fig. 8. Evaluation of ecological compound on DCRA of treated fabric at various concentrations, in two steps BW and AW.

The effect of the compound "GALASOLV NF 62"is more important than GALACID XT 88 ones on the DCRA before and after washing, particularly if the sample was subject to a higher concentration. In spite of compound type, DCRA values BW are higher than those obtained AW. It can be clarified by the effect of washing additives that reduces the ability of fabric to wrinkle, because of the elimination of the rest of non-fixing compound on the surface of cellulosic fibers. Additionally, it is possible that water leads to the relaxation of the strength of the fiber and inter- and intra-fiber that it assists to decrease the capability to crumpling GALASOLV NF 62 has a more significant effect on DCRA compared to GALACID XT 88 (Figs. 7). DCRA can reach higher values 180 (°C), when applying higher concentrations 150 (g/L) and especially with a higher curing temperature. Possibly, the size of GALASOLV NF 62molecules play asignificant function after having cross linked with cellulose chains of cotton; this promotes the ability of finished fabric to wrinkle at the two steps: AW and BW Consequently GALASOLV NF 62 reduces the mechanical property (explained previously).

4 Conclusion

The compound "GALACID XT 88" based on lactic acid which is a renewable compound, like most bio compounds, known for its hydrophilic nature which allows it to interact with cellulosic fibers, this property is due to the presence of polar hydroxyl functions which have a high reactivity. The application of this compound has enabled us to achieve these highly sought-after properties of touch, shine, transparency, stability, thanks to its plasticizing character, but also to its biodegradable, biocompatible and bioresorbable character, since it is a well tolerated product and does not present any toxicity towards the human body as a handler or also a user. This limits its use, its ability to withstand high temperatures, which is decisive for its application. The compound "GALASOLV NF 62" based on ethyl lactate "LE" as being a green solvent synthesized from lactic acid which is generally used in mixtures with other solvents, mainly in the industry where its presence in varnishes increases their fluidity and gloss. Following the study of substitution as a textile finishing product allowed us to achieve the same results obtained by that of "GALACID XT88" but with a greater improvement in the mechanical characteristics. The minority significant effects are expected in these studies apart from irritation of the respiratory tract, following its application, its smell, detectable at low concentration. We conclude that the ecological compounds gives a suitable quality of 3 D effects, as well, they can be used in the finishing of denim articles because they preserve the mechanical properties of treated fabrics and at the same time a slight improvement in the quality of the structure, distinct the usual toxic resins which generate excessive losses in the mechanical properties of treated fabrics and the desired final quality. Compared to the conventional resins, these ecological compounds provide a safer textile finishing alternative. These green finishing compounds make denim more fashionable, durable and sustainable without risky chemicals. They present a best solution for eco-friendly textile production.

Acknowledgements. The authors would like to thank, especially of Chimitex Plus Company (Tunisia), and the team of Sitex Company (Tunisia) for providing materials for investigation. Finally, sincere thanks are also addressed to the scientific team of Textile Physics and Mechanics Laboratory LPMT (Mulhouse, France) for their support.

References

1. Horrocks, A.R., Anand, S.C.: Handbook of Technical Textiles, pp. 540–548. Woodhead Publish. Limit. in assoc. with Textile Institute, Cambridge, England (2000)
2. Naeem, F., Asim, F., Tufail, M.: Investigation into the effect of resin finish on the functional characteristics of plain fabrics using different curing methods. Pigment Resin Technol. **52**(2), 192–202 (2021)
3. Beyene, K.A., Gebeyehu, E.K., Adamu, B.: The effects of pretreatment on the surface roughness of plain-woven fabric by the Kawabata Evaluation System. Text. Res. J. **93**(9-10) (2023)
4. JE. Opre, EP. Bergemann and M. Henneberry., U. S. Patent 6191087, (2001).
5. M. Henneberry, JA. Snively, GJ. Vasek, and R. Datta., U. S. Patent 6797684, (2004).
6. Chang, J.: Chem. Mark. Rep. **267**, 4 (2002)
7. Udomkichdecha, W., Kittinaovarat, S., Thanasoonthornroek, U., Potiyaraj, P., Likitbanakorn, P.: J. Text. Inst. **73**, 401 (2003)
8. Ho, L.-Y., Chi-Wai, K.: Effect of Resin Finishing on Some Properties of 100% Cotton Light Weight Woven Fabric. Coatings. **12**(11), 1791 (2022)
9. J. L. WERTZ, ValBiom Gembloux Doc. Agro-Bio. Tech., pp. 12–17, Note de synthèse, (2011).
10. Litim, N., Baffoun, A., Khoffi, F., Hamdaoui, M., Abdessalem, S.B., Durand, B.: Effect of finishing resins on mechanical and surface properties of cotton Denim fabrics. J. Text. Inst. **108**(5) (2017)
11. Litim, N., Baffoun, A., Abdessalem, S.B.: Impact of modified DMDHEU and copolymer acrylic resin using spraying treatment before and after an enzymatic washing on the mechanical properties of Denim cotton fabric. J. Polym. Text. Eng.., IOSR-JPTE. **3**(4), 24–34 (2016)
12. Litim, N., Baffoun, A.: Investigation of acrylic resin treatment and evaluation of cationic additive quality impact on the mechanical properties of finished cotton fabric. Int. J. Eng. Adv. Technol. IJEAT. **5**(6), 76–84 (2016)
13. Schindler, W.D., Hauser, P.J.: Chemical Finishing of Textiles, Chapitre 5: Easy Care and Durable Press Finishes of Cellulosics, pp. 54–70. Woodhead Publishing Limited and The textile Institute (2004)
14. Tomasino, C.: Chemistry and Technology of Fabric Preparation and Finishing, pp. 106–115. NC State University, Coursebook, Raleigh, U. S. A (1996)
15. Dannie, S., Stylios, G.K.: Cotton fabric mechanical properties affected by post-finishing processes. Fiber Polym. **13**(8), 1050–1057 (2012)
16. Naujokaityte, L., Strazdiene, E.: The effect of finishing upon textile mechanical properties at low loading. Mater. Sci. (Medžiagotyra). **13**(3) (2007)
17. Ibrahim, N.A., Abo-Shosha, M.H., Elnagdy, E.I., Gaffar, M.A.: Eco-friendly durable press finishing of cellulose-containing fabrics. J. Appl. Polym. Sci. **84**, 22–43 (2002)
18. Wang, C., Chen, J.C., Yao, W.H., Chen, C.C.: Cross-linking of cotton cellulose in the presence of Alpha-aminoacids, Part III: Pore structures. Text. Res. J. **76**(4), 336–342 (2006)

Textile Industry 4.0

The Effects of Digital Preparation Before the Exploitation of AI in Tunisian Clothing Companies

Slim Boussaadoun[1(✉)], Sejri Nejib[2], and Mohamed Hamdaoui[1]

[1] Université de Monastir, ENIM, Unité de Recherche Matériaux et Procédés Textiles (MPTex), UR17ES33, 5019 Monastir, Tunisie
slimbinz@gmail.com

[2] Textile Engineering Laboratory (LGTex), ISET of Ksar Hellal, Monastir University, Monastir, Tunisia

Abstract. Today, the evolution of the textile and clothing industry has several organizational, social, qualitative and quantitative challenges arising from economic, industrial and technological changes and draconian market demands. These challenges have an impact on the implementation of an artificial intelligence system. In this context, this study aims to use the gambling and stratified sampling methods to evaluate the digital performance of five Tunisian clothing companies, with the aim of selecting the most efficient ones. We reached the following results: E1 = 49.25%, E2 = 44.25%, E3 = 45.75%, E4 = 33.75%, and that company E5 had the best score of approximately 62%, revealing a DP < 50%. While E5 achieved a DP ratio of approximately 62%, this enabled us to prepare the company to address AI technologies, such as Industry 4.0, the Internet of Things, and others, in a simple and sustainable manner. Not only that, but in this research, we also focused on six pillars (leadership, culture and organization, technology management, data management, measurement system, and customer experience). Finally, for a good digital preparation of the company, as well as to combat risk factors such as absenteeism, overall company performance, sources of non-performance, errors, omissions, defects, organizational factors, etc., this method offers the potential, along with other organizational tools, to verify their impact on the preparation of companies before implementing digital systems such as Industry 4.0 and the Internet of Things (IoT).

Keywords: Gamache method · digital performance · Industry 4,0 · Internet of Things · artificial intelligence

1 Introduction

The swift development of digital technologies is currently driving major changes in products, services, innovation processes, business models, and the nature of commercial activities in industrial ecosystems [1]. Furthermore, the production, quality, and financial information system plays a pivotal role in providing the necessary information to

M. Hamdaoui et al. (Eds.): ITDM'3 2025, PTT, pp. 221–230, 2026.
https://doi.org/10.1007/978-981-95-7704-0_20

company managers. Therefore, the successful implementation of a digital evaluation system requires management commitment, and a strong organizational culture is crucial [2]. Including AI such as data management technologies, communication and connection techniques [3]. According to a Gartner report published in October 2024, by 2028, 15% of daily business decisions will be made autonomously by AI agents. Similarly, according to a McKinsey study conducted in November 2024, business units related to customer service, marketing, sales, software engineering, and research and development will represent 75% of the total annual value generated by studied AI use cases. On the other hand, Gartner predicts that 90% of companies will adopt a hybrid cloud approach by the end of 2027. Edge computing, an extension of cloud computing, is also gaining importance. This technology, which brings data closer to its source, will be critical for applications that require low latency, such as autonomous machines and the Industrial Internet of Things [4, 5]. The impact of digital transformation on businesses is significant. Technological innovation is an important means for businesses to gain competitive advantages [6]. In addition, digital transformation can also impact technological innovation by integrating these different dimensions. It also impacts important characteristics such as technology and the efficiency of information systems use within small and medium-sized enterprises [7]. Furthermore, Al-Mutawa and al studied the adoption of cloud computing by SMEs and its impact on their sustainability, and their research identified several factors, such as cost reduction and ease of use, that affect the sustainability of SMEs through the adoption of this technology [8, 9]. While Zhang and al explored the key factors and interaction mechanisms that influence the success of digital transformation of SMEs in China, highlighting the importance of technological, organizational, and environmental factors in this process. Digital technologies, such as artificial intelligence and the Internet of Things (IoT), play a vital role in improving business performance [10]. Integrating these technologies helps automate tasks, improve internal and external communications, and optimize operational processes. For example, companies that adopt AI solutions for customer service often experience increased customer satisfaction and, consequently, increased economic performance [11]. In this context, the importance of choosing an assessment method is highlighted. There are several assessment methods, such as the MIT-Capgemini (Managed Infrastructure Services) method for studying the digital maturity of Swiss industrial companies, which focused on digital intensity (technology) and digital culture (organizational skills) [12], and the Digital Maturity Model (DMM) from Berghauss & Back in 2022, which is based on customer experience, innovation, and operations [13]. The Gamache model is an interesting tool, especially in the context of small and medium-sized enterprises (SMEs), for assessing a company's digital maturity. It is generally based on an analytical grid that examines several dimensions of a company's digital technology: strategy, technology, skills, digital culture, and more [14]. This method covers several important areas (not only technological tools, but also culture and processes). It is designed to be used even by non-digital specialists. Moreover, it is suitable for SMEs: it takes into account the reality of small structures. For this reason, company selection is crucial, which is carried out by assessing digital maturity and performance, according to the organizational changes necessary for the successful deployment of digital technologies [15]. And according to a 2019 Gamache study, the digital maturity of textile companies in 3 countries such as Canada, Germany and France is around 49.5% [16]. To meet these challenges, the study proposes the adoption of the 6S principles (Sort, Put in order, Shine, Standardize, Maintain and Secure) [17]. The use of new technologies such as IOT (Internet of Things), the intervention of This research aims to analyze the influence of management commitment and organizational culture on the implementation of a hybrid approach to apply it and improve the conditions of digital performance of companies. Ultimately this cross-sectional exploratory study also requires: integrating Lean tools such as: 5S, 5 M, 6S, visual management, Andon…) [18]. In order to prepare the ground for the application of artificial intelligence in this study.

2 Methods of Work and Materials Used

In this study, we started with the preparation of the working team which is multidisciplinary according to the challenges of AI. Then we used a hybrid participatory approach with the Gamache method was designed to implement new artificial intelligence techniques. This work was based on two phases: In the first phase, we administered a digital maturity quiz to the company's employees. This survey contained 109 questions, divided into six axes based on 28 criteria. n this study, we used a hybrid collaborative approach using leadership, culture and organization, technology management, data management, measurement system, and customer experience to identify the target company with the highest digital maturity scores. See the Table 1.

In this study, we also applied six important determinants for the evaluation of the digital performance (DP) of companies in several dimensions such as: (LDS: Leadership, CLS: Culture and Organization, TES: Technology Management, DMS: Data Management, MSS: Measurement System, EXS: Customer Experience). We also implemented this tool (Gamache quiz) in an anonymous, professional and practical manner. For the evaluation method, we used several factors such as: (age, gender, seniority, age brackets, degree of versatility, etc.). In addition, for the weighting of the scores with this model, we used a scale of 1 to 4, depending on the degree of appreciation, as shown in Table 1.

After defining the assessments and results, we will move on to explaining the equations related to companies' digital performance. In addition, the global numerical average formulas of Eq. (1) provide a general assessment for comparing companies' numerical performance [19]. In fact, the indicators of Eqs. 2 to 8 focus on the potential improvement factors for each dimension, as they facilitate the development of an action plan for the efforts made. In particular, it is plausible that a company masters certain dimensions more than others. This makes it possible to assess and refine the level of digital performance in order to avoid focusing on efforts or gains that have a limited impact on the organization as a whole.

The Equations: Formulas for Digital Maturity According to the Axes
In this study we applied the following formulas) according to the Gamaches method [15].

$$\text{The company}'\text{s digital performance} = \text{PN} = \sum_{i=1}^{n} \frac{Ci}{Nci}$$

Table 1. Gamache Quiz according to the 6 axis

Axis	N°	Criteria Code	Evaluation criteria digital performance	Score	Average Score
Leadership	**1**	**C1**	Strategic Vision		
	2	**C2**	Technology Monitoring		
	3	**C3**	Commitment and Exemplary Leadership		
	4	**C4**	New Business Models		
Technology Management	**5**	**C5**	Technology Mastery		
	6	**C6**	Cybersecurity		
	7	**C7**	Automation		
	8	**C8**	Digital Architecture and Ecosystem		
Customer Experience	**9**	**C9**	E-Commerce		
	10	**C10**	Service Orientation and Loyalty		
	11	**C11**	Personalization		
	12	**C12**	Co-creation		
Culture and Organization	**13**	**C13**	Lean and Work Organization		
	14	**C14**	Skills Acquisition and Development		
	15	**C15**	Change Management		
	16	**C16**	Culture and Organization		
	17	**C17**	Internal Communication		
	18	**C18**	Agility and Innovation		
	19	**C19**	Investment and Resources		

(*continued*)

Table 1. (*continued*)

Axis	N°	Criteria Code	Evaluation criteria digital performance	Score	Average Score
Measurement System	**20**	**C20**	External Outreach		
	21	**C21**	Operational Use of Data		
Culture and Organization	**22**	**C22**	Strategic Use of Data		
	23	**C23**	Data Collection		
	24	**C24**	Data Integrity and Quality		
	25	**C25**	Connectors, APIs, and other web services have been developed to promote data sharing		
	26	**C26**	Systems/tools are implemented to capture data in real time		
	27	**C27**	Interoperability is taken into account, particularly with IoT data		
	28	**C28**	Data Delivery		

Table 2. Digital Maturity Score according to Gamache

Assessment	Score
Non-existent	**0**
Beginner (in the initialization phase)	**1**
Advanced (in the process of being implemented)	**2**
Undergoing improvement)	**3**
Confirmed (operational device and/or	**4**
NA = Not applicable / Don't know	**/**

With Ci: The criteria
Nci: the number of criteria

$$\text{PN} = \frac{(\text{ LDS} + \text{CLT} + \text{TEC} + \text{DNE} + \text{MSR} + \text{EXP})}{6}$$

$$\text{LDS} = \frac{(\text{ C1} + \text{C2} + \text{C3} + \text{C4})}{\text{4 compoments of leadership}}$$

$$\text{TES} = \frac{(\text{ C5} + \text{C6} + \text{C7} + \text{C8})}{\text{4 compoments of technologies}}$$

$$\text{EXS} = \frac{(\text{ C9} + \text{C10} + \text{C11} + \text{C12})}{\text{4 compoments of costumer experience}}$$

$$\text{CLS} = \frac{(\text{ C13} + \text{C14} + \text{C15} + \text{C16} + \text{C17} + \text{C18} + \text{C19})}{\text{7 compoments of culture and organization}}$$

$$\text{MSS} = \frac{(\text{ C20} + \text{C21})}{\text{2compoments of measuring system}}$$

$$\text{DMS} = \frac{(\text{ C22} + \text{C23} + \text{C24} + \text{C25} + \text{C26} + \text{C27} + \text{C28})}{\text{7 data management compoments}}$$

Explanation of abbreviations:
DP: Digital Performance
LDS: Score associated with the Leadership dimension
CLS: Score associated with the Culture and Organization dimension
TES: Score associated with the Technology Management dimension
DMS: Score associated with the Data Management dimension
MSS: Score associated with the Measurement System dimension
EXS: Score associated with the Customer Experience dimension
C: From C1 to C28, these are the evaluation criteria

We also used the statistical method of stratified sampling [19] in this study which included a population of 300 people from a total workforce of 1200 people in 5 companies where the number of women is 1100 (91.66%) and men is 100 (8.34%) with a confidence level of 95% and an accuracy rate of 0.05. In the second step after selecting the company, we built a list of indicators with respect to confidentiality: absenteeism rate 11.40%, work schedule 8 h/day, expected defect rate < 5%, company average performance 75%, supervision rate about 10%, employee training rate 67.3%, applicable to evaluation during the working day (up to 8 h).

For further explanation, if the company has a DP score that is ≥2 or a percentage ≥ 50% which presents the minimum threshold to apply AI in this case we can say that it has an advanced digital maturity in this area. In addition, we deployed a continuous improvement tactic and measured the results according to tangible indicators [20] such as scores and percentages of the different axes. In addition, we used IT tools such as: XLSTAT-Premium, and Excel office 365, the statistical calculation application "statistic calculator", computers and smartphones to display the measured results.

3 Results et Discussion

Tunisian companies face numerous challenges, such as product diversity, constant model changes, defects at production sites, the use of advanced technologies, customer requirements, customer experience, etc. In this context, choosing the method of preparing companies is important for the subsequent application of artificial intelligence and improving business processes. After studying the average digital performance (DP) of companies, we arrived at the following results for the five companies (E1, E2, E3, E4, and E5) (see Table 3).

Table 3. Average digital performance of the 5 companies

Compagnies	E1	E2	E3	E4	**E5**
Average DP	1,97	1,77	1,83	1,35	**2,48**

Table 2 presents the average digital performance (DP) of the companies in this study. We note that the (PN) of the first 4 companies <2: This explains why they are in the initiation phase and are not yet ready to apply artificial intelligence. Although company E1 has made some progress in score 1.97; but it still has gaps and shortcomings to rectify so that it can optimize its (DP). While company E5 has an average score (DP) of around 2.48, which reflects its progress. In addition, E5 is quite developed in terms of monitoring its production and quality indicators in real time with Industry 4.0. This justifies our choice of company E5 in this study. All this leads us to say that company E5 was well prepared to welcome the application of artificial intelligence systems. In addition, Fig. 1 clearly shows the percentages of variations in digital performance for each of the companies.

Figure 1 clearly shows that the percentages of variations in the digital performance of companies are as follows: E1 = 49.25%, E2 = 44.25%, E3 = 45.75%, and E4 = 33.75%. These are percentages lower than the minimum threshold accepted for digital maturity (which is: 50%). This means that they are not yet prepared to apply artificial intelligence. This reveals their (DP) < 50%. While company E5 has a percentage of (DP) of around 62%, which allowed us to apply AI in a simple way. Not only that, but we also focused this research on the 6 axes (Leadership, Culture and Organization, Technology Management, Data Management, Measurement System, Customer Experience) in order to obtain the results below according to Fig. 2.

According to the radar chart in Fig. 2, we obtained the following results: E5 company has a good digital maturity, which is reflected in the following values: 75% Leadership, 68.75% Technologies, 62.5% Measuring System, 58.33% Data management, 57.14% Culture and organization and 50% customer experience. These results clearly show that we can move forward in this study since the ground is favorable to apply artificial intelligence. Moreover, the first three highest percentages highlight their relevance for the successful application of AI with Industry 4.0. This includes the number transformation as the introduction of Industry 4.0 in textile enterprises or other industrial sectors Knowing that the six key levers of the Gamaches method have shown great effectiveness

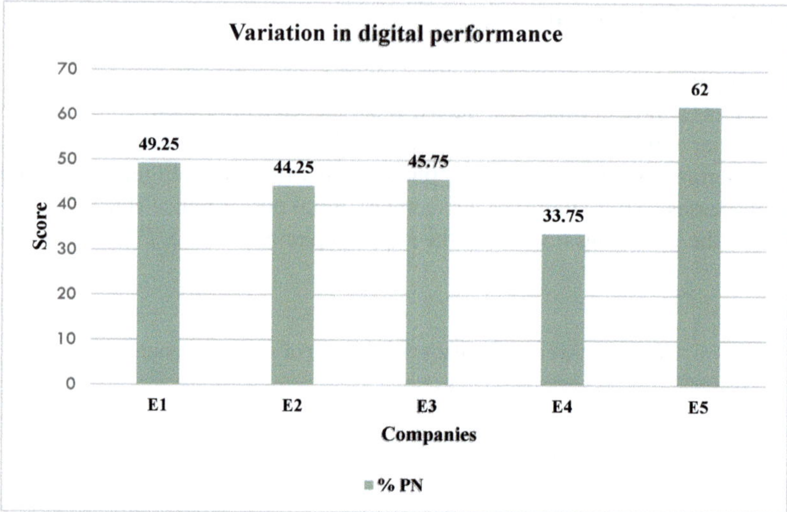

Fig. 1. Digital Variations in the performance of companies

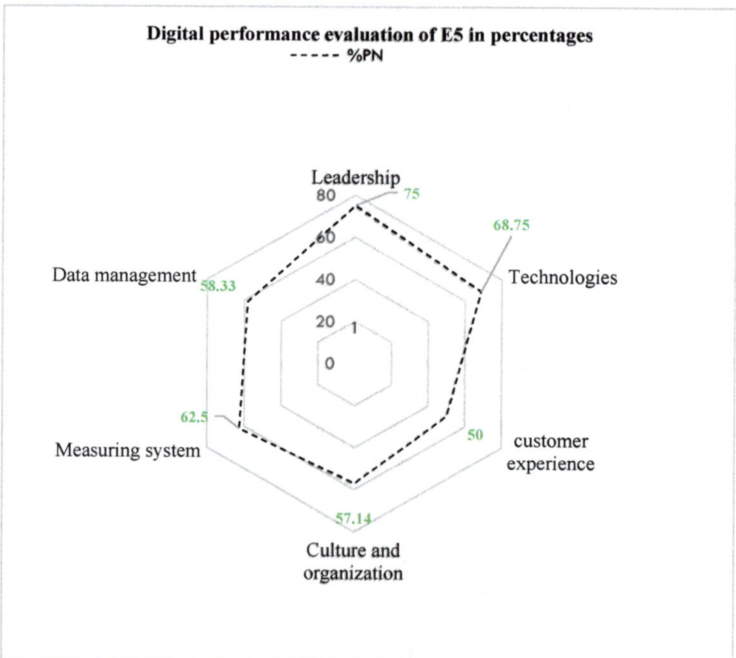

Fig. 2. Digital performance evaluation of E5 in percentages

if we compare them to other tools such as the MIT-Capgemini (Managed Infrastructure Services) method, which is a method for digital maturity and focuses on digital intensity (technology) and digital culture (organizational skills), or the Digital (DMM) method, which is based on customer experience, innovation, and processes. While our method takes care of everything for the good of the company and human beings. Finally, this text mentions the evaluation of digital performance, by associating organizational performance with digital performance, based on original ideas.

4 Conclusion

This study validated the effectiveness of the Gamache method to assess the maturity and performance of companies such as the E5 which has a digital performance of 2.48 and a percentage of 62%, as well as the application of the hybrid approach with artificial intelligence. It demonstrated its advantages for the measurements and results obtained by evaluating the actions of organizational factors according to six axes such as: 75% Leadership, 68.75% Technologies, 62.5% Measuring System, 58.33% Data management, 57.14% Culture and organization and 50% customer experience. These significant results at different percentages clearly prove the effectiveness of the Gamache method for preparing for the application of artificial intelligence systems such as industry 4.0. As it aims to bridge the dissonance between research and practical implementation in digital performance [21] in other sectors. Finally, this study showed that the Gamache method offers potential with other organizational tools such as Lean to validate its impact on the preparation of companies before the implementation of digital systems.

References

1. Kolagar, M., Parida, V., Sjodin, D.: Ecosystem transformation for digital servitization: A systematic review, integrative framework, and future research agenda. J. Bus. Res. **146**, 176–200. Published by Elsevier Inc. This is an open access article under the CC BY-NC-ND licens. https://doi.org/10.1016/j.jbusres.2022.03.067
2. Achmad Fadjar, Karhi Nisjar Sardjudin, Factors that influence decision making through the application of management accounting information systems, Int. J. Psychosoc. Rehabil., date 2023, Vol. 24 No. 2 https://doi.org/10.61841/hf1swx46
3. Jain, S.: Bibliometric analysis of internet of things (IoT) in the construction industry using Scopus database: Trends, characteristics, and future directions. Build. Eng. (BE) belongs to Academic Publishing Pte Ltd. **2**(2), 1503 (2024). https://doi.org/10.59400/be.v2i2.1503
4. Brabara Ware, Rebecca Mercury, Cloud computing 2025 and beyond: trends, obstacles and new possibilities. March 2025 doi: https://doi.org/10.20944/preprints202503.0724.v1 CC BY 4.0 License
5. Das, A., Thampi, M.P., Shaik, K., Kashyap, C.M.: Serverless cloud computing: addressing challenges and exploring future opportunities. In: 2nd International Conference on Advances and Key Challenges in Green Energy and Computing (AKGEC), pp. 1–6. IEEE (November 2024)
6. Bingbing Liu, How does digital transformation enhance enterprise technological innovation? Evidence from Chinese manufacturing listed companies Author links open overlay panel Show mor Add to Mendeley Share journal *Technology in Society* Volume 82, September 2025, 102884 March 2025, Accepted 21 March 2025, Available online 22 March 2025, Version of Record 24 March 2025. doi. https://doi.org/10.1016/j.techsoc.2025.102884

7. Alsurayyi, A., La Alsughayer, S.: relation entre la gouvernance d'entreprise et la performance de l'entreprise: l'effet de l'audit interne et de la planification des ressources de l'entreprise (ERP). Open J. Account. **10**, 56–76. https://doi.org/10.4236/ojacct.2021.102006
8. ID Baha Mohamed, Eddani Mohamed. Impact of Digital Technologies on the Performance of Moroccan SMEs: A Case Study of Companies in the " Souss Massa" Region. Afr. Sci. J., Vol : 3, Numéro 23, Avril 2024 doi: https://doi.org/10.5281/zenodo.11048428
9. Al-Mutawa, B., Al Mubarak, S., M. M.: Impact of cloud computing as a digital technology on SMEs sustainability. Compet. Rev.: Int. Bus. J. **34**(1), 72–91 (2024)
10. Makovoz, O., Lysenko, S.: Evolution of digital transformations in IT companies. Lond. J. Soc. Sci. **8**, 1–7 (2024). https://doi.org/10.31039/ljss.2024.8.202
11. Zhang, X., Qi, Z., Ma, L., Zhang, G.: Assessing the Curvilinear Relationship in Employee Digital Performance: A Task-Technology Fit Perspective. Int. J. Hum.-Comput. Interact. **41**, 2615–2633 (2024). https://doi.org/10.1080/10447318.2024.2327181
12. Salih, A.A., Alsalhi, L., Abou-Moghli, A.: Entrepreneurial orientation and digital transformation as drivers of high organizational performance : Evidence from Iraqi private banks. Uncertain Supply Chain Manag. (2024). https://doi.org/10.5267/j.uscm.2023.10.022
13. Thordsen, T., Bick, M.: Une décennie de modèles de maturité numérique: beaucoup de bruit pour rien ? Inf Syst E-Bus Manage. **21**., journal springer Nature Link Volume 21, 947–976 (2023). https://doi.org/10.1007/s10257-023-00656-w
14. D'Souza, R.C., Marti, I.: Organization as spaces for Carning: A case of an Antitrafficking Organization in india. J. Bus. Ethics. **177**, 829–842 (2022). https://doi.org/10.1007/s10551-02205102-4
15. Charbonneau-Genest, M., Gamache, S.: The importance of digital readiness for the performance of manufacturing SMEs towards Industry 4.0: a case study. In: Kim, K.Y., Monplaisir, L., Rickli, J. (eds.) Flexible Automation and Smart Manufacturing: The Nexus of Humans, Data, and Technology. FAIM 2022 Lecture Notes in Mechanical Engineering. Springer, Cham (2023). https://doi.org/10.1007/978-3-031-17629-6_66
16. Thesis, G.'s., Sébastien, G.: Industry 4.0 Implementation Strategies in Small and Medium-Sized Manufacturing Enterprises in Quebec, Canada (2019)
17. Wodajeneh, S., Kitaw, D.: Addis Ababa University, Kassu Jilcha Synergy of Ergonomic Principles with Lean Philosophy Model to Improve Productivity in Shoe Manufacturing Industry. Int. J. Workplace Health Manag. (December 2024). https://doi.org/10.1108/IJWHM-10-2023-0157
18. Ankesh Mittal, Pardeep Gupta, Vimal Kumar, Ali Al Owad, Seema Mahlawat, Sumanjeet Singh. The performance improvement analysis using Six Sigma DMAIC methodology: A case study on Indian manufacturing companyAvailable online 18 March 2023 2405–8440/© 2023 The Authors. Published by Elsevier Ltd. This is an open access article under the CC BY-NC–/di.org/10.1016/j.heliyon. 2023.e14625
19. Samsuden, N.S., Kohar, U.H.A., Saleh, F.A., Khatib; and Alhamzah F.: Abbas Digital Capabilities and Business Performance: A Systematic Literature Review by Malaysia. Sustainability. **16**(24), 11108 (2024). https://doi.org/10.3390/su162411108
20. Diallo, B.: Study of the factors of satisfaction with the use of the general surgery department of the Gabriel Touré University Hospital in Bamako. Mali Public Health Journal. **11**(1), 44–49 (2021). https://doi.org/10.53318/msp.v11i1.1892
21. Muhammad Saleem Ullah Khan, Performance Measurement in the Digital Age: A Literature Review the Intersection of Management Accounting, Big Data Analytics and Impact of Emerging Technologies on Performance Evaluation, December 2024 Int. J. Res. Innov. Soc. Sci. (IJRISS) (ISSN 2454–6186) doi: https://doi.org/10.51244/IJRSI.2024.1108009

Optimization of Textile Cutting Workshop Scheduling Through Fuzzy-Enhanced Simulated Annealing

Ichrak Trabelsi[1](✉), Faten Fayala[2], and Adel Ghith[1]

[1] Université de Monastir, ENIM, Unité de Recherche Matériaux et Procédés Textiles (MPTex), UR17ES33, 5000 Monastir, Tunisie
ichrak.trabelsi@enim.u-monastir.tn

[2] University of Monastir, Laboratory of Thermal and Energetic Systems Studies (LESTE), Monastir Textile Science & Technology Research Center, 5000 Monastir, Tunisia

Abstract. Optimizing production scheduling in textile cutting workshops is crucial for improving efficiency and reducing lead times. Traditional scheduling methods frequently exhibit limited adaptability to fluctuating workloads, leading to suboptimal system performance.

To address this challenge, this study integrates Fuzzy Logic (FL) with the Simulated Annealing (SA) algorithm to enhance scheduling adaptability. FL dynamically adjusts SA key parameters, including the initial temperature, the cooling rate, and the number of iterations per temperature level, based on three input variables: System Load (SL), Makespan Variability (MV), and Job Urgency (JU).

A simulation is conducted in a textile cutting workshop where five cutting orders are processed through three stages: spreading, cutting, and labeling. Machines operate under resource constraints, and scheduling decisions aim to minimize makespan while adapting to workload fluctuations. The results demonstrate that the proposed approach effectively adjusts SA parameters in response to system conditions, leading to an improved scheduling performance. Comparative analysis with a conventional SA approach highlights the benefits of integrating FL, particularly in reducing makespan variability.

These findings underline the potential of hybridizing metaheuristic algorithms with intelligent decision-making techniques in complex manufacturing environments. The proposed methodology provides a foundation for further research on adaptive scheduling solutions in textile and other production industries.

Keywords: Scheduling · Simulated Annealing · Fuzzy Logic · Textile production system

1 Introduction

The textile industry operates within a complex framework of production processes that require efficient planning and scheduling to optimize resource use and maintain high productivity. Production planning in this sector is challenged by multiple constraints,

M. Hamdaoui et al. (Eds.): ITDM'3 2025, PTT, pp. 231–242, 2026.
https://doi.org/10.1007/978-981-95-7704-0_21

including machine availability, workforce scheduling, and strict order deadlines, making it a highly complex combinatorial optimization problem. Traditional planning methods often struggle to adapt to the dynamic and uncertain nature of real-world textile manufacturing environments.

To overcome these challenges, various optimization techniques have been explored. Among them, Simulated Annealing (SA) has emerged as an effective metaheuristic method for solving complex scheduling problems. However, the performances of SA are heavily influenced by its parameter settings which are often determined empirically [1].

To enhance the adaptability and efficiency of SA, the integration of Fuzzy Logic (FL) offers a promising way. FL's capacity to handle uncertainty and model imprecise informations makes it particularly suitable for dynamic decision-making scenarios in manufacturing [2]. By incorporating FL, SA can dynamically adjust its key parameters based on real-time system conditions, thereby increasing the flexibility and resilience of the scheduling process.

This paper proposes a hybrid approach that combines SA and FL to optimize production planning in textile industry. The main objective is to improve scheduling efficiency, reduce production delays, and enhance overall process flexibility.

2 Problem Modeling

In a textile cutting workshop, efficient scheduling ensures optimal machine utilization, minimal idle time, and overall production efficiency. The scheduling problem can be defined using the following variables:

- $T = \{T_1, T_2, \ldots, Tn\}$ represents the set of tasks, including spreading, cutting, and labeling;
- $M = \{M_1, M_2, \ldots, Mm\}$ represents the set of available machines (spreading tables, cutting machines, labeling stations);
- Si denotes the expected start time of task Ti;
- di is the expected duration of task Ti;
- Li represents the expected completion time of task Ti, given by:

$$L_i = S_i + d_i \tag{1}$$

- $x_{i,j}$: a binary variable indicating whether task T_i is assigned to machine M_j

The objective function is to minimize the makespan:

$$\text{Min}\ (\max\ (L_i))$$

To ensure feasible scheduling, the following constraints must be satisfied:

- The spreading task for each fabric layer must be completed before the cutting task starts:

$$S_{cut,i} \geq L_{spread,i}, \forall i \in T \quad (2)$$

- Each task is assigned to exactly one machine:

$$\sum_{j \in M} x_{i,j} = 1, \forall i \in T \quad (3)$$

- A machine cannot process two tasks simultaneously. If two tasks Ti and Tk are assigned to the same machine Mj, their execution times must not overlap:

$$S_i > L_k \text{ or } S_k > L_i, \forall i, k \in T, i \neq k \text{ and } x_{i,j} = x_{k,j} = 1 \quad (4)$$

3 Simulated Annealing Method

The SA algorithm follows an iterative approach to explore the solution space until a predefined stopping criterion is achieved. It begins with an initial state S0 characterized by an energy E0 and temperature T0. At each iteration, a new state S' is generated with an associated energy E' and temperature T'. Consequently, the system's energy variation is given by $\Delta E = E0 - E'$ and two cases can arise:

- If $\Delta E < 0$: the new state S' has a lower energy. The SA algorithm accepts this solution as it leads to an immediate benefit in minimizing the makespan and it becomes the new current state;
- If $\Delta E \geq 0$: the new state S' presents a higher energy and does not improve the makespan. The SA can accept this solution with a probability P that is defined by the following Boltzmann distribution:

$$P(\text{current state} = S') = e^{(-\Delta E/(T.K))} \quad (5)$$

where T denotes the temperature of the state and K is the Boltzmann constant ($K = 1.38 \times 10–23/\text{Kelvin}$) [3, 4].

The acceptance rule described above is known as the Metropolis criteria.

In order to effectively implement the SA algorithm, it is essential to consider several key parameters that significantly influence its performance such as the initial temperature (T_0), the temperature decay rate (α) and the number of iterations per temperature level (N_{iter}).

3.1 The Initial Temperature T_0

The initial temperature T_0 is a critical parameter in simulated annealing, as it governs the algorithm's exploration capacities. A sufficiently high initial temperature increases the probability of accepting worse solutions, which enables the algorithm to escape local optimum during the early stages of the search. However, setting T0 too high can result in excessive exploration and extend the cooling process, while a low T0 may lead to premature convergence toward suboptimal solutions [3, 5]. Several strategies have been proposed to determine an appropriate initial temperature. One common approach is to set T0 such that the initial acceptance ratio of worse solutions lies between 0.8 and 0.9, ensuring a good balance between exploration and exploitation at the beginning of the search [6]. Empirical calibration based on preliminary runs or problem-specific characteristics is also widely used [7].

3.2 The Temperature Decay Rate

The temperature decay rate controls how rapidly the temperature decreases during the search process. Initially, the high value of T0 increases the probability of accepting solutions that degrade the objective function. As the algorithm progresses, the temperature is multiplied by α, which typically ranges between 0.8 and 0.99 [8]. Consequently, the temperature gradually decreases with each iteration. This decrease allows only solutions that improve the objective or lead to a low objective degradation to be accepted.

A higher α slows down the cooling process, which allows the algorithm to perform a more exhaustive exploration of the solution space but increases computational cost. In contrast, a lower α accelerates convergence but may lead to insufficient exploration, increasing the risk of getting trapped in local minimum [9]. Adaptive cooling schedules have also been explored to improve robustness in dynamic environments, where the decay rate is adjusted based on search feedback or environmental variations [10]. Such adaptive mechanisms are particularly relevant in manufacturing applications, where process conditions can fluctuate rapidly.

3.3 The Number of Iterations Per Temperature Level

In the SA cooling process, the number of iterations per temperature level refers to the number of times the algorithm explores the solution space at a given temperature. This parameter must be carefully selected to ensure a balance between the exploration of the solution space and the execution time. In fact, a large Niter provides more extensive exploration but can significantly increase computational time. However, a small Niter reduces the runtime but might result in premature convergence to suboptimal solutions.

N_{iter} is often chosen empirically or based on the size and the complexity of the problem. In this context, the more complex the problem is, the higher the number of iterations is set.

In a traditional SA algorithm, these parameters are often fixed or heuristically determined. However, real-world manufacturing environments, particularly in textile cutting workshops, are subject to uncertainties such as variations in machine availability, fluctuating production demands, and operator efficiency. A rigid parameter setting may lead to suboptimal solutions when facing dynamic production conditions [11].

4 Fuzzy Logic System

To avoid the limitations of SA in dynamic production environments and improve its adaptability, an adaptive approach based on FL is incorporated. Unlike classical SA, which relies on fixed chosen parameters, a fuzzy logic- enhanced SA can dynamically adjust its parameters based on the system's current state.

Zadeh [2] introduced FL as a robust mathematical framework for dealing with uncertainty and imprecise information. Its capability to model human reasoning using linguistic variables makes it particularly well-suited for real-world manufacturing problems. In recent years, FL has been increasingly applied in production scheduling, control systems, and optimization problems to improve system flexibility and robustness [12].

In the context of SA, FL serves as a decision-support layer that continuously monitors key performance indicators and adjusts algorithmic parameters like temperature or acceptance probability. This hybridization allows the algorithm to maintain a balance between exploration and exploitation throughout the search process.

4.1 Fuzzification and Rule Base Construction

The choice of input variables is critical, as they must reflect key operational conditions that impact the scheduling system's performance. In this paper, three main input variables are used: System Load (SL), Makespan Variability (MV), and Job Urgency (JU).

These indicators were selected as they effectively capture essential uncertainties and operational dynamics, such as machine breakdowns, setup time fluctuations, and varying job priorities in textile cutting workshops. SL represents resource saturation and indirectly accounts for factors such as machine breakdowns and capacity fluctuations. MV addresses the variability in job processing times, which may arise from variations in setup times, technical failures, and unexpected delays. JU reflects the system's responsiveness to delivery deadlines and changes in tasks priorities.

By combining these three indicators, it is possible to account for significant sources of operational uncertainty without explicitly modeling each disturbance individually.

The fuzzy intervals and the fuzzy rules are based on industry experience and scheduling stability requirements, ensuring a balanced trade-off between exploration and exploitation in the SA algorithm.

System Load System load reflects the level of machine use, indicating how much work is assigned to the available machines. It is calculated as:

$$\sum\nolimits_{(j \in M)} A_j \tag{6}$$

Where:

- M_j is a machine in the cutting workshop such as a spreading table, cutting machine or labeling station
- A_j is the time when the machine is available to execute a task.

To adjust the SA parameters based on SL value, we have classified it into three linguistic categories:

- Low ($SL < 30\%$): there is greater flexibility to explore different solutions;
- Medium ($30\% \leq SL \leq 70\%$): the algorithm must balance exploration and exploitation;
- High ($SL > 70\%$): production deadlines become critical, requiring rapid convergence.

The fuzzy sets for SL are represented by triangular membership functions:

- Low: Triangle function with the peak at 0% and range from 0% to 30%;
- Medium: Trapezoidal function, overlapping between 30% and 70%;
- High: Triangle function with the peak at 100% and range from 70% to 100%.

The rule base used to adjust the SA parameters based on SL is built from industrial experience and scheduling practice. A non-exhaustive list of fuzzy rules is as follows:

- Rule 1: If SL is low then T_0 is high and α is low and Niter is high;
- Rule 2: If SL is medium then T_0 is medium and α is medium and Niter is medium;
- Rule 3: If SL is high then T_0 is low and α is high and Niter is low.

Makespan Variability Makespan variability represents the fluctuation in estimated completion times due to unexpected delays or inefficiencies. A high MV suggests scheduling instability, while a low MV suggests a more predictable scheduling process.

To measure MV, we first determine the average makespan across multiple scheduling scenarios. These scenarios are generated during the execution of the SA algorithm, which explores different scheduling solutions.

For each temperature level in SA, the algorithm performs multiple iterations, generating a set of possible schedules. Each schedule corresponds to a different makespan value based on task assignments and sequencing.

The average makespan $\overline{C}$ is determined as follows:

$$\overline{C} = \frac{\sum_{i=1}^{N_{scenarios}} C_i}{N_{scenarios}} \tag{7}$$

Where:

- Ci is the completion time of the ith scheduling scenario
- Nscenarios is the total number of scheduling scenarios considered across all temperature levels.

Once $\overline{C}$ is determined, the MV is computed using the standard deviation formula:

$$MV = \frac{\sqrt{\sum_{i=1}^{N_{scenarios}} (C_i - \bar{C})^2}}{N_{scenarios}} \tag{8}$$

To adjust the SA parameters based on MV, we classify it into three linguistic categories:

- Low MV ($0 \leq MV \leq 10\% \overline{C}$): the schedule is stable, requiring minimal adjustment;
- Moderate MV ($10\% \overline{C} < MV \leq 25\% \overline{C}$): a balance between exploration and exploitation is necessary;
- High MV ($25\% \overline{C} < MV$): the system presents high instability requiring greater solution diversity.

The fuzzy sets for MV are represented by trapezoidal membership functions:

- Low: Triangle function with the peak at 0% and range from 0% to 10%;
- Moderate: Trapezoidal function overlapping between 10% and 25%;
- High: Trapezoidal function extending beyond 25%.

The rule base for SA parameter adjustment based on MV are as follows:

- Rule 4: If MV is low then T0 is low and α is high and Niter is low;
- Rule 5: If MV is moderate then T0 is medium and α is medium and Niter is medium;
- Rule 6: If MV is high then T0 is high and α is low and Niter is high.

Job Urgency It quantifies the criticality of a task based on its proximity to the delivery deadline. It is defined as follow:

$$JU_i = \frac{D_i - L_i}{D_i} \tag{9}$$

Where Di is the predefined completion deadline of task i within the workshop. It represents the latest time the task should exit its designated station to ensure smooth workflow and meet overall production deadlines.

To effectively integrate JU into the optimization process, we define three fuzzy linguistic terms: low urgency, medium urgency and high urgency. Here we present the range of each term and its effect on SA parameters

- Low urgency ($JU > 0.5$): the task has significant remaining time allowing for more scheduling flexibility;
- Medium urgency ($0 \leq JU \leq 0.5$): the task is approaching its deadline, requiring balanced prioritization;
- High urgency ($JU < 0$): the task is overdue and must be prioritized.

The fuzzy sets for JU are represented by triangular membership functions:

- Low Urgency: Triangle function with the peak at 0 and the range from 0.5 to 1;
- Medium Urgency: Trapezoidal function, overlapping between 0 and 0.5;
- High Urgency: Triangle function with the peak at 0 and the range from 0 to 0.5.

The corresponding fuzzy rules are:

- Rule 7: If JU is low then T0 is high and α is moderate and Niter is high;
- Rule 8: If JU is medium then T0 is medium and α is medium and Niter is medium;

– Rule 9: If JU is high then T0 is low and α is high and Niter is low.

In cases where multiple fuzzy rules are activated simultaneously, potentially leading to overlapping or conflicting outputs, a weighted aggregation is performed using the centroid method. This ensures a consistent and smooth adjustment of SA parameters, even under complex or ambiguous scheduling conditions.

4.2 Defuzzification

Defuzzification is the final step of the fuzzy inference process, where fuzzy output sets are transformed into precise values that can be used in optimization algorithms for decision-making.

Among the various defuzzification techniques, the Center of Gravity (CoG) method, known also as the centroid method, is widely used due to its simplicity and effectiveness in producing balanced outputs. The CoG method is particularly effective in environments where fuzzy outputs need to be mapped to actionable parameters, as it computes a weighted average of all possible output values. These weights correspond to the degrees of membership of each value within the fuzzy output set [13].

This method was selected in this study because it offers a favorable balance between computational efficiency and decision stability, which is critical in dynamic optimization environments such as SA. Unlike simpler methods such as maxima averaging, which can be sensitive to minor variations in membership degrees, the CoG technique offers enhanced robustness particularly when fuzzy sets overlap significantly or when multiple rules contribute conflicting information. Its ability to deliver a nuanced output reflecting the aggregate FL makes it especially suitable for real-time adaptive control [14].

In the context of SA, the CoG method contributes meaningfully to the stability of SA key parameters. By avoiding abrupt shifts in parameter values the CoG method supports a more reliable and consistent optimization process.

The formula for the CoG is given as follows

$$\mu_{\mathrm{defuz}} = \frac{\sum_{i=1}^{n} \mu_i * x_i}{\sum_{i=1}^{n} \mu_i} \tag{10}$$

Where:

– μdefuz: the value resulting from the defuzzification process
– μi: the membership degree of the fuzzy output value xi
– xi: the actual value corresponding to the fuzzy output
– n: the total number of output values considered.

5 Implementation of Fuzzy Logic Within Simulated Annealing

This section describes how FL is integrated into the SA algorithm to adjust key parameters dynamically based on operational conditions. The main objective is to enhance the search process by incorporating fuzzy decision- making to control the temperature settings, cooling rate, and the number of iterations per temperature level.

The fuzzy inference system is applied iteratively at each temperature level of the SA process. The hybrid algorithm is implemented as follows:

1. Initialize an initial solution S_0
2. Initialize an initial temperature T_0
3. Initialize fuzzy logic input variables
4. Use fuzzy inference to determine T0, α and Niter
5. While the stopping condition is not achieved do:

   While the temperature has not reached a stable state do:

   For i: = 1 to Niter:

   Generate a random solution S' in the neighborhood of S_0

   Calculate $\Delta = f(S') - f(S0)$

   If $\Delta < 0$ then:

   Update S0: S0 := S′

   Else:

   Accept S′ with the probability $P = e^{-\Delta/T.K}$

   End if

   Update fuzzy input variables (SL, MV, JU) based on the new scheduling state

   Recalculate T_0, α and N_{iter} using fuzzy inference

   End for

   Update the temperature: $T_0 = T_0 * \alpha$
6. Return the best solution found: S0

6 Simulation

To demonstrate the effectiveness of the proposed approach, numerical simulations were conducted on a 5.50 GHz microprocessor, and both algorithms were implemented using Python. The primary objective was to evaluate the impact of integrating FL into the SA algorithm, particularly in terms of reducing makespan.

The simulation considers a cutting workshop where five cutting orders must sequentially pass through three processing stages: spreading, cutting, and labeling. Each order must follow this sequence without skipping or reordering the steps. The workshop is equipped with four spreading machines, three cutting machines, and three labeling stations. The objective is to schedule these tasks efficiently to minimize the makespan while ensuring adaptability to dynamic workload conditions.

For this study, some industrial data were designed to highlight the key aspects of the proposed approach. Tasks durations were assigned in a way that creates variations in system load across different processing stages. Tasks duration for each cutting order, in minutes (min), are listed in Table 1.

The parameters for the standard SA used in the simulation were set as follows:

- $T_0 = 1000$, a commonly used value in similar configurations to ensure sufficient exploration of the initial search space;
- $\alpha = 0.95$, chosen to allow gradual cooling while maintaining a good exploration of the annealing process;
- Niter = 500, selected after preliminary trials to strike a balance between solution quality and computation time.

Table 1. The tasks durations (min)

Order	Spreading	Cutting	Labeling
O1	5	15	5
O2	5	15	5
O3	7	20	10
O4	7	25	15
O5	10	30	15

These parameters were set empirically, based on common practices in similar problems.

In order to evaluate the effectiveness of the SA and the combination of FL within SA (FL_SA) in reducing makespan, Table 2 presents the results of these two methods.

Table 2. Comparison of scheduling performance

Method	Initial makespan (min)	Final makespan (min)	Improvement
SA	82	77	6.09%
SA_FL	82	62	24.39%

The results show that integrating FL with SA leads to a more effective optimization of the production scheduling process. The reduction in makespan is significantly higher with the hybrid method (24.39%) compared to SA (6.09%). This indicates that FL enhances the algorithm's adaptability to workload variations and leads to more effective and flexible scheduling outcomes.

To further illustrate the performance of the algorithms, convergence plots were generated to show the progression of the makespan over iterations. These plots highlight the convergence speed and stability of each method and reveal the superior convergence behavior of FL_SA method (Fig. 1).

In terms of computational time, the SA algorithm took approximately 42 s to complete the optimization process, while the FL_SA required 58 s. This increase is attributed to the additional computational load introduced by the fuzzification and rule-based adaptation mechanisms.

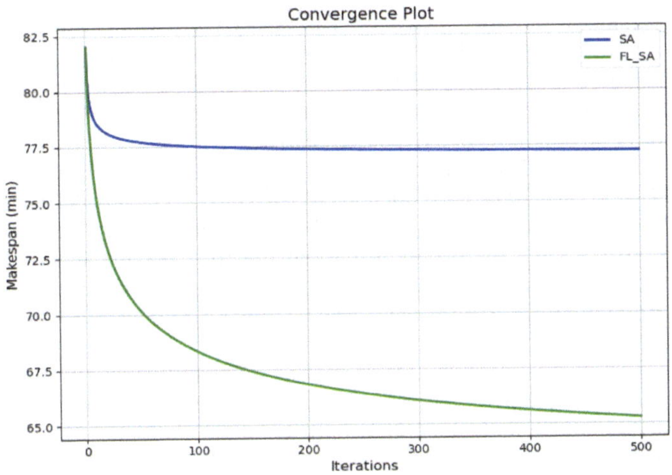

Fig. 1. Makespan reduction over iterations for SA and FL_SA algorithms

7 Conclusion

This study presents an adaptive approach that integrates FL on SA algorithm for optimizing production scheduling in a textile cutting workshop. By dynamically adjusting key SA parameters based on workload conditions, the proposed method enhances both efficiency and adaptability.

The simulation results demonstrated that the fuzzy-controlled SA algorithm reduces the makespan while responding dynamically to variations in system load, job urgency, and makespan variability. Unlike traditional SA implementations with fixed parameters, this approach enables real-time adaptation, leading to more robust and efficient scheduling solutions.

References

1. R. Heckmann et T. Lengauer, A simulated annealing approach to the nesting problem in the textile manufacturing industry, Ann. Oper. Res., vol. 57(1), p. 103–133, déc. 1995, https://doi.org/10.1007/BF02099693
2. L. A. Zadeh, The concept of a linguistic variable and its application to approximate reasoning—II, Inf. Sci., vol. 8(4), p. 301–357, janv. 1975, https://doi.org/10.1016/0020-0255(75)90046-8
3. P. J. M. van Laarhoven et E. H. L. Aarts, Simulated annealing, in Simulated Annealing: Theory and Applications, P. J. M. van Laarhoven et E. H. L. Aarts, Éd., Dordrecht: Springer Netherlands, 1987, p. 7–15. https://doi.org/10.1007/978-94-015-7744-1_2
4. D. Delahaye, S. Chaimatanan, et M. Mongeau, Simulated annealing: From basics to applications. In: M. Gendreau et J.-Y. Potvin, Éds., Handbook of Metaheuristics. International Series in Operations Research & Management Science, vol. 272, p. 1–35. Springer International Publishing, Cham (2019). https://doi.org/10.1007/978-3-319-91086-4_1
5. S. Kirkpatrick, C. D. Gelatt, et M. P. Vecchi, Optimization by simulated annealing, Science, vol. 220(4598), p. 671–680, mai 1983, https://doi.org/10.1126/science.220.4598.671

6. L. Ingber, Simulated annealing: Practice versus theory, Math. Comput. Model., vol. 18(11), p. 29–57, déc. 1993, https://doi.org/10.1016/0895-7177(93)90204-C
7. M. Mascagni, E. Aart, et J. Korst, Simulated annealing and boltzmann machines: A stochastic approach to combinatorial optimization and neural computing., Math. Comput., vol. 55(191), p. 393, juill. 1990, https://doi.org/10.2307/2008816
8. R. W. Eglese, Simulated annealing: A tool for operational research, Eur. J. Oper. Res., vol. 46(3), p. 271–281, juin 1990, https://doi.org/10.1016/0377-2217(90)90001-R
9. Nourani, Y., Andresen, B.: A comparison of simulated annealing cooling strategies. J. Phys. Math. Gen. **31**(41), 8373–8385 (1998). https://doi.org/10.1088/0305-4470/31/41/011
10. V. Černý, Thermodynamical approach to the traveling salesman problem: An efficient simulation algorithm, J. Optim. Theory Appl., vol. 45(1), p. 41–51, janv. 1985, https://doi.org/10.1007/BF00940812
11. Simulated annealing overview, ResearchGate. Consulté le: 18 avril 2025. [En ligne]. Disponible sur: https://www.researchgate.net/publication/238690391_Simulated_annealing_overview
12. Petrovic, D., Roy, R., Petrovic, R.: Modelling and simulation of a supply chain in an uncertain environment. Eur. J. Oper. Res. **109**(2), 299–309 (1998). https://doi.org/10.1016/S0377-2217(98)00058-7
13. Ross, T.J.: Fuzzy Logic with Engineering Applications, 1st éd. edn. Wiley (2010). https://doi.org/10.1002/9781119994374
14. Zadeh, L.A., Klir, G.J.: Fuzzy Sets and Fuzzy Logic: Theory and Applications. Prentice Hall, Upper Saddle River, NJ (1995)

Management, Quality, and Textile Processes

Study of the Influence of Relative Humidity and Storage Modes on Knitted Surface Properties

Maroua Sayahi[1](✉), Hassen Hedfi[2], Mohamed Hamdaoui[1], and Omar Harzallah[3]

[1] University of Monastir, National Engineering School of Monastir, Textile Materials and Processes Research Unit MPTex, UR17ES33, 5019 Monastir, Tunisia
sayahimaroua09@gmail.com

[2] University of Monastir, National Engineering School of Monastir, Mechanical Engineering Laboratory (LGM), LR99ES32, 5019 Monastir, Tunisia

[3] Laboratory of Textile Physics and Mechanics–LPMT, National School of Engineering of Southern Alsace, University of Haute-Alsace, Mulhouse, France

Abstract. The concept of surface roughness is intuitive: it can be associated with a sensory perception describing the adhesion and sliding characteristics of a surface. Roughness is an essential element to understand the physical characteristics of various surfaces and their evolution. Nowadays, in many application fields, we use a variety of textile surfaces. These are distinguished by their geometric and structural parameters at three levels: the microscopic level (fiber morphology, fibrous level), the mesoscopic level (yarn level), and the macroscopic level (knit level). The evaluation of surface roughness is crucial in many areas: surface finish, knitwear quality, and perception of garment comfort. The objective of this article is to investigate the factors affecting the surface of knitwear during storage in order to better understand their impact on subsequent production phases. To achieve this objective, we carried out the evaluation of the surface properties of knitwear by the Kawabata warp surface module "KES- FB4", focusing on the effect of relative humidity, storage modes, as well as the type of knit structure (jersey, rib and fleece knits) and its composition (cotton, viscose and polyester) on this phenomenon. Cellulose-based knitwear stored in a high humidity environment becomes rougher (less smooth), due to the increase in the linear thickness deviation (SMD). It was found that the textile composition, environmental conditions and storage conditions and configurations have an impact on the surface properties of the knitwear studied. This study could be of great use in predicting the behavior of knitwear in the different phases of production, marketing and use; thus, allowing to predict and define the optimal conditions to be given to the knitwear in these phases to ensure: dimensional stability, comfort and durability.

Keywords: Knit · Storage Modes and conditions · Roughness · KES-FB4

M. Hamdaoui et al. (Eds.): ITDM'3 2025, PTT, pp. 245–257, 2026.
https://doi.org/10.1007/978-981-95-7704-0_22

1 Introduction

Crossing one or more threads to form stitches is the craft of knitting. Whereas the rows are oriented perpendicular to the production direction, the columns are oriented parallel to it. The lay of a knitted fabric is the fundamental interlacing pattern of the yarns that comprise a stitch (Fig. 4). A knitted fabric is identified by its assembly and the length of yarn integrated per stitch, or LFA, in addition to the characteristics of the yarns that comprise it [1].

The interaction between the surfaces of two items is critical when they come into contact with one another, highlighting the significance of surface finish. The recurring departures from the nominal surface that make up the surface's three-dimensional configuration are referred to as micro-geometry or surface topography. Surface texture includes surface waviness (macro), the main alignment of surface features, surface defects, and roughness (nano and micro) [1].

It is challenging to come up with a thorough description of roughness because it is a complicated notion. This idea spans a wide range of possible geometric arrangements and changes based on the circumstance. Smoothness and roughness are frequently compared. Only a small portion of the whole of continuous surfaces is studied mathematically through the ideas of regularity, continuity, and differentiability in the context of smooth surfaces. In a non-differentiable notion, roughness is instead called "irregularity" [2].

At times, it can be challenging to connect roughness to a surface's desired feature. Microgeometric variations in the theoretical geometric surface are often linked to roughness. First, short-range anomalies with different amplitudes and spacings are caused by physical and chemical changes in the material's microscopic structure. Despite being focused on a narrow portion of the surface, these alterations might degrade the component's quality or, in certain cases, render the surface unsatisfactory [2].

This work is devoted to the presentation of the results obtained during the study of the surface condition of the knitted fabrics object of the study, with emphasis on the analysis of the influence of relative humidity, storage method as well as weave and composition on this phenomenon.

2 Materials and Methods

2.1 Textile Supports

Natural fibers, like cotton, have long been utilized to meet a basic human need. However, new fibers are being introduced to the globe and their share in the production of textile structures is growing due to scientific and technological advancements (e.g., viscose and polyester) [3].

Thinness, remarkable extensibility relative to its average diameter, specific strength, and flexibility are the characteristics that define the textile fiber. Fibers can be produced industrially or naturally. The length, form, and strength of these fibers are the traditional characteristics that set them apart. Additionally covered are technical characteristics including configuration, flexibility, robustness, and moisture absorption [4].

Cotton fiber. Natural fibers have unique qualities that should be emphasized, such as their renewable source, their durability and biodegradability, and the notable mechanical properties of some fibers. The usage of natural fibers is not restricted to replacing synthetic fibers [5]. The fiber (see Fig. 1 (a)) has a length of 1 to 7 cm. Its diameter ranges from 20 to 40 μm. The fiber is curled and empty. The middle channel is broad. The part has a bean-like form. The material is cellulose. Its polymerization level ranges from 2500 to 3000 [6].

A cotton fiber's high hygroscopicity is due to its 87% cellulose content. In addition to the typical recovery rate of 8.5%, water infiltration in the fiber can result in a transverse expansion of 16% and a longitudinal expansion of 1%. Surface structures are deformed by these values, which are impacted by meteorological factors as temperature and relative humidity [7]. About 90% of cotton is made up of cellulose, which is its primary constituent. Wax, pectins, water, proteins, potassium, magnesium, calcium, sodium, and colorful particles are among the numerous substances that surround it [8].

Viscose fiber. Pre-existing natural materials are chemically processed to form a "pasty" substance that may be spun through a die to create artificial fibers [9]. The most widely used fibers are viscose fibers, which can be produced using a variety of methods (such as the xanthate process or the NMMO process, which involves N- methylmorpholine-N-oxide). Furthermore, they can be altered by adding active ingredients straight to the solution prior to solvent spinning [10].

In terms of production volume, viscose is the most common type of regenerated cellulose fiber. Cellulose pulp is first aged in an alkaline solution (NaOH) to make it soluble by producing cellulose xanthate (cellulose-O-CSSNa), a derivative of cellulose that contains carbon sulfide. This paste is then extruded through a die after being molded to the proper viscosity texture. After being immediately coagulated in an aqueous solution of sodium zinc sulfate salts and sulfuric acid, which permits xantate to regenerate into cellulose, the extracted filaments (see Fig. 1 (b)) are released from the bath at a regulated pace [11].

There are significant physical and chemical differences between cellulose and cotton since cellulose is a polymer that is formed from cellulose but has a lower degree of polymerization. Despite being more brittle, viscose has a lower thermal conductivity than cotton but is more resistant to heat, moisture, and abrasion [11].

Polyester fiber. All synthetic fibers are created chemically using recycled plastic or oil. These were created in response to the worldwide scarcity of natural fibers and to offer fibers with certain qualities that satisfy customer demands at a reasonable cost [9]. Because of its fast growing applications, virgin polyester is regarded as one of the most important engineering polymers of the past 20 years. It is frequently utilized in the textile sector and is regarded as the preferred material for a variety of applications. Excellent tensile and impact resistance, chemical strength, transparency, colorability, ease of handling, and good thermal stability are all provided by this material [12].

Organic molecules arranged in ester repetitions (COO) produce linear polyester, also known as polyethylene terephthalate (PET). When terephthalic acid was incorporated into the polymer molecule's repeat group, polyester's use in fibers was discovered [13]. Polyethylene terephthalate, or PET (see Fig. 1 (c)), is a semi-crystalline thermoplastic that is a member of the polyester group. The crystallinity of PET can reach 40% by volume. Its mechanical properties are significantly influenced by this degree of crystallinity [14].

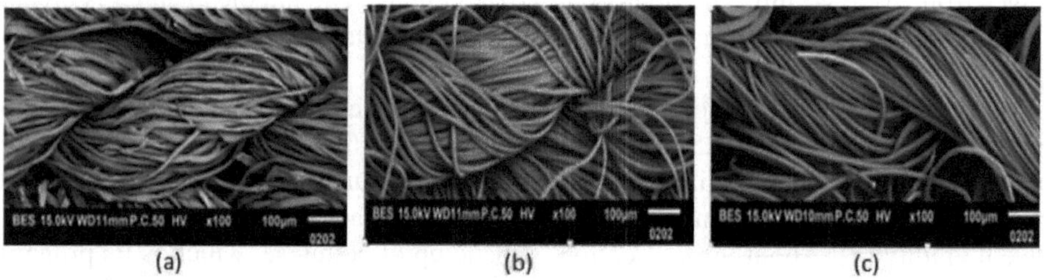

Fig. 1. Fibers observed at SEM (a) Cotton fibers, (b) Viscose fibers and (c) Polyethylene terephthalate (PET) fibers

Knitted fabrics. A mesh is created when many points are combined. As a result, the size of the exposed stitch components—such as the legs, head, or feet—determines how the knit looks. The easiest knitting is the jersey (see Fig. 2(a)), which has simply knitted meshes on one side and stitches on the other [15].

A knitted garment made of jersey with extra yarns tied by loops based on the rows, each of which has a linear mass greater than the foundation yarn. The names visible hook point 1 and 3 or single fleece 1 and 3 (see Fig. 2(b)) are used to identify knitting in which the pattern is attached to the base of a load for a floating on three threads [16].

Ribbed Weft Knitting is a general word for weft knitting in which all of the stitches in a column are the same, either all of the spot stitches or all of the reverse stitches, but there are both spot stitches and reverse stitches throughout the rows. Only the stitches on the right side are visible on both sides when the rib knit is not extended [16].

Typically, an interlock calage technique is used to create ribbed pattern knitting, which consists of two columns of spot meshes next to two columns of reverse meshes (see Fig. 2(c)).

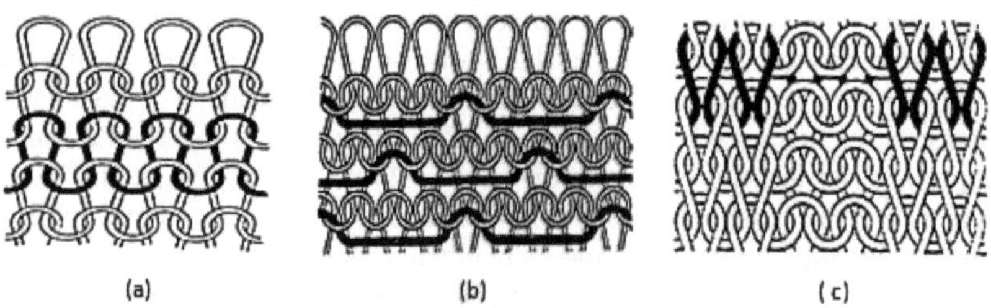

Fig. 2. Modeling a knitted fabrics [16]

Using a circular knitting machine, the experiment makes use of delicate jersey, fleece, and seam stitches. Depending on the production belt, this kind of machine may knit many rows of meshes continuously in each cycle to produce a knit with different inclination angles.

The metrological characteristics of knitted fabrics are described in Table 1. Experiments are applied to different samples for the same knitted fabric.

Table 1. Characterization of knitted fabrics

Reference	TEc	TBm	TVr	TVl	TRc
Material	Cotton	Cotton/ PET	Viscose	Cotton/ PET	Cotton
Type	Jersey	Jersey	Jersey	fleecy fabric	Rib fabric
Density (g/m2)	169.91	147.21	262.22	246.97	330.72
Thickness (mm)	0.486	0.604	0.741	0.922	1.172
Stitches number/ Wale (cm^{-1})	20	20	20	21	19
Stitches number/ Course (cm^{-1})	14	14	16	16	12
Yarn number (Tex)	24	20	20	20	26

2.2 Methods

Conditioning In our metrology laboratory, we use three storage methods for our samples: rolled, folded and suspended (see Fig. 3). During this activity, the samples are placed in a controlled environment (relative humidity of 63% and temperature set at 21 °C) for 48 h.

(a) (b) (c)

Fig. 3. Different storage configurations

For each test, take five square samples (200 mm x 200 mm) with column and row orientation: the different knitted fabrics are sampled for the three storage configurations (rolled (a), folded (b) and suspended (c)).

KES-FB4-A Surface Tester. Surface interactions are a significant factor in friction between two objects, underscoring the significance of the surface condition of those items. Roughness, surface ripples, the prevailing direction of the components on the surface, and surface anomalies are some of the surface's features [1].

The profile of a surface is experimentally determined using a profilometer. A probe is used to horizontally scan the surface in order to obtain a contact measurement. A motion sensor captures the detector's vertical movements as it interacts with the surface under study. In order to guarantee point contact at an infinite angle and avoid material distortion, the probe is typically fitted with a spherical head [1].

Using a roughness measurement equipment and the coefficient of friction of the Kawabata chain KES-FB4 (see Fig. 4), we have conducted a comprehensive investigation into the surface condition of knitted fabrics in this article. We have also examined the impact of various parameters on this phenomenon.

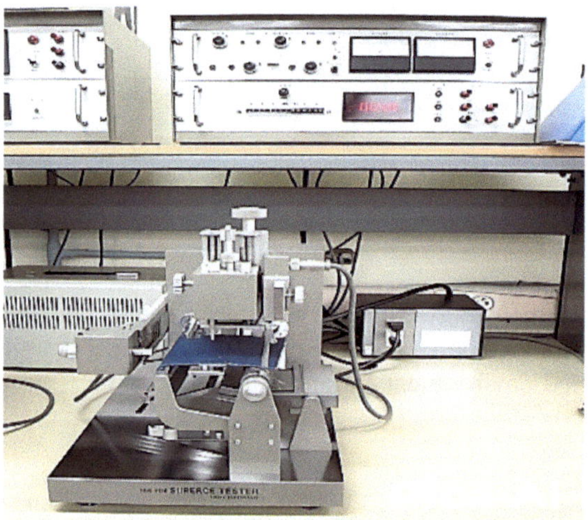

Fig. 4. KES-FB4 from Kawabata

A needle that moves steadily across the surface and whose vertical movement makes it possible to record the profile defines the instrument. A CPU, a mechanical progression unit, an amplifier/conditioner, and perhaps recording devices are also included [17] (see Fig. 5).

It is not possible to characterize roughness by a single value. Various statistical indicators can be determined from a profile to describe it [18]. The main statistical parameters that can be used to characterize a surface profile (see Fig. 6) are:

- Geometric roughness (SMD) measurement
- The coefficient of friction measurement (MIU).
- The average friction coefficient deviation (MMD).

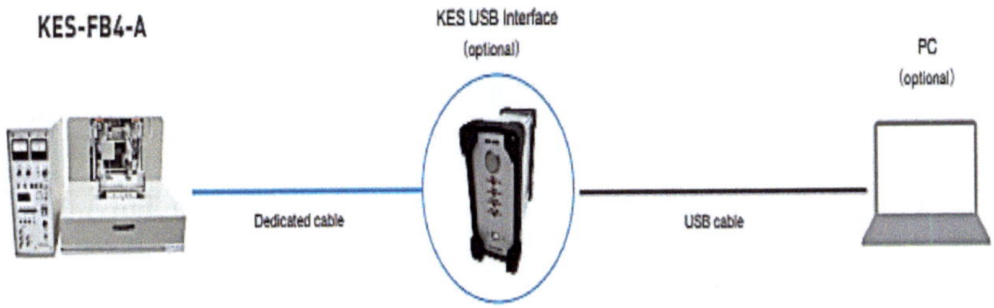

Fig. 5. System configuration diagram/measurement DATA

Roughness is represented by the mean deviation SMD (average linear deviation of thickness) of y (with: y an arbitrary starting position of vertical displacement of the sensor):

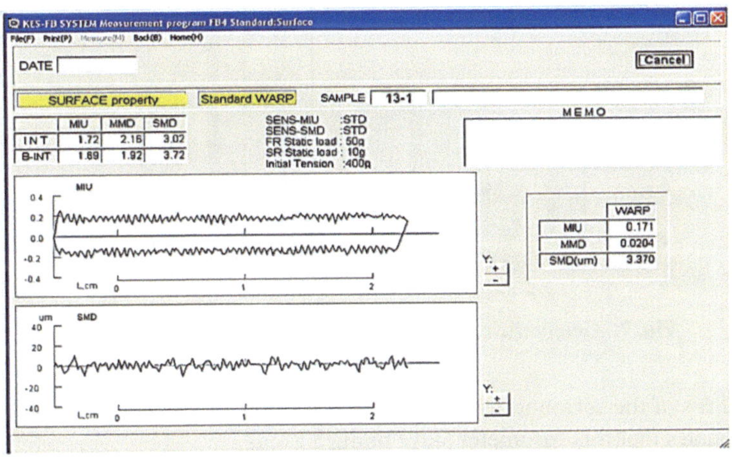

Fig. 6. Surface properties

To have correct measurements it is mandatory to make the electronic and mechanical adjustments of the instrument. The roughness of the samples to be tested is measured under a preload weight: 200 g.

The test is repeated several times at different locations in the sample along the wale and course to ensure that the measurement is representative.

3 Results and Discussions

In this section, we will present the results of measuring the roughness and coefficient of friction of different knitted fabrics. This study includes surface irregularities which is influenced by the following parameters:

3.1 Influence of Relative Humidity

Because polymers contain functional groups, they have certain chemical polarities. By definition, they are drawn to water. Two factors contribute to moisture transport: water/polymer attraction brought on by the presence of hydrophilic sites and diffusion in the free volume, which includes microvolumes situated in between polymer chains.

The following graphs illustrate the change in the average roughness value of the various knitted fabrics within the humidity range [40%, 80%] at 21 °C.

When the sensor travels in the direction of the columns in a moist environment, the average geometric roughness of a cotton jersey fluctuates considerably (see Fig. 7). In a dry environment, the latter is shown to have significant values of about 5 μm for a capture movement in the direction of the rows.

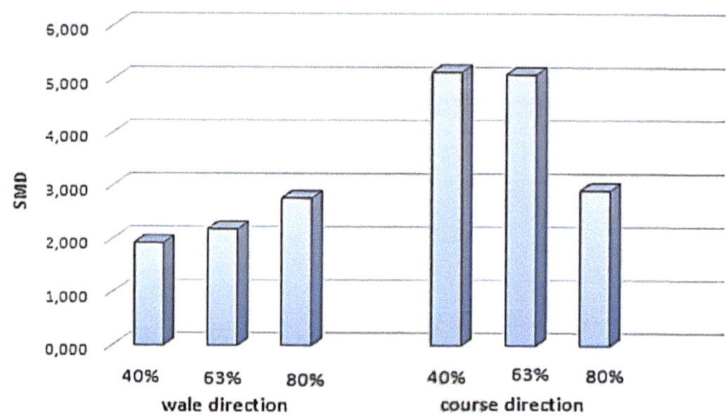

Fig. 7. Geometric roughness measurement in relation to HR%

Regardless of the surrounding humidity, the friction coefficient progress graph (see Fig. 8) indicates that this parameter stays high.

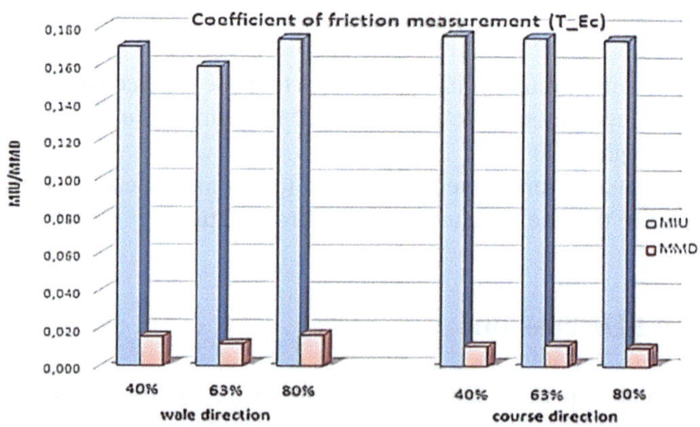

Fig. 8. Changes in the coefficient of friction with HR%

3.2 Influence of Storage Method

We examined the impact of storing knitted fabrics in similar conditions of relative humidity around 63% at 21 °C. The figures below present the results of this research (Fig. 9):

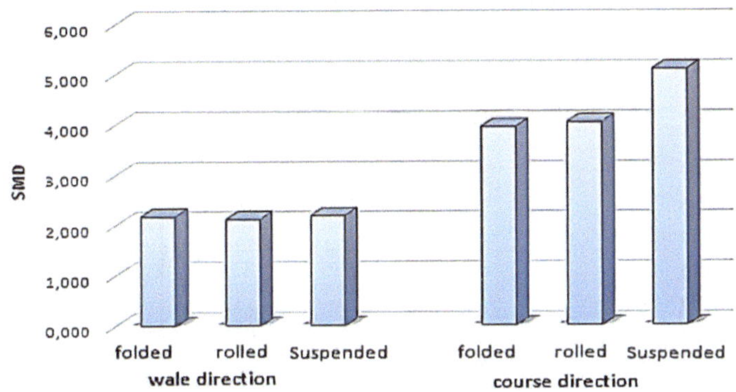

Fig. 9. Geometric roughness measurement in relation to storage method

It may be inferred from the device sensor's movement against the columns that the storing technique has no effect on geometrical roughness.

The mean roughness deviation for various storage techniques rises as the sensor moves in a direction perpendicular to its trajectory; for hanging knitwear, this maximum value is approximately 5 µm.

For all kinds of storage, the coefficients of friction are important, and they are at their highest when folded fabrics are folded (see Fig. 10).

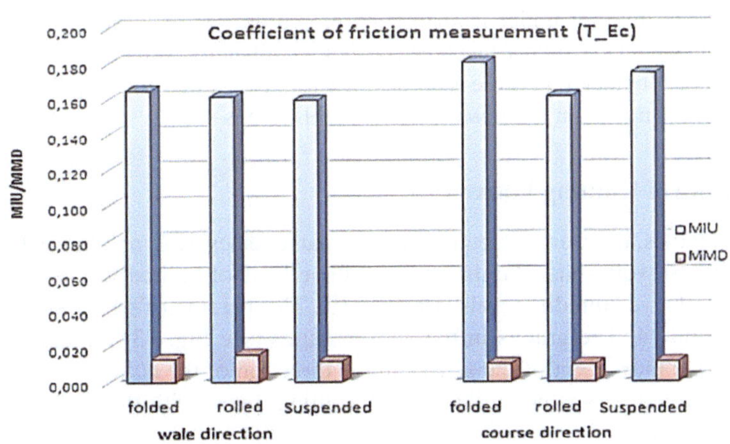

Fig. 10. Changes in the coefficient of friction with storage method

3.3 Influence of Composition

The many fiber types from natural to synthetic and regenerated fibers used to make knitted fabrics are examined in this section. It also examines how the origin of the fibers utilized affects the texture of tissues.

We evaluated the effect of knitted fabric composition under similar relative humidity conditions, about 63%, at a temperature of 21 °C. Fig. 11 shows the findings of this study:

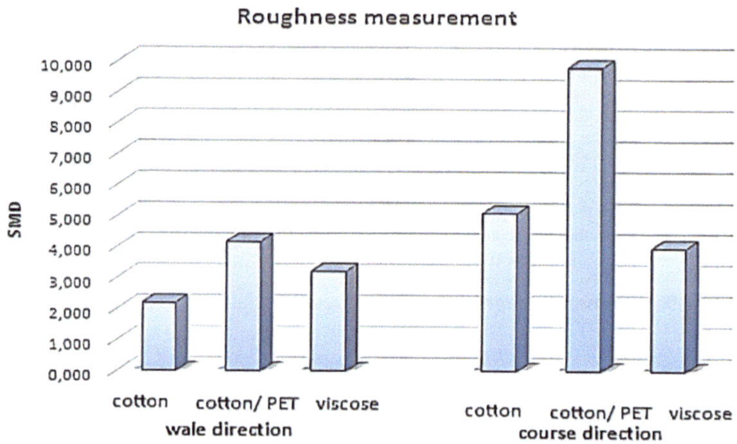

Fig. 11. Geometric roughness measurement in relation to composition

In general, textile fibers have a distinct feel to the touch. In our study, we found that the roughness value for a knitted fabric made from mixed yarns of cotton and polyester is significant, reaching about 10 μm, irrespective of sensor orientation.

The different textiles under investigation are composed of yarns containing fibers from different sources, which affects the coefficient of friction value (see Fig. 12). When compared to cotton fiber, this value is noteworthy for viscose and cotton/polyester blends. When the sensor moves in the rows' direction, this rise is seen.

3.4 Influence of Knitting Weave

Regarding textiles, there are structures; we can name jersey, ribs, and fleece knitwear as the fundamental weft knits. We investigated the effects of knitted weave at 21 °C and comparable relative humidity levels, around 63%. This parameter's analysis is displayed in Figs. 13 and 14. The graphs in the above figures (see Fig. 13) indicate that, when considering the movement of the sensor in the direction of the columns, the difference in geometric roughness is negligible for all knitting types, with the rib knitted having the most variation.

When the sensor is moved in the direction of the rows, the coefficient of friction in rib knitting increases to its maximum value. His own armor's reliefs are connected to this ascent (see Fig. 14).

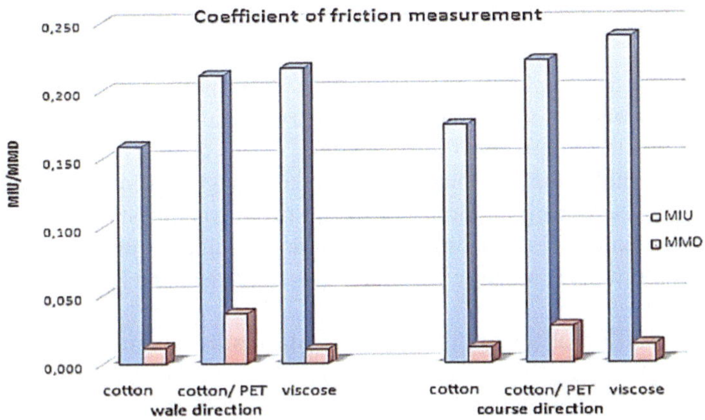

Fig. 12. Changes in the coefficient of friction with composition

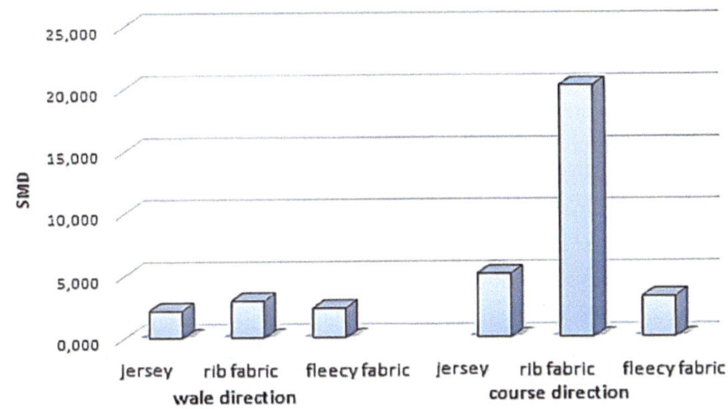

Fig. 13. Geometric roughness measurement in relation to knitting weave

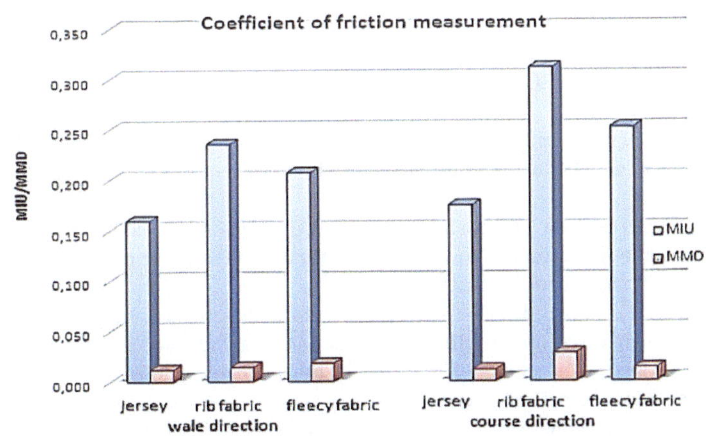

Fig. 14. Changes in the coefficient of friction with knitting weave

4 Conclusion

The average geometric roughness of a cotton jersey varies noticeably in humid conditions before reaching a value of more than 5 µm in a dry medium. Regardless of the level of ambient humidity, the friction coefficient analysis shows that this number stays high.

It may be concluded that the measured geometric roughness in the column direction is unaffected by the storing method. If not, the mean roughness deviation rises with different storage techniques, reaching a maximum of roughly 5 µm for knitwear that is hung. Friction coefficients are crucial for all types of storage and are at their highest for folded textiles.

The unique texture of textile fibers upon contact is its defining characteristic. Our research has shown that, for knitted fabrics created from yarn combining cotton and polyester, roughness measurement is crucial, regardless of the angle of measurement, reaching nearly 10 µm. The friction coefficient value is affected by the fact that the various fabrics under analysis are composed of fiber yarns from diverse suppliers. In contrast to cotton fiber, this value is significant for viscose and cotton/polyester blends. When measurements are made in the row direction, this rise is seen.

The change in geometric roughness is minimal regardless of the style of knitting. When measured in the other direction, it varies and reaches its maximum for knits in slope (20 µm). When measured in the direction of the rows, the coefficient of friction is large and reaches its maximum. His own armor's reliefs are the cause of this rise.

References

1. Romain Bocquet, Doctoral thesis: Study of adhesion and deformation mechanisms causing friction of textile surfaces, University of Haute Alsace-Mulhouse, France (2013)
2. Poull, C., et al.: Surface Roughness and its Applications. French Computer Graphics Days (2022)
3. Sundaresan Jayaraman, W. W. CarrL. And Howard Olson: Textile Engineering, Georgia Institute of Technology
4. Ben Hassine, L.: Mechanical and physical study of creasing and uncreasing of cellulosic textiles based on cotton and linen. Doctoral School: Mathematics and Information and Engineering Sciences (ED 269), University of Haute-Alsace & University of Strasbourg (2022).
5. Christophe, B.: Natural reinforcement fibers for composite materials. Eng. Techniques. (2020)
6. Atelier du Papetier: The textile fibres and their classification, Journal Ouestfrance, published 2020
7. Bhouri N., Badel E., Ben Nasrallah S., Perré P.: Comparison between the free deformations of a Canvas and a Jersey under the effect of variations in their water conditions. International Congress of Applied Textile Research, (2006)
8. Bhouri Naoufel, Doctoral thesis: Thermodynamic and dimensional behavior of textile materials subjected to variations in climatic conditions. The National School of Engineers of Monastir, in cooperation with the University Henri Poincaré. (2009).
9. The fiber lab: Chemical fibre, Federation, mesh, lingerie and spa. (XXXX)
10. Laurence Caramaro: Fibers and yarns for technical use, This article is from Materials, Industrial Textiles, Journal Engineering Techniques, Publication date: November 10, 2005
11. Roux, C., Robertson, J.: Chapters FIBERS Types, pp. 838–854. Encyclopedia of Forensic Sciences (2000)

12. Firas Awaja and Dumitru Pavel: Recycling of Pet, European Polymer Journal, Available online 16 March 2005
13. Lin, C.A.: Manufacture of Polyester Fibres. Feng Chia University, Taiwan, Woodhead Publishing Series in Textiles (2008)
14. Asma, D.: Doctoral thesis: Experimental study of the pyrolysis of poly(ethylene terephthalate) (PET): Fine characterization of the products and their formation kinetics. University of Lorraine (2019)
15. Sylvia Piotin: Towards a CAO tool for knitwear: Doctorate from the University of Reims Champagne-Ardenne (URCA). (2008).
16. International standard, ISO 8388:1998 (fr): Mesh Fabrics, Types and Vocabulary
17. Fateh Makhloufi: Three-dimensional Metrology and Surface Conditions, Institute of Applied Sciences and Techniques. University of the Brothers Mentouri Constantine. 1, (2021)
18. International standard, ISO 4287: Geometrical Product Specifications (GPS), surface texture: Profile method-Terms, definitions and surface texture. (1997)

Comparative Study: Agile Kanban Method and Traditional Method in Two Tunisian Clothing Companies

Slim Boussaadoun[1](✉), Sejri Nejib[2], and Mohamed Hamdaoui[1]

[1] Université de Monastir, ENIM, Unité de Recherche Matériaux et Procédés Textiles (MPTex), UR17ES33, 5019 Monastir, Tunisie
slimbinz@gmail.com

[2] Textile Engineering Laboratory (LGTex), ISET of Ksar Hellal, Monastir University, Monastir, Tunisia

Abstract. In recent years, Tunisian textile and clothing SMEs have experienced strong competitive intensity in terms of deadlines, productivity, quality, performance, efficiency and customer satisfaction. Indeed, in this confusing global context, we think agile before being digital and innovative in the face of rapid changes and customer demands. In this context, the purpose of this comparative study between two Tunisian textile and clothing companies with different organizational structures. We applied in this study the criteria of the agile kanban method on both companies E1 and E2 in order to evaluate and identify the most efficient according to the percentages of agile scores. We found the following results: company E1, which applies agile kanban and 4.0, has an overall Agile score percentage 38.17% higher than company E2. This gap clearly reflects its impact on the performance and efficiency criteria of both companies. The results of our research show that the Agile method improves the productivity and efficiency of companies, and that they provide fertile ground for exploring other agile indicators. This gap clearly reflects its impact on performance and efficiency metrics at both companies. Our research findings also demonstrate that the Agile methodology improves companies' productivity and efficiency, providing fertile ground for exploring other Agile indicators.

Keywords: SMEs · Agile Kaban · Agile Crieria · Performance

1 Introduction

Today, high competitive intensity, respect for deadlines, productivity, quality, performance [1], efficiency and customer satisfaction are becoming key factors in the excellence of textile and clothing companies [2]. In this chaotic global context, we think agile before being digital and innovative in order to bring about the necessary changes that small and medium-sized enterprises in the textile and clothing sector need to evolve [3]. The objective of this study aims to compare two Tunisian textile and clothing companies

M. Hamdaoui et al. (Eds.): ITDM'3 2025, PTT, pp. 258–263, 2026.
https://doi.org/10.1007/978-981-95-7704-0_23

in the first E1 applies industry 4.0 and the approach (Agile Kanban) in the intermediate phase while the second company E2 which uses traditional classical methods, in order to identify the most efficient company. The Agile approach has undergone several variations: such as Scrum which is characterized by short cycles of 2 to 4 weeks (Sprints) [4]. In addition, the Kanban flow management method which has as its principle the visualization of tasks in a whiteboard with columns such as: "To Do, Doing, Done". This method aims to limit work in progress and optimize production flows, as well as the Extreme Programming XP method: it is a practical and technical method… [5]. Several studies conducted in different countries have demonstrated that the application of the agile Kanban method is particularly effective in textile and clothing. In fact, companies that adopt this method note an increase in productivity of around 32% compared to traditional methods, often criticized for their rigidity and the delays they cause. Other approaches, such as the SAFe (Scaled Agile Framework) method, also aim to promote agile adoption across the entire company [6]. Regarding the quality of products exported to customers, the application of this tool has been able to reduce the percentage of defects by 45% thanks to its self-control system; because it is based on continuous improvement tests with iterations. As a result, traditional methods have been throughout the work process have the same problems and the same percentages of errors. In particular, organizational agility increases the performance of SMEs (+37% productivity) and improve the quality (+55%). It optimizes the use of scarce resources, reduces costs, and strengthens sustainability. Performance acts as a partial mediator in this relationship, opening up avenues for research [7]. In addition, companies that have implemented Kanban with good organizational flexibility have had their capacity increased by 35% in the face of changes and fluctuations in the market. Additionally, companies that implemented Kanban with good organizational flexibility increased their ability to cope with market changes and fluctuations by 35% [8]. The expertise of the company's workforce also has a decisive impact on total unit production costs by 8% in the textile and clothing sector [9], which can subsequently impact companies' productivity and efficiency. Finally, note that customer satisfaction with a Kanban project through Agile, according to recent studies such as this one, showed that customers benefited from a 42% higher satisfaction rate than companies that apply the traditional approach [10]. Note that these two companies were similar in many aspects: size, types of products, company culture, and customer relationships, but in other aspects they were different in: defect rate, performance, communication, and decision-making [10]. All this leads us to make a comparative study between the two clothing companies E1 and E2, which deal with two different organizational approaches. To know their impact on deadlines, productivity, quality, efficiency and customer satisfaction. In order to identify the most efficient and effective company and to approve and validate the best approach [11].

2 Methods Used and Materials

We used a comparative methodology based on Agile benchmarks between the two companies. To ensure the success of our project, we took into account the following steps: We defined the company's objectives, which are performance through Agile indicators [12], and we also defined the actors (Agile team) for the Agile project in two stages. In the first phase (6 months) we formed two teams of 12 people, according to agile principles: such as: (Deadlines, productivity, customer satisfaction, collaboration, this team is made up of: Team members, developers, designers, testers, service delivery Manager; Product Owner, stakeholders (Customers/Management), Kanban Coach (agile consultant/internal coach). In the second phase (6 months) we expanded the team size to around 50 people for each team which is worth 40% of the overall objective. In addition, we used the following criteria: We applied the following formulas to calculate the following indicators: [12–17]

$$\text{Delay of} = \frac{\big(\text{Need in min of an order in min}}{\text{Capacity of a company in min /days}} \quad \text{days} \tag{1}$$

$$\text{Delay de livraison} = \sum\nolimits_{k=1}^{n} \frac{(\text{Date de livraison i} - \text{Date de commande i})}{n} \tag{2}$$

$$\text{Productivity} = \frac{\text{Production (quantity)}}{\text{Number of hours worked}} \quad \text{Pieces/days} \tag{3}$$

$\%\text{Defects} = \frac{\text{defective quantity}}{\text{Quantity produced}} * 100$ (4)

$$\text{Efficiency\%} = \frac{\text{actual results}}{\text{optimal results}} * 100 \tag{5}$$

$$\text{CAST\%} = \frac{\text{Number of satisfied customers}}{\text{Total number of responses}} * 100 \tag{6}$$

CAST %: this is the customer satisfaction rate

Explaining agile kanban and hardware tools in detail [13, 18]

To calculate the agile score, we used this formula:

$$\text{Score Agile} = \sum\nolimits_{k=1}^{n} \text{P1} * \text{C1} + \text{P2} * \text{C2} \ldots \text{Cn} * \text{Pn} \tag{7}$$

P1 to… P_n: This is the weight used to weight the agile scores as a percentage.

Example: $P_1 = 5\% = 0.05$.

C_n: Agile criteria from 1 to n

C_n criterion score: Rate from 1 to 10

We applied the agile kanban approach for continuous flow in the clothing sector with the work team. We also used a computer, Office 365, Excel, and E1's internal ERP software called GCP (General Control of Production), as well as E2's data from documents recorded using traditional methods and ERP software with limited capabilities and options.

3 Results and Discussion

Before citing and analyzing the results obtained, it is important to highlight the multiple challenges faced by Tunisian companies: such as the criteria below indicated in Fig. 1, where we obtained the following results.

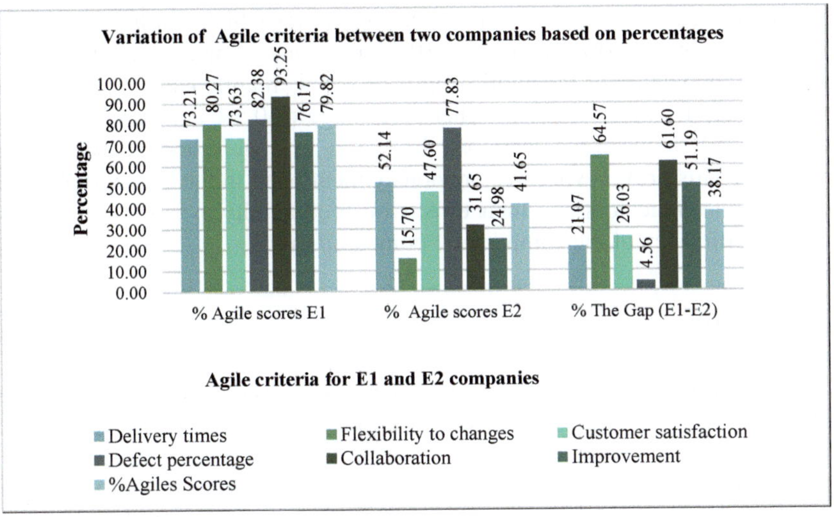

Fig. 1. Variation of Agile Criteria between two compagnies E1 and E2 based on percentages

Figure 1 shows the variations of the Agile criteria of the two companies according to the percentages of agile scores. We note that the percentages of E1 are higher than those of company E2, and that the sum of % Agile overall scores are 79.82% while the percentage of % Agile score of company E2 is 41.65%. This significant difference of 38.17% between the two companies allows us to conclude that company E1 is 38.17% more efficient than company E2. In addition, we note that company E1, which applies agile in the intermediate phase (for 12 months), has exceeded E2 in all agile criteria by a percentage that varies between 15.70% and 77.83%. And that E1 has strong points compared to E2 concerning its flexibility to changes which is of the order of 80.27 and the collaboration of personnel of 93.25%. As well as its capacity to satisfy the customers which the high percentage 77.63% these results lead us to say that the agile kanban in the company E1 has confirmed what was said previously cited in this study. While we deduce that these criteria influence the performance, quality and productivity of the company. In addition, companies E1 and E2 affirmed in particular the results through Table 1.

Table 1. Indicator results of two companies E1 and E2

N	Indicator	E1	E2	The Gap (E1-E2)
1	Delivery times	10 Days	11 Days	1 Day
2	Productivity	2550 P/Day	1950 P/Day	600 P/Day
3	Quality	5.5%	5%	0.5%
4	Performance	85%	65%	20%
5	Efficiency	85%	75%	10%

From this table, we see that the gaps in delivery times which is 1 day, productivity is 600 pieces/day, yield is 20% efficiency is 10% shows that company E1 is more efficient thanks to the application of the hybrid approach 4.0 with agile kanban [18]. While for quality company E2 has a slightly higher percentage of 0.5% compared to E1, this is thanks to the experience of the operators, the quality managers, and the extensive training in quality. This proves what was previously mentioned in the bibliographic studies concerning agile indicators and their effects on productivity, performance, efficiency [19] and the efficiency of the company E1 compared to E2.

4 Conclusion

This article proposes an applied research framework between two Tunisian companies E1 and E2 similar in terms of size, type of article to be made, region, and climate... In addition, they have certain differences in terms of organizational structure, one (E1) adopts agile kanban with industry 4.0 while (E2) applies a classic traditional method. In particular, this study also showed us the importance of the new hybrid approach (Industry 4.0 and Agile Kanban) in improving the performance indicators of a textile clothing company by a significant percentage of about 38.17%, with a 20% improvement in performance and a 10% increase in efficiency. These interesting results prompt us to recommend that E2 adopt the Eagle Kanban model to improve its position. Finally, given these positive results, we are committed to developing and working on other Agile standards, while simultaneously deploying the Agile system within the company and completing the third phase of the Agile process, thereby integrating other areas into the future of both companies.

References

1. Homayoun, S., Salehi, M., Arminkia, A., Novakovic, V.: The mediating effect of innovative performance on the relationship between the use of information technology and organizational agility in smes. Sustainability. **16**, 9649 (2024). https://doi.org/10.3390/Su16229649
2. Lee, H.: Agile quality 4.0 in textile industry. J. Manuf. Syst. **72**, 1–14 (2024). https://doi.org/10.1016/J.Jmsy.2024.01.005
3. Fan, Q., Abbas, J., Zhong, Y., Pawar, P.S., Adam, N.A., Alarif, G.B.: Role of organizational and environmental factors in firm green innovation and sustainable development: Moderating role Of knowledge absorptive capacity. J. Clean. Prod. **411**, 137262 (2023)
4. Anoir, L., Khaldi, M., Erradi, M.: Use of the agile scrum method to manage an e-learning platform in the face of the application of the addie model in the pedagogical engineering of a training course. Kotebe J. Educ. **1**(2), 66–82 (2024). https://doi.org/10.61489/30053447.1(2.66)
5. Martinez, et al.: Kanban Vs Scrum in software maintenance teams. J. IEEE Trans. Software Eng. (2024). https://doi.org/10.1109/Tse.2024.3367759
6. Tamanda, J.: Impact of agile methodologies on the success of development projects. **6** (2024). https://doi.org/10.47672/ijpm.2140
7. wafa, R., Khan, M.Q., malik, F., Abdusalomov, A.B., Cho, y.I., Odarchenko, R.: The impact of agile methodology on project success, with a moderating role of person-job fit in the IT sector of Pakistan. MDPI J. **12**(21) (2022). https://doi.org/10.3390/app122110698

8. Mbonigaba Celestin, S. Sujatha, A. Dinesh Kumar, The rise of agile methodlogies in managing complex collaboration, Indo Americain Journal Of Multidisciplinary Resarch And Review (Iajmrr). Volume 8, Issue 2, 2024, https://doi.org/10.5281/Zenodo.13871832
9. Taylor And Al: Strateg. Manag. J. (2023).
10. Giboing, B.: The Strategy Toolbox, 2nd edn. Dunod, Spain (2017)
11. Muller, J., Schneider, E., Weber, R.: Impact of agile practices on customer satisfaction in digital transformation projects. Int. J. Inform. Manag. Mars. (2024). https://doi.org/10.1016/J.Ijnfomgt.2024.102765
12. Martinez and al: J. Mod. Project Manage. (2023). https://doi.org/10.1016/J.Jmp.2023.04.012
13. Krajewski and al Operations Management Processes And Supply Chains. (2019). DOI: Person Edication
14. Christopher M Logistic & Supply chain Management journal Person Uk. (2016). DOI:https://doi.org/10.1057/978-1-137-58410-2
15. Souti, H., Malainine, C., Mansouri, M.: Organizational agility and organizational performance in small and medium enterprises: A systematic literature review. J. Altern. Manageriales Economiques. (2023). https://doi.org/10.48374/Imist.Prsm/Ame-V5i.37692
16. Fâre, R., et al.: Productivity growth, technical progress, and effeiciency change. J. Prod. Anal. (1994). https://doi.org/10.1007/Bf01073474
17. Cooper, R.G., Sommer, A.: Agile stage gate hybrids: The next stage for product developement. J. Prod. Innov. Manage. **33**(5), 563–572 (2016)
18. Sickles, R.C., Zelenyuk, V.: Measurement of Productivity and Effeciency: Theory and Practice. Cambridge University Press. https://doi.org/10.1017/978113956981
19. Tavana, M., Di Caprio, D., Rostamkhani, R.: A model for evaluating the total quality management action plan in supply chain management using Lean and Agile scores. J. Innov. Knowl. **10**(1) (2024). https://doi.org/10.1016/j.jik.2024.100633
20. Al-Refaie, A., Lepkova, N.: A fuzzy larg index for assessing the lean, agile, resilience, and green paradigms in industrial companies. J. Mdpi Sustain. **17**, 1863 (2025). https://doi.org/10.3390/Su17051863

Kinetics of Capillary Rise in Single-Sided and Double-Sided Woven Fabrics

Hassen Hedfi[1(✉)], Sabra Hamzaoui[2], and Mohamed Hamdaoui[3]

[1] University of Monastir, National Engineering School of Monastir, Mechanical Engineering Laboratory (LGM), LR99ES32, 5019 Monastir, Tunisia
hassen.hedfi@enim.u-monastir.tn
[2] University of Monastir, National Engineering School of Monastir, 5019 Monastir, Tunisia
[3] University of Monastir, National Engineering School of Monastir, Textile Materials and Processes Research Unit MPTex, UR17ES33, 5019 Monastir, Tunisia

Abstract. Moisture and liquid transport through textile fabrics is one of the critical factors affecting the physiological comfort of the individual, particularly for sportswear, underwear, workwear, and protective clothing. Wettability plays an important role in determining the comfort and moisture transport of a fabric.

The objective of this work is to study the influence of the texture of woven fabrics on their wetting behavior or ability. To do this, the construction parameters of the fabrics are varied and the wetting characterization was studied using the capillary rise method.

In this context, an experimental study of the vertical wetting properties of double-sided (by weft) fabrics is investigated. The influence of the characteristics of double-sided fabrics (composition of each face, Picking and weave patterns) on the capillary rise, by determining the kinetics of the rise and the heights reached at equilibrium, is studied and analyzed. A comparison with the wetting behavior of a simple woven structure having the same characteristics is also undertaken in this work.

It has been found that the kinetics of the capillary rise of the wetting liquid is inversely proportional to the hydrophilicity of the fibers constituting the textile fabric. The capillary rise and the maximum height reached at equilibrium are strongly influenced by the structural parameters of the fabrics, namely the picking and the weave patterns. The composition of the wetting liquid has a significant influence on capillary rise through woven fabrics: the addition of electrolyte (NaCl) leads to a reduction in the maximum height reached at equilibrium and accelerates capillary rise.

Keywords: Woven texture · Capillary rise · Wettability

1 Introduction

Our modern lifestyle has created an increased demand for new functional textiles, ranging from simple comfort to multi-protective functions in hazardous or extreme environments. Today's textile products are not simply clothing, or traditional home furnishings

M. Hamdaoui et al. (Eds.): ITDM'3 2025, PTT, pp. 264–278, 2026.
https://doi.org/10.1007/978-981-95-7704-0_24

and textiles, but should also be advanced shields for the human body, capable of preventing attacks from pathogenic microorganisms, toxic chemicals, flames, UV rays, and potentially any environmental threat or vulnerability. In the medical sector, for example, textile supports meet multiple functional criteria thanks to their barrier effect, absorption capacity, biocompatibility, compression capacity and their capacity and potential in Tissue engineering (TE) [1].

The structuring of textile surfaces during manufacturing operations is possible by working on new textile architectures. One example is 3D fabrics, which are three-dimensional textile textures made of staple fibers or continuous yarns arranged in three spatial directions.

In the literature, several approaches have been adopted to functionalize textiles by modifying the texture. Among these approaches, we can cite waterproof-breathable fabrics. The first technique for producing waterproof-breathable fabrics is the creation of high-density fabrics obtained from conventional yarns/filaments or those obtained from specialty fibers and filaments. Tight-woven fabrics characterized by high warp and weft yarn densities are made either from absorbent and hydrophilic yarns or from synthetic microfiber yarns. The resulting fabric has small pores, providing maximum protection against wind and rain. For this type of fabric, a compromise must be made between waterproofing and water vapor permeability. In other words, the inter-thread spaces must be as large as possible to maximize the transmission of water vapor through the fabric. At the same time, to ensure maximum protection against wind and rain, the inter-thread spaces must be as small as possible and the outer surface of the garment must be hydrophobic to minimize wettability by rain, snow, and slush [2, 3].

The wetting behavior of a liquid is influenced by several parameters. For example, the temperature which acts on the properties of liquid as well as solid (density, structural parameters, etc.), it modifies the surface tension (generally decreases with the increase in temperature [4], the oxidation behavior, etc. We also cite the effect of time, or the duration of wetting. The properties of the wetting liquid also represent a crucial parameter controlling the wetting phenomenon. Among these properties, we cite surface tension, the density of the liquid, its viscosity, etc. For example, for the viscosity of liquid, we intuitively note that a drop of high viscosity spreads less quickly than a less viscous droplet that reaches an equilibrium state in a short time.

Capillary rise can be defined as a consequence of the spontaneous wetting of a porous medium by a wetting liquid. Indeed, the capillary rise of a liquid in a porous structure is driven by the capillary forces resulting from the wetting of the solid's surface (in the absence of any external forces) [5].

Textile fabrics are composed of several capillaries located between the threads that intertwine to form them and between the fibers that form the threads. Generally speaking, like all fibrous structures, textile materials contain empty spaces or pores that can occupy a greater or lesser portion of their total volume, allowing them to be considered a porous medium characterized by its porosity and permeability. These interstices (pores) can be likened to capillary tubes that are not truly straight and parallel and can overlap. Furthermore, the cross-section of the threads shows that the shapes are not circular and the sizes vary. These voids allow the capillary rise of a fluid in the fibrous material. Thus, the size and shape of the interstices can influence the amount of liquid absorbed. In the case of textile materials, we distinguish three major scales according to pore size [6]. The microscopic scale referring to the accessible sorption sites in the amorphous zones of the fibers that constitute it (relating to pores whose diameter does not exceed 2 nm. In the second level we find the mesoscopic scale referring to the intra-thread volume and concerning the micro-pores located between the fibers in the threads constituting the fabric (relating to pores whose diameter is between 2 and 50 nm). Finally, we find the macroscopic scale, referring to the inter-thread volumes and concerning the macro-pores located between the threads constituting the fabric (relating to pores whose diameter is greater than 50 nm). Indeed, for textile materials, capillary impregnation is the replacement of the fiber-air interface by that of the fiber-liquid. Within textiles, the spontaneous propagation of the liquid is guided by the capillary forces created by the phenomenon of Wetting can be considered as the initial behavior of the fabric, yarn, or fiber when they are in contact with a wetting liquid. This spontaneous migration of fluids in capillary systems depends essentially on the properties of the fluids (density, surface tension, etc.), fluid-solid interactions, and the geometry of the network forming the porous structure [7].

In this paper, we study the effect of the structure of single-sided and double-sided textile fabrics and their texture characteristics (nature of the yarns composing these fabrics, the density of weft yarns) on the capillary rise of wetting liquids. Three wetting media are used: distilled water (standard bath), bath containing sodium chloride and bath containing acid hystidine. This paper is structured as follows. An introduction presents the phenomenon studied and a brief overview of studies on this phenomenon. Then, we present the materials and methods implemented. We then present the results obtained as well as their discussions.

2 Methods and Materials

2.1 Textile Substrates

All the fabrics used in this study were produced in the weaving workshop at ENIM (National Engineering School of Monastir), on the DORNIER weaving machine, which has the following characteristics:

- DORNIER weaving machine equipped with a BONAS Jacquard mechanism with electronic reading;
- Weft insertion system: rigid rapier with positive grippers;
- One warp: 100% white pinched polyester, 18 tex, and warp count 24 threads/cm;
- Two wefts:
- *Weft 1: 100% cotton, 31 tex*, for right side (in case of double-sided fabric: called side-cotton)
- *Weft 2: 30/70 Cotton/PES, 29 tex*; for back side (in case of double sided fabric: called side-C-PES)
- Reed width: 180 cm;
- Reed stitching, 2 threads/spindle.

The study focused on double-sided fabrics (by weft) and single-sided fabrics (weft 100% cotton and warp 100% PES). These fabrics have the same construction parameters.

Only the weave and the picking (number of weft threads) were varied. The weaves used and the picking of different fabrics studied are listed in following table (see Table 1).

Table 1. The weaves used and the picking of different fabrics

	Fabrics	**Weave patterns**		**Picking (picks/cm)**	**Fabric Code**
		Right side	**Back side**		
Simple fabric	Simple Fabric : Weave: 3-ply twill	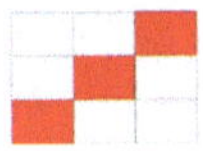		10	SF-T-P10*
				20	SF-T-P20
				30	SF-T-P30
Complex fabrics	Double-sided weft: 3-ply twill weave	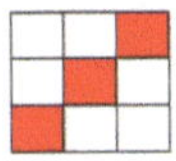		20	DF-TT-P20**
				40	DF-TT-P40
				60	DF-TT-P60

(*) SF: Simple Fabric, T: indicates the fabric weave pattern, T for Twill fabric and P10: indicates piking: 10 picks/cm for this case.

(**) DF: double sided fabric, TT: indicates fabric weave pattern, T for Twill (right side) and T for Twill (back side) and P20 indicates picking: 20 picks/cm for two sides.

2.2 Experimental Device and Wetting Liquids

Experimental device for measuring capillary rise. At ENIM's textile finishing laboratory, a device was developed to ensure the textile remains vertically held by touching the water surface in order to study the capillary rise of water through woven or knitted fabrics.

In a dark room, the sample, 20 cm wide and 40 cm long, is held vertically from the top using a steel rod, partially submerged in distilled water, which causes it to wet. In fact, it is only under the effect of its own weight (see Fig. 1).

Wetting liquids used. The liquid used to characterize the wetting behavior by capillary rise is colored distilled water. We chose an acid dye that has no affinity for either cotton or polyester. The acid dye is chosen because it has no dyeing affinity, neither for cotton nor for polyester. It is simply used to color the water and make it easier to identify the height reached at equilibrium by the water during capillary rise. Table 2 gives the composition of wetting baths used in this study.

2.3 Porous Media Related Characteristics of Textile Fabrics

We consider the following characteristics: porosity, air permeability, surface wetting resistance and water absorption rate.

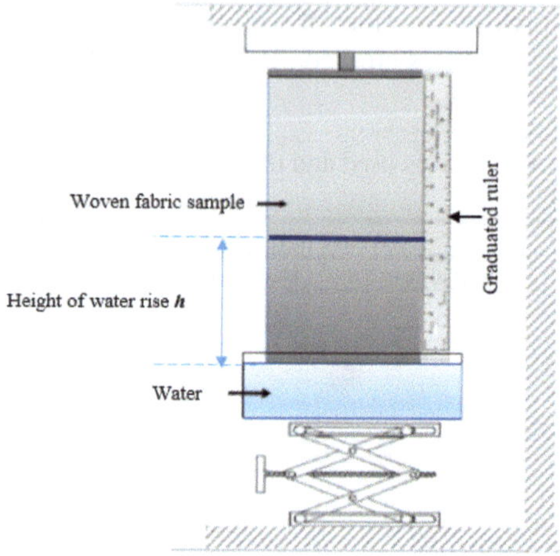

Fig. 1. Experimental device used to measure the height of water capillary rise in different samples of woven fabrics.

Table 2. Composition of wetting baths

Type of wetting baths	Acid dye[a] ($g\,L^{-1}$)	Sodium chloride NaCl ($g\,L^{-1}$)	Hystidine acid ($ml\,L^{-1}$)
1	1	0	0
2	1	0	5
3	1	30	0

[a]The acid dye is chosen because it has no dyeing affinity, neither for cotton nor for polyester

Porosity. It represents the volume fraction of the void; it is defined as the ratio of the volume of the pore space to the total volume. For a textile fabric, we can calculate the porosity using the following relationship:

$$\varepsilon = 1 - \frac{m_s}{\rho_s e} \tag{1}$$

Where e is the thickness of the fabric, ρ_s the density of the material that constitutes the fabric and m_s is the surface mass [8].

Water absorption rate. The water absorption rate of a fabric is its ability to absorb water and the speed at which it does so (its kinetics), which depends on its fiber type, weave structure, thickness and any water-repellent finishes. In this study we investigate a static absorption test which involves immersing the fabric (in distilled water for 30 min and draining for 5 min) and calculating difference between the weight of the sample after immersion and the weight of the dry sample to find the absorption ratio using the following equation.

$$Water\ absorption\ rate(\%) = \frac{M_f - M_i}{M_f} \times 100 \tag{2}$$

With M_i is the initial mass of the dry sample and M_f is the mass of the sample after immersion in distilled water for *30 min* and draining for *5* min.

Air permeability. The TEXTEST (FX3300) air permeability tester was used to measure air permeability according to ISO 9237:1995.

Surface wetting resistance. The Spray Test, also known as the "resistance to surface wetting test," evaluates a fabric's capacity to resist water absorption during surface spraying, in accordance with ISO 4920 standard. It does not quantify water penetration through the fabric, but rather its resistance to surface wetting, by spraying water onto the sample and visually assessing its condition using standards (Grades ranging from Grade 1 to Grade 5).

2.4 Mathematical Modeling of Capillary Rise

A double exponential mathematical model (DEM) was used by Wilezak et al. [9] to describe the adsorption of lead and copper on activated carbon. Hamdaoui et al. [6] validated this same model to study the flow of matter through simple fabrics, Achour et al. [5] also used this DEM model to describe the transfer of water within knitted structures. Recently, Said et al. [10] carried out the application of direct dyes on pre-treated cotton jersey knitted fabric using an experimental system based on the capillary rise technique.

Based on the experimental results of capillary rise found in our study, it was observed that the rise curves can be modeled with two independent processes: one representing the fast phase and a second process for the slow phase. Therefore, we propose to use the double exponential mathematical model (DEM) which describes, during a vertical capillary rise test of water through complex and simple woven fabrics.

The double exponential mathematical model (DEM) is described by the following expression:

$$h_t = h_e - h_1\exp(-K_1 t) - h_2\exp(-K_2 t) \tag{3}$$

With, h_t *(cm)*: the height reached by the water at time *t*; h_e *(cm)*: the height reached by the water at equilibrium; h_1 *(cm)* and h_2 *(cm)* are the water heights reached, respectively, during the fast and slow phases of capillary rise; and K_1 *(min⁻¹)* and K_2 *(min⁻¹)* are the capillary diffusion parameters of water during the fast and slow phases, respectively.

We also model each capillary rise curve using a DEM model. We thus identify the parameters of the corresponding DEM model for each textile fabric by experimental/numerical recalibration. This identification is done using MatLab® software and using the two parameters: the coefficient of determination R^2 and the sum of squared residuals (Sum of squared Errors SSE) to judge the relevance of the model.

$$R^2 = 1 - \left[\frac{\sum_{i=1}^{N}\left(X_{th,i} - X_{\exp,i}\right)^2}{\sum_{i=1}^{N}\left(\overline{\overline{X}}_{\exp,i} - X_{th,i}\right)^2}\right] \tag{4}$$

$$SSE = \frac{\sum_{i=1}^{N}\left(X_{\exp,i} - X_{th,i}\right)^2}{N} \tag{5}$$

Where $X_{\exp,\, i}$ and $X_{th,\, i}$ are respectively the value of the height of the wetting liquid determined experimentally and that determined theoretically by the proposed model, and N is the number of experimental points.

3 Results and Discussions

3.1 Characterization of Manufactured Fabrics

The physical characteristics of fabrics. Once the fabrics are manufactured, we proceed to characterize them.

In the following table (see Table 3) we present the characteristics of the manufactured fabrics. These characteristics relate to picking, surface mass, thickness, shrinkage and crimp.

Table 3. Picking, surface mass density, thickness, Shrinkage and crimp of single and double-sided fabrics

Fabric Type	Fabric code	Picking *Picks/cm*	Surface mass density $g\ m^{-2}$	Thickness *mm*	Shrinkage %	Crimp %
SF	SF-T-D10	10	78.30	0.94	8.26	3.83
SF	SF-T-D20	20	112.30	0.99	10.19	6.72
SF	SF-T-D30	30	153.40	0.99	12.11	7.33
DF	DF-TT-D20	20	110.30	0.81	9.11	4.45
DF	DF-TT-D40	40	177.80	0.84	10.38	4.84
DF	DF-TT-D60	60	258.00	0.89	11.65	6.74

The porous media characteristics of fabrics. In the following table (see Table 4) we indicate the results of this characterization.

When developing a functional textile for clothing, parameters influencing comfort, such as air permeability, must be taken into account. From the histograms presented in Fig. 2, it is clear that fabric manufacturing parameters have a considerable effect on the variation in air permeability.

The results confirm that permeability depends on the physical characteristics of the fabric: higher the weft count, lower the permeability. Indeed, for low weft counts, the fabric has numerous inter-thread and inter-fiber pores, thus achieving very high porosity, whereas for high weft counts, the structure is denser and the porosity is lower (see Table 4, porosity column).

The weave also influences this characteristic; for the same weft count, permeability is higher for simple sided fabrics than for double sided ones.

We make the same observations when considering the measurements obtained for water absorption rates: the higher the number of picks, the lower the water absorption rates (see Table 4, water absorption rate column)

Table 4. Porosity, water absorption rate, air permeability and surface wetting resistance

Fabric Type	Fabric code	Porosity %	Water absorption rate %	Air permeability		Surface wetting resistance	
				Right side $mm\ s^{-1}$	Back side $mm\ s^{-1}$	Right side	Back side
SF	SF-T-D10	89.88	79.00	1740.00	1740.00	Grade-4	Grade-4
SF	SF-T-D20	84.14	66.30	1082.50	1082.50	Grade-4	Grade-4
SF	SF-T-D30	79.33	65.00	536.30	536.30	Grade-4	Grade-4
DF	DF-TT-D20	93.80	76.00	961.70	951.00	Grade-5	Grade-4
DF	DF-TT-D40	91.29	65.00	62.07	61.73	Grade-5	Grade-3
DF	DF-TT-D60	88.47	62.00	28.43	26.70	Grade-5	Grade-3

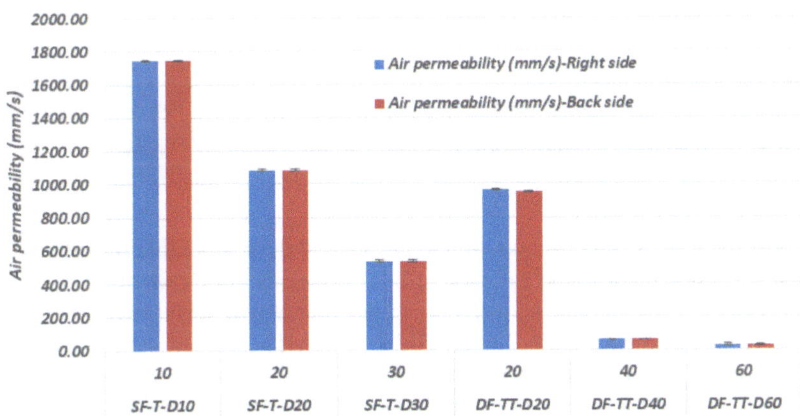

Fig. 2. Air permeability of in right side and back side of fabrics

For the samples studied, the observed evaluations (ratings) of surface wetting resistance are:

- *Grade-3*: wetting of the upper surface at watering points;
- *Grade-4*: slight adhesion or light dispersed wetting of the upper surface;
- *Grade-5*: no adhesion or wetting of the upper surface.

Overall, we note good surface wetting resistance of the various fabrics manufactured, with pronounced wetting resistance for double-sided fabrics, specifically the right side.

3.2 Kinetics of Capillary Rise

In this section, we present the capillary rise results obtained for the different fabrics.

Influence of fabric texture parameters. As previously presented, the manufactured double-sided fabrics (DF) consist of a 100% polyester warp and two wefts: the first is

100% cotton (side-cotton) and the weft of the second side is composed of 70% polyester and 30% cotton (side-C-PES).

The experimental and theoretical values of the height of the wetting liquid reached at equilibrium, $h_{e,exp}$ and he, th, the values of the different diffusion coefficients and constants of the theoretical model used, as well as the values of the coefficient of determination and the sum of squares of the residuals (SSE) are grouped in Table 5.

Table 5. The results of capillary rise measurements and the coefficients of the developed theoretical models

Fabric Type	Fabric code	Fabric side code	$h_{e,exp}$ (cm)	$h_{e,th}$ (cm)	h_1 (cm)	h_2 (cm)	K_1 (min^{-1})	K_2 (min^{-1})	R^2	SSE
SF	SF-T-D10	SF-T-D10	23.70	22.96	8.92	10.21	0.351	0.036	0.998	3.893
SF	SF-T-D20	SF-T-D20	25.00	23.50	6.08	13.17	0.302	0.032	0.998	2.916
SF	SF-T-D30	SF-T-D30	18.00	18.20	2.91	34.02	0.282	0.029	0.990	4.839
DF	DF-TT-D20	DF-TT-D20-Side-Cotton	23.50	23.40	6.73	14.29	0.660	0.048	0.999	1.061
		DF-TT-D20-Side-C-PES	24.00	24.00	7.51	14.59	1.194	0.050	0.999	2.523
DF	DF-TT-D40	DF-TT-D40-Side-Cotton	23.10	22.57	4.27	16.28	0.304	0.031	0.996	5.102
		DF-TT-D40-Side-C-PES	23.70	23.31	4.32	16.73	0.650	0.031	0.997	3.560
DF	DF-TT-D60	DF-TT-D60-Side-Cotton	15.80	16.81	5.93	9.03	0.131	0.011	0.997	2.318
		DF-TT-D60-Side-C-PES	16.00	16.97	6.12	9.35	0.132	0.013	0.997	2.447

According to the results found with the DEM model, we note that the value of R^2 is very close to 1 (> 0.99) and the value of SSE is between 1 and 5, we can say, therefore, that there is a good agreement between the experimentally determined values, h_{exp} (t), and those predicted theoretically by the DEM, $h_{th}(t)$. The values indicated in Table 5 (see $h_{e,exp}$ and $h_{e,th}$ columns) confirm this observation: The experimental ($h_{e,exp}$) and theoretical ($h_{e,th}$) values of this are very close. We also note that the height reached by the water during the second phase is greater than that absorbed during the first phase of capillary rise: $h_1 < h_2$ (see h_1 and h_2 columns in Table 5).

The results presented in the table above confirm the effect of the weft composition on the height reached at equilibrium. Indeed, the experimental and theoretical deviations depend on the weave and the weft. For example, an incorporation of 70% polyester in the weft yarn results in an increase of approximately of 2.08%, 2.53%, and 1.25% of the height of the liquid reached at equilibrium for twill fabric with 20, 40, and 60 picks/cm, respectively. We do not observe a big difference since the two sides of the fabric are interconnected (in the case of double-sided fabrics), and the composition of the weft yarn used for second side is not made of 100% PES, but it contains about 30% cotton. For a comparison between the kinetics of capillary rise of two sides of each fabric, see Fig. 5, and also K_1 and K_2 columns in Table 5.

We examine the evolution of the kinetics of the rise of the wetting liquid during capillary rise. The curves presented in Figs. 3 and 4 show that the kinetics of the rise are influenced by the composition of the fabric, the weave: single sided fabrics (see Fig. 4) or double sided fabrics and the number of weft threads.

We note that the kinetics of capillary progression during phase 1 is much higher than that of phase 2: $K_1 > K_2$ (see Fig. 5 and K_1 and K_2 columns in Table 5).

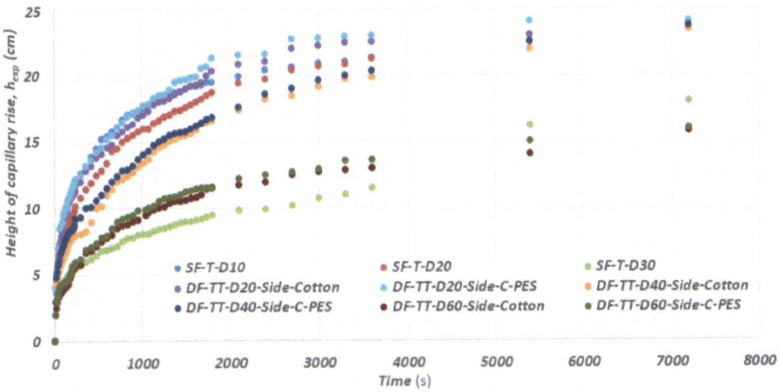

Fig. 3. Capillary rise curves of different textile fabrics: single-sided and double-sided fabrics

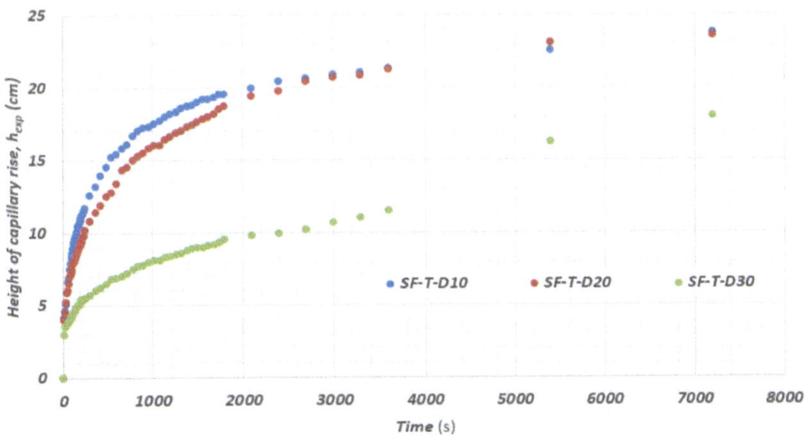

Fig. 4. Capillary rise curves of different textile fabrics: single-sided fabrics

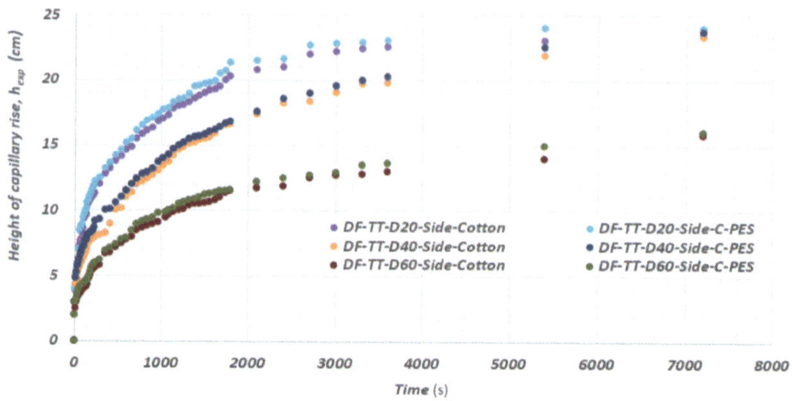

Fig. 5. Capillary rise curves of double-sided textile fabrics: comparison between the two sides of each fabric

Concerning the effect of the composition on the kinetics of the rise, we observe that the speed of the capillary rise in *side 1*, during the two phases of slow and fast progression, is lower than *side 2*. This can be explained by the fact that polyester is a synthetic fiber whose crystallinity is very high (*80–90%*), its micropores occupy a small volume which leads to a poor absorption capacity. This accelerates the kinetics of water rise and promotes the advancement of liquids in the pores and large interstices: the macropores. However, cotton fiber has a high-water absorption capacity and a large number of amorphous regions, where water molecules can penetrate and bind very well to hydroxyl groups (these act as primary adsorption centers due to the formation of hydrogen bonds between water molecules and OH functional groups in cotton fibers). Therefore, it has a better absorption capacity than blended polyester fiber and slower progression kinetics.

These results lead us to conclude that the kinetics of capillary rise of the wetting liquid is inversely proportional to the hydrophilicity of the fibers constituting the textile fabric.

Examining the results from the point of view of the structure of textile fabrics, we see that the weave significantly impacts the height h_e and the capillary rise kinetics for complex structures (double sided fabrics). This is explained by the modification of the interweaving of the threads in the fabric, whether simple or complex.

For the DF fabric in 3-twill weave, when the weft count is increased, the kinetics of the capillary rise of two phases as well as the heights $h_{e,exp}$ and $h_{e,th}$ decrease. This can be explained by the fact that when the weft count increases, the fabric will be more compact, the number of threads per centimeter increases, the amount of fibrous material increases and the pores between the threads become smaller. Therefore, the rise of the liquid will be more difficult and slower. In addition, a 3-twill weave fabric with a large weft count has a higher value of thread tortuosity. This reduces the macropores of the structure, slows down the progression of water and limits the maximum height reached at equilibrium.

Influence of the composition of the wetting liquid. It has been shown in the literature [11] that the properties of the liquid determine the wetting behavior of the textile. Therefore, in this section, we will study the effect of adding sodium chloride salt and the effect of adding hystidine on the kinetics of capillary rise.

As for the effect of sodium chloride salt on the kinetics of capillary rise presented in Fig. 6 (for slow phase) and in Fig. 7 (for rapid phase), we note that they are influenced by the addition of sodium chloride salt: for the kinetics of the rapid phase (see Fig. 7), we observe that when the number of weft yarns increases, the kinetics of the rise of the wetting liquid containing sodium chloride salt increases. Indeed, for different fabrics with a low number of weft yarns, around *20 picks/cm*, the rise rate of the liquid without sodium chloride salt is greater than that of the liquid containing salt. For fabrics with *40 picks/cm*, the two kinetics are similar. Whereas for strong wefting, the most significant kinetics are detected with the wetting liquid containing sodium chloride salt.

Regarding the kinetics of the rise in the slow phase, we note that for DF twill fabrics, the addition of sodium chloride salt leads to an increase in the kinetics as the wefting increases.

The addition of electrolytes, sodium chloride (Na^+, Cl^-), has the effect of reducing the negative potential of the cotton fiber. This addition does not have a significant impact on the polyester fiber. The potential barrier is more easily crossed by the dye, and there is an acceleration of the adsorption process on the surface and the fixation of the water molecule within the cellulosic fiber. The mass of absorbed water will be greater, while the maximum height reached at equilibrium decreases.

As for the capillary rise kinetics of the fast and slow phases, we observe that the effect of adding hystidine depends on the weft yarn density (see Figs. 6 and 7). Indeed, in the case of twill fabrics, we note that for a low picking (20 picks/cm), the kinetics of the fast phase K_1 of the liquid without hystidine are higher. For a fabric with 40 picks/cm, the two kinetics without and with hystidine are similar. However, for maximum number of weft yarns, we observe that the kinetics of the liquid containing hystidine are much higher than those without hystidine. During the slow phase, we observe that the rise kinetics of the liquid containing hystidine are lower than those without hystidine for fabrics with picks densities of 20 and 40 picks/cm. However, for maximum wetting the effect of hystidine is reversed, the kinetics of the wetting liquid containing hystidine is slightly higher.

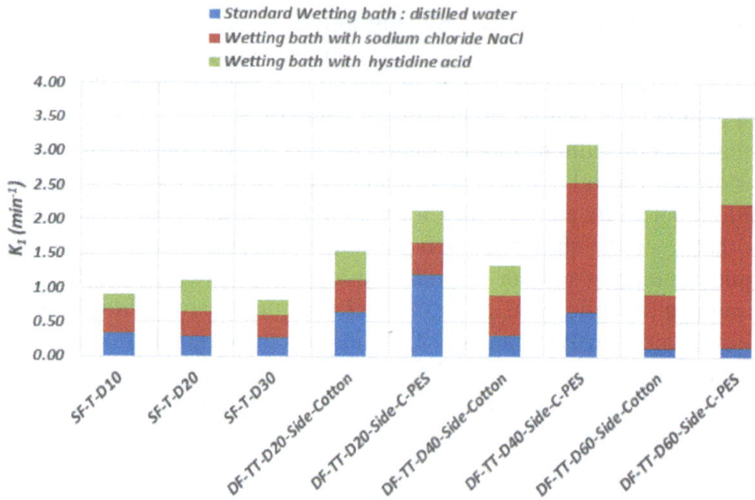

Fig. 6. Variation of kinematic parameters as a function of the wetting liquid: slow phase K_1

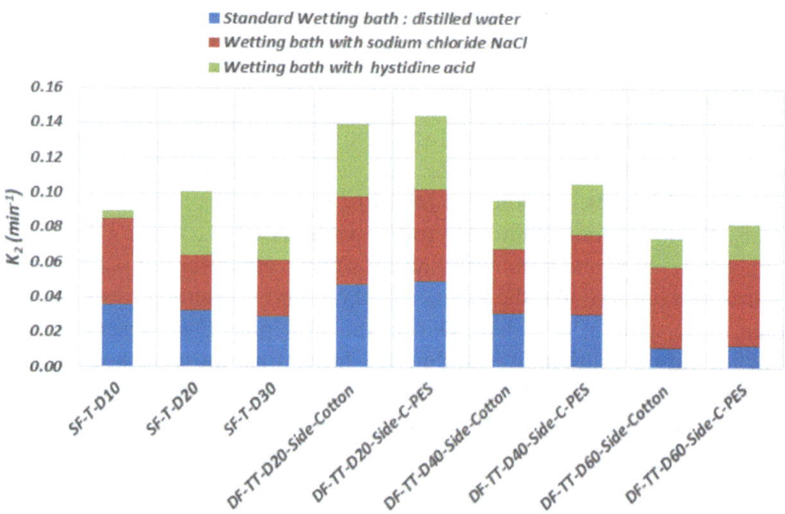

Fig. 7. Variation of kinematic parameters as a function of the wetting liquid: rapid phase K_2

4 Conclusion

This paper presents the experimental and theoretical results of the study on the kinetics of capillary rise for single-sided and double-sided textile fabrics. To this end, the effects of the texture characteristics (weave, fiber type, and weft yarn density) of single-sided and double-sided fabrics and the composition of the wetting liquid on capillary rise (the kinetics of capillary rise and the maximum height reached at equilibrium) were investigated. The experimental results obtained were then used to model the phenomenon using a double exponential *DEM* mathematical model.

The experimentally determined and theoretically predicted results show that the texture characteristics of the textile fabrics and the wetting liquid significantly influence the kinetics of capillary rise and the maximum height reached at equilibrium.

These results confirm that the capillary rise phenomenon has two independent phases: fast and slow. The analysis of the results leads us to note that the kinetics of the capillary rise of the wetting liquid is inversely proportional to the hydrophylie of the fibers constituting the textile fabric; the face whose weft is composed of *100%* cotton has the lowest kinetics and height. The capillary rise and the maximum height reached at equilibrium are strongly influenced by the structural parameters of the fabrics, namely the weave and the weave. Indeed, the kinetics of the capillary rise as well as the maximum height reached at equilibrium are inversely proportional to the weave in the case of twill fabrics.

The composition of the wetting liquid has a great influence on the capillary progression within the woven fabrics, the addition of salt leads to a decrease in the maximum height reached at equilibrium and accelerates the capillary rise. On the other hand, the addition of acid hystidine leads to an increase in the maximum height reached at equilibrium. This study could be extended and generalized to other woven or knitted textile structures, especially complex structures, such as other types of double-sided textile fabrics, with compositions that vary significantly from one side to the other.

Another parameter, that of the surface roughness of double-sided textile fabrics, could be examined to study its influence on the wetting behavior of these surfaces, and work toward designing surfaces with different wetting behaviors from one side to the other.

References

1. Scholpp, S., Hoffmann, L.A., Schätzlein, E., Gries, T., Emonts, C., Blaeser, A.: Interlacing biology and engineering: An introduction to textiles and their application in tissue engineering. Mater. Today Bio. **31** (2025)
2. Mukhopadhyay, A., Midha, V.K.: A Review on Designing the Waterproof Breathable Fabrics Part I: Fundamental Principles and Designing Aspects of Breathable Fabrics. J. Ind. Text. **37**(3), 225–262 (2008)
3. Mukhopadhyay, A., Midha, V.K.: A review on designing the waterproof breathable fabrics part II: Construction and suitability of breathable fabrics for different uses. J. Ind. Text. **38**(1), 17–41 (2008)
4. Satish, G.K., Mark, E.S.: Contact angles and interface behavior during rapid evaporation of liquid on a heated surface. Int. J. Heat Mass Transf. **45**(8), 3771–3780 (2022)
5. Achour, N.S., Hamdaoui, M., Ben Nasrallah, S.: Wicking & sorption ability on knitted fabrics: Experimental and theoretical studies. TEKSTİL ve KONFEKSİYON. **27**(1), 46–52 (2017)
6. Hamdaoui, M., Ben Nasrallah, S.: Capillary rise kinetics on woven fabrics – Experimental and theoretical studies. Indian J. Fibre & Textile Res. **40**(6), 150–156 (2015)
7. Kissa, E.: Wetting and wicking. Text. Res. J. **66**(10), 660–668 (1996)
8. Özdil, N., Marmaralı, A., Kretz, S.D.: Effect of yarn properties on thermal comfort of knitted fabrics. Int. J. Therm. Sci. **46**, 1318–1322 (2007)
9. Wilczak, A., Keinath, T.M.: Kinetics of sorption and desorption of copper (II) and lead (II) on activated carbon. Water Environ. Res. **65**(3), 238–244 (1993)

10. Said, S., Hamdaoui, M., Sahraoui, W.: Mathematical models to follow the dyeing of cotton knitted garment. Fibers Polym. **24**, 45–56 (2023)
11. Hamdaoui, M., Fayala, F., Ben Nasrallah, S.: Dynamics of capillary rise in yarns: Influence of fiber and liquid characteristics. J. Appl. Polym. Sci. **104**, 3050–3056 (2007)

Correction to: Piezoelectric Smart Textiles for Energy Harvesting

Ichrak Trabelsi, Adel Ghith, and Faten Fayala

Correction to:
Chapter 11 in: M. Hamdaoui et al. (Eds.): *Proceedings of the Third International Conference of Innovative Textiles and Developed Materials*, PTT, https://doi.org/10.1007/978-981-95-7704-0_11

The original publication of this chapter was accidentally issued without the appropriate permissions credit line for figure 1. The phrase "Reproduced from Ref. [4] with permission from the publisher." has been added.

The updated version of this chapter can be found at
https://doi.org/10.1007/978-981-95-7704-0_11

M. Hamdaoui et al. (Eds.): ITDM'3 2025, PTT, p. C1, 2026.
https://doi.org/10.1007/978-981-95-7704-0_25

Author Index

M. Hamdaoui et al. (Eds.): ITDM'3 2025, PTT, p. 279, 2026.
https://doi.org/10.1007/978-981-95-7704-0